Y0-AAF-306

MATH-
STAT.
LIBRARY

SPECTRAL THEORY OF LINEAR OPERATORS

L.M.S. MONOGRAPHS

Editors: P. M. Cohn *and* G. E. H. Reuter

1. Surgery on Compact Manifolds *by* C. T. C. Wall, F.R.S.
2. Free Rings and Their Relations *by* P. M. Cohn
3. Abelian Categories with Applications to Rings and Modules *by* N. Popescu
4. Sieve Methods *by* H. Halberstam and H.-E. Richert
5. Maximal Orders *by* I. Reiner
6. On Numbers and Games *by* J. H. Conway
7. An Introduction to Semigroup Theory *by* J. M. Howie
8. Matroid Theory *by* D. J. A. Welsh
9. Subharmonic Functions, Volume 1 *by* W. K. Hayman and P. B. Kennedy
10. Topos Theory *by* P. T. Johnstone
11. Extremal Graph Theory *by* B. Bollabás
12. Spectral Theory of Linear Operators *by* H. R. Dowson

Published for the London Mathematical Society
by Academic Press Inc. (London) Ltd.

SPECTRAL THEORY OF LINEAR OPERATORS

H. R. DOWSON

Department of Mathematics
University of Glasgow
Glasgow, Scotland

1978

ACADEMIC PRESS

LONDON NEW YORK SAN FRANCISCO

A Subsidiary of Harcourt Brace Jovanovich, Publishers

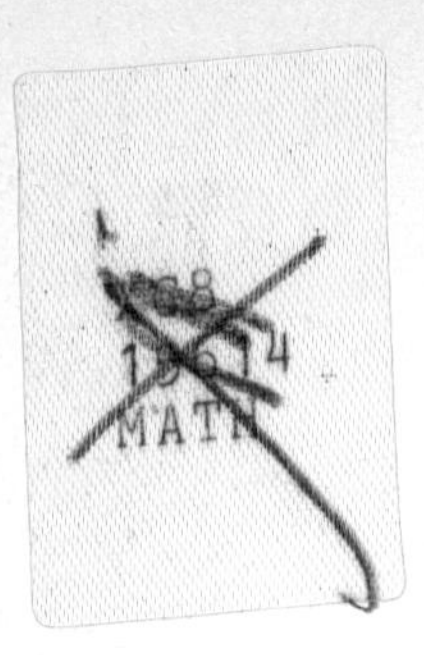

ACADEMIC PRESS INC. (LONDON) LTD.
24/28 Oval Road
London NW1

United States Edition published by
ACADEMIC PRESS INC.
111 Fifth Avenue
New York, New York 10003

Library of Congress Catalog Card Number: 77-79301
ISBN: 0-12-220950-8

PRINTED IN GREAT BRITAIN BY
PAGE BROS (NORWICH) LTD, NORWICH

Preface

During the period 1958–1971 there appeared the three volumes of the monumental treatise "Linear Operators" by N. Dunford and J. T. Schwartz. This work contains a wealth of material. The purpose of my book is to develop this research further in one particular direction; namely the study of various classes of linear operators on a complex Banach space which possess a rich spectral theory.

General spectral theory is developed in Part One. The material presented is very much influence by the aforementioned work of N. Dunford and J. T. Schwartz and by the book "Introduction to Functional Analysis" by A. E. Taylor.

Part Two contains two chapters. The first relies heavily on ideas of F. F. Bonsall and V. I. Lomonosov to simplify the usual presentation of the spectral theory of compact operators and of the theory of superdiagonal forms for compact operators due to J. R. Ringrose. In Chapter 3 we present the theory of Riesz operators initiated by A. F. Ruston and later developed by T. T. West.

Part Three is a very brief section containing all the properties of hermitian operators on a Banach space required for later sections of the book.

Part Four is the longest section of the book and is devoted to the theory of prespectral operators. It includes brief sections on spectral operators and normal operators on Hilbert space that contain proofs of the basic theorems significantly simpler than those found in existing textbooks.

In Part Five we develop the theory of well-bounded operators initiated by D. R. Smart and developed by J. R. Ringrose and others.

My thanks are due to Professor John Ringrose on two counts. First, as my research supervisor in the early 60s he first aroused my interest in many of the topics discussed in this book and second because Chapters 2, 15 and 16 are heavily dependent on his research papers on compact and well-bounded operators. I should also like to express my gratitude to Professor Earl

Berkson for many valuable discussions on hermitian operators, prespectral operators and well-bounded operators. I am indebted to Dr Philip Spain for submitting to me several of his unpublished manuscripts and for reading and criticizing an earlier version of Part Three of this book.

My thanks are due to Dr Trevor West for reading and criticizing an earlier version of Parts One and Two of this book. Also, I am very grateful to Dr Alastair Gillespie for reading and criticizing an earlier version of Part Four of this book and for helping me very considerably in writing Chapters 18 and 20 by sending to me several of his unpublished manuscripts.

Finally, I am much indebted to Miss Daphne Davidson for her patient and very careful work in producing the typescript of this volume.

University of Glasgow H. R. DOWSON
March, 1977

Contents

Note to the Reader

It will be assumed that the reader of this book has a basic knowledge of functional analysis as could be acquired from "Elements of functional analysis" by A. L. Brown and A. Page. We assume also a knowledge of Gelfand theory in commutative Banach algebras and of elementary spectral theory in general Banach algebras. Our standard reference for these topics is "Complete Normed Algebras" by F. F. Bonsall and J. Duncan. For the results on measure theory that we shall need, the reader is referred to "Measure Theory" by P. R. Halmos. It will be assumed also that the reader is familiar with the theories of vector-valued holomorphic functions and of integration of vector-valued functions presented in Chapter III of "Linear Operators" by N. Dunford and J. T. Schwartz. However, the deeper results, derived from this theory, on the representation of weakly compact linear mappings are specifically recalled.

Theorems, propositions, lemmas, corollaries, definitions and notes are numbered consecutively. For example, Theorem 4.17 refers to the seventeenth item in Chapter 4.

We have divided the bibliography into two sections. The works in "General background reading" are referred to in the text by a number alone. Other references are cited by giving the author's name followed by a number.

Some Terminology, Notation and Conventions Used Throughout This Book

Throughout, X denotes a non-zero complex Banach space, otherwise arbitrary unless the contrary is explicitly stated. H denotes a non-zero complex Hilbert space, otherwise arbitrary unless the contrary is explicitly stated. Operator means "bounded linear operator". The Banach algebra of operators on X is denoted by $L(X)$. The dual space of X is denoted by X^*. We write $\langle x, \phi \rangle$ for the value of the functional ϕ in X^* at the point x of X. In a Hilbert space setting this notation is also used for the inner product of two vectors. The adjoint of an operator T on X is denoted by T^*. A similar notation is used in a Hilbert space setting for the Hilbert adjoint of an operator. The *annihilator* $Y^\perp$ of a closed subspace Y of X is the set

$$\{\phi \in X^*: \langle y, \phi \rangle = 0 \quad \text{for all } y \text{ in } Y\}.$$

If Y is a closed subspace of H, then $Y^\perp$ denotes the orthogonal complement of Y.

R denotes the set of real numbers
C denotes the set of complex numbers
Z denotes the set of integers
N denotes the set of positive integers

Throughout, scalars and functions are complex-valued unless the contrary is explicitly stated. $\subseteq$ is used for "is contained in", while $\subset$ is reserved for "is strictly contained in". $\{a\}$ denotes the set consisting of the point a alone. $\chi(\tau; z)$ denotes the characteristic function of the set τ evaluated at the point z. Occasionally the characteristic function of τ is denoted by χ_τ. $C(K)$ denotes the Banach algebra of continuous complex-valued functions on the compact set K under the supremum norm. Σ_p denotes the σ-algebra of Borel subsets of the complex plane.

Let $A \subseteq X$. We denote the norm closure of A in X by $\bar{A}$ or $\operatorname{cl} A$. If Y is a closed subspace of X, the quotient space of X by Y is denoted by X/Y. Let

$T \in L(X)$ and let $TY \subseteq Y$. The restriction of T to Y is denoted by $T|Y$ and the operator induced on X/Y by T is denoted by T_Y.

If $\mathscr{A} \subseteq L(X)$, then $\mathscr{A}'$ denotes the commutant of $\mathscr{A}$ and $\mathscr{A}''$ denotes the bicommutant of $\mathscr{A}$. Let $T \in L(X)$. We write $N(T)$ for the null-space of the operator T and $R(T)$ for the range space TX. $\sigma(T)$ and $\rho(T)$ denote respectively the spectrum and resolvent set of T. $\mathscr{F}(T)$ denotes the family of complex functions analytic on some open neighbourhood of $\sigma(T)$. $\sigma_a(T)$, $\sigma_c(T)$, $\sigma_p(T)$, $\sigma_r(T)$ denote respectively the approximate point spectrum, the continuous spectrum, the point spectrum and the residual spectrum of T. $v(T)$ denotes the spectral radius of T.

Let $J = [a, b]$ be a compact interval of $\mathbf{R}$. Let $BV(J)$ be the Banach algebra of complex-valued functions of bounded variation on J with norm $|||\ |||$ defined by

$$|||f||| = |f(b)| + \operatorname{var}(f, J) \qquad (f \in BV(J)),$$

where $\operatorname{var}(f, J)$ is the total variation of f over J.

Let $AC(J)$ be the Banach subalgebra of $BV(J)$ consisting of absolutely continuous functions on J. For f in $AC(J)$

$$|||f||| = |f(b)| + \int_a^b |f'(t)|\,dt.$$

Let $NBV(J)$ be the Banach subalgebra of $BV(J)$ consisting of those functions f in $BV(J)$ which are normalized by the requirement that f is continuous on the left on $(a, b]$.

Part 1

GENERAL SPECTRAL THEORY

1. General Spectral Theory

In this chapter, we develop the spectral theory of a bounded linear operator on a non-zero complex Banach space. The concepts of spectrum and resolvent set are introduced and the various subdivisions of the spectrum are studied. A functional calculus for such operators is introduced. The theory of ascent and descent of linear operators is developed. Various results on the spectra of restrictions of operators to closed invariant subspaces are proved.

Let $T \in L(X)$.

Definition 1.1. The *resolvent set* $\rho(T)$ *of* T is the set of complex numbers λ for which $\lambda I - T$ is invertible in the Banach algebra $L(X)$.

Definition 1.2. The *spectrum* $\sigma(T)$ *of* T is defined to be $\mathbf{C} \backslash \rho(T)$.

Definition 1.3. The function

$$\lambda \to (\lambda I - T)^{-1} \qquad (\lambda \in \rho(T))$$

is called the *resolvent* of T.

THEOREM 1.4. *Let $T \in L(X)$. The resolvent set $\rho(T)$ is open. Also the function $\lambda \to (\lambda I - T)^{-1}$ is analytic in $\rho(T)$.*

Proof. Let λ be a fixed point in $\rho(T)$ and let μ be any complex number with $|\mu| < \|(\lambda I - T)^{-1}\|^{-1}$. We show that $\lambda + \mu \in \rho(T)$. Consider the series $\sum_{k=0}^{\infty} (-\mu)^k (\lambda I - T)^{-(k+1)}$. Since $\|\mu(\lambda I - T)^{-1}\| < 1$ this series converges in the norm of $L(X)$. Denote its sum by $S(\mu)$. Then

$$[(\lambda + \mu) I - T] S(\mu) = (\lambda I - T) S(\mu) + \mu S(\mu) = I,$$

$$S(\mu)[(\lambda + \mu) I - T] = S(\mu)(\lambda I - T) + \mu S(\mu) = I.$$

It follows that $\lambda + \mu \in \rho(T)$ and the function $\mu \to S(\mu) = [(\lambda + \mu) I - T]^{-1}$ is analytic at the point $\mu = 0$.

COROLLARY 1.5. *Let $T \in L(X)$. If $d(\lambda)$ is the distance from λ to the spectrum $\sigma(T)$ then*

$$\|(\lambda I - T)^{-1}\| \geqslant \frac{1}{d(\lambda)} \qquad (\lambda \in \rho(T)).$$

Therefore $\|(\lambda I - T)^{-1}\| \to \infty$ as $d(\lambda) \to 0$, and the resolvent set is the natural domain of analyticity of the resolvent.

Proof. In the course of proving Theorem 1.4 it was shown that if $|\mu| < \|(\lambda I - T)^{-1}\|^{-1}$ then $\lambda + \mu \in \rho(T)$. Hence $d(\lambda) \geqslant \|(\lambda I - T)^{-1}\|^{-1}$, from which the statements follow.

THEOREM 1.6. *Let $T \in L(X)$. Then $\sigma(T)$ is compact and non-empty.*

Proof. For $|\lambda| > \|T\|$, the series $\sum_{n=0}^{\infty} T^n/\lambda^{n+1}$ converges in the norm of $L(X)$. Let $S(\lambda)$ denote its sum. Then

$$(\lambda I - T) S(\lambda) = S(\lambda)(\lambda I - T) = I.$$

Hence

$$S(\lambda) = (\lambda I - T)^{-1} \qquad (|\lambda| > \|T\|).$$

It follows that $\sigma(T)$ is bounded. By Theorem 1.4, $\sigma(T)$ is closed. Hence $\sigma(T)$ is compact. It remains to show that the spectrum is non-empty. If $\sigma(T) = \varnothing$, then the resolvent of T is an entire function. Since $(\lambda I - T)^{-1}$ is readily seen to be analytic at infinity, it follows from Liouville's theorem that $(\lambda I - T)^{-1}$

is a constant. Hence the coefficient of λ^{-1} in the Laurent series $\sum_{n=0}^{\infty} T^n/\lambda^{n+1}$ vanishes, so that $I = 0$, which contradicts the assumption $X \neq \{0\}$. This completes the proof.

Definition 1.7. Let $T \in L(X)$. The *spectral radius* $v(T)$ *of* T is defined by

$$v(T) = \sup\{|\lambda| : \lambda \in \sigma(T)\}.$$

PROPOSITION 1.8. *Let* $T \in L(X)$. *The spectral radius of* T *has the properties*

$$v(T) = \lim_{n \to \infty} \|T^n\|^{1/n} \leqslant \|T\|.$$

Proof. In the course of proving Theorem 1.6 it was shown that

$$(\lambda I - T)^{-1} = \sum_{n=0}^{\infty} T^n/\lambda^{n+1} \quad (1\lambda| > \|T\|)$$

and so $v(T) \leqslant \|T\|$. By Theorem 1.4, the resolvent is analytic in $\rho(T)$. Let $x \in X$, $y \in X^*$. Then the function $\lambda \to \langle(\lambda I - T)^{-1} x, y\rangle$ is analytic for $|\lambda| > v(T)$. Hence the singularities of this function all lie in the disc $\{\lambda : |\lambda| \leqslant v(T)\}$. Thus the series $\sum_{n=0}^{\infty} \langle \lambda^{-n-1} T^n x, y\rangle$ converges for $|\lambda| > v(T)$ and for such λ we have

$$\sup_n \left| \frac{\langle T^n x, y\rangle}{\lambda^{n+1}} \right| < \infty.$$

The principle of uniform boundedness shows that there is M_λ such that

$$\|T^n \lambda^{-n-1}\| \leqslant M_\lambda < \infty$$

and hence that

$$\limsup_n \|T^n\|^{1/n} \leqslant |\lambda|.$$

Since λ is an arbitrary number with $|\lambda| > v(T)$ it follows that

$$\limsup_n \|T^n\|^{1/n} \leqslant v(T).$$

To complete the proof we show that $\nu(T) \leqslant \liminf_n \|T^n\|^{1/n}$. Observe that if $\lambda \in \sigma(T)$ then $\lambda^n \in \sigma(T^n)$; for the factorization

$$(\lambda^n I - T^n) = (\lambda I - T)\,P(T) = P(T)(\lambda I - T)$$

shows that if $(\lambda^n I - T^n)$ has an inverse in $L(X)$ so will $\lambda I - T$. Thus $|\lambda|^n \leqslant \|T^n\|$, and hence

$$\sup\{|\lambda| : \lambda \in \sigma(T)\} \leqslant \|T^n\|^{1/n},$$

$$\nu(T) \leqslant \liminf_n \|T^n\|^{1/n}.$$

Definition 1.9. Let $T \in L(X)$. T is said to be *quasinilpotent* if and only if $\lim_{n\to\infty} \|T^n\|^{1/n} = 0$.

PROPOSITION 1.10. *Let $T \in L(X)$.*

(i) *T is quasinilpotent if and only if $\nu(T) = 0$.*

(ii) *T is quasinilpotent if and only if $\sigma(T) = \{0\}$.*

Proof. These results follow at once from Definition 1.7 and Proposition 1.8.

PROPOSITION 1.11. *Let $T \in L(X)$. The following identity, known as the resolvent equation, is valid for every pair of points λ, μ in $\rho(T)$.*

$$(\lambda I - T)^{-1} - (\mu I - T)^{-1} = (\mu - \lambda)(\lambda I - T)^{-1}(\mu I - T)^{-1}.$$

Proof. Observe that

$$(\mu I - T)(\lambda I - T)\{(\lambda I - T)^{-1} - (\mu I - T)^{-1}\} = (\mu I - T) - (\lambda I - T),$$

$$(\mu I - T)(\lambda I - T)\{(\lambda I - T)^{-1} - (\mu I - T)^{-1}\} = (\mu - \lambda)\,I.$$

Multiply both sides of this equation by $(\lambda I - T)^{-1}(\mu I - T)^{-1}$ to complete the proof.

Let $T \in L(X)$. There is an operator T^* in $L(X^*)$ called the *adjoint* of T such that

$$\langle Tx, y\rangle = \langle x, T^*y\rangle \qquad (x \in X,\ y \in X^*).$$

The map $T \to T^*$ is an isometric linear map of $L(X)$ into $L(X^*)$ with the additional property

$$(AB)^* = B^*A^* \qquad (A, B \in L(X)).$$

PROPOSITION 1.12. *Let $T \in L(X)$. The spectrum of T^* is equal to the spectrum of T. Moreover*

$$((\lambda I - T)^{-1})^* = (\lambda I^* - T^*)^{-1} \qquad (\lambda \in \rho(T)).$$

Proof. If T^{-1} exists and is in $L(X)$ then

$$T^*(T^{-1})^* = (T^{-1}T)^* = I^* = (TT^{-1})^* = (T^{-1})^*T^*.$$

Thus $(T^*)^{-1}$ exists, is in $L(X^*)$, and $(T^*)^{-1} = (T^{-1})^*$. Conversely, if $(T^*)^{-1}$ exists and is in $L(X^*)$ then, by what has already been proved, $(T^{**})^{-1}$ exists and is in $L(X^{**})$. Thus T^{**} is a homeomorphism of X^{**} onto itself. It is also an extension of T. Hence T is one-to-one and TX is closed. It only remains to show that $TX = X$. If $TX \neq X$, there is f in X^* with $f \neq 0$ and

$$\langle Tx, f \rangle = \langle x, T^*f \rangle = 0 \quad (x \in X).$$

Hence $T^*f = 0$, contradicting the assumption that T^* is one-to-one. The theorem follows easily.

We now introduce some important subsets of the spectrum.

Definition 1.13. Let $T \in L(X)$. Define

$$\sigma_p(T) = \{\lambda \in \mathbf{C}: \lambda I - T \text{ is not one-to-one}\};$$

$$\sigma_c(T) = \{\lambda \in \mathbf{C}: \lambda I - T \text{ is one-to-one}, \; \overline{(\lambda I - T)X} = X \text{ but } (\lambda I - T)X \neq X\};$$

$$\sigma_r(T) = \{\lambda \in \mathbf{C}: \lambda I - T \text{ is one-to-one but } \overline{(\lambda I - T)X} \neq X\}.$$

$\sigma_p(T)$, $\sigma_c(T)$ and $\sigma_r(T)$ are called respectively the *point spectrum*, the *continuous spectrum* and the *residual spectrum* of T. Clearly $\sigma_p(T)$, $\sigma_c(T)$ and $\sigma_r(T)$ are disjoint and

$$\sigma(T) = \sigma_p(T) \cup \sigma_c(T) \cup \sigma_r(T).$$

If $\lambda \in \sigma_p(T)$ we say that λ is an *eigenvalue* of T. If this is the case there is a non-zero vector x in X such that $Tx = \lambda x$. Such a vector is called an *eigenvector* corresponding to the eigenvalue λ of T.

PROPOSITION 1.14. *Let $T \in L(X)$. Then*

$$\sigma_r(T) \subseteq \sigma_p(T^*) \subseteq \sigma_r(T) \cup \sigma_p(T).$$

Proof. Let $\lambda \in \sigma_r(T)$. Then $\lambda I - T$ is one-to-one but

$$\overline{(\lambda I - T)X} \neq X.$$

Hence there is f in X^* with $f \neq 0$ such that

$$\langle(\lambda I - T)x, f\rangle = 0 \quad (x \in X);$$

$$\text{i.e.} \quad \langle x, (\lambda I^* - T^*)f\rangle = 0 \quad (x \in X).$$

Therefore $T^*f = \lambda f$ and so $\lambda \in \sigma_p(T^*)$.

Now let $\lambda \in \sigma_p(T^*)$. There is a non-zero f in X^* such that

$$\langle x, (\lambda I^* - T^*)f\rangle = 0 \quad (x \in X);$$

$$\text{i.e.} \quad \langle(\lambda I - T)x, f\rangle = 0 \quad (x \in X).$$

If $\lambda I - T$ is not one-to-one then $\lambda \in \sigma_p(T)$. If $\lambda I - T$ is one-to-one the last equation shows that $f(y) = 0$ for all y in $\overline{(\lambda I - T)X}$, and so $\overline{(\lambda I - T)X} \neq X$. In this case $\lambda \in \sigma_r(T)$. This completes the proof.

Another subset of the spectrum proves to be useful.

Definition 1.15. Let $T \in L(X)$. Define

$$\sigma_a(T) = \{\lambda \in \mathbf{C}: \text{there is a sequence } \{x_n\} \text{ in } X \text{ with } \|x_n\| = 1 \text{ and } \lim_{n \to \infty} \|(\lambda I - T)x_n\| = 0\}.$$

$\sigma_a(T)$ is called the *approximate point spectrum of* T.

The following result summarizes the main properties of the approximate point spectrum.

THEOREM 1.16. *Let $T \in L(X)$.*

(i) *The set $\sigma_a(T)$ is a closed non-empty subset of the spectrum of T.*
(ii) *The boundary of $\sigma(T)$ is contained in $\sigma_a(T)$.*
(iii) $\sigma_p(T) \subseteq \sigma_a(T)$.
(iv) $\sigma_c(T) \subseteq \sigma_a(T)$.

Proof. We show first that $\sigma_a(T) \subseteq \sigma(T)$. Suppose $\lambda \in \rho(T)$. Then $\lambda I - T$ has a bounded inverse and so

$$\begin{aligned}\|x\| &= \|(\lambda I - T)^{-1}(\lambda I - T)x\| \\ &\leqslant \|(\lambda I - T)^{-1}\|\,\|(\lambda I - T)x\| \quad (x \in X).\end{aligned}$$

This means

$$\|Tx - \lambda x\| \geqslant \varepsilon\|x\| \quad (x \in X)$$

where $\varepsilon = \|(\lambda I - T)^{-1}\|^{-1} > 0$, and so $\lambda \in \mathbf{C}\backslash\sigma_a(T)$. Hence $\sigma_a(T) \subseteq \sigma(T)$.

If $\lambda_0 \in \mathbf{C}\backslash\sigma_a(T)$, then we have shown that there is $\varepsilon > 0$ such that $\|(T - \lambda_0 I)x\| \geqslant \varepsilon$ for all unit vectors x in X. Consequently if $\|x\| = 1$ and $|\lambda - \lambda_0| < \varepsilon/2$ then

$$\|Tx - \lambda x\| \geqslant \|Tx - \lambda_0 x\| - |\lambda_0 - \lambda| \geqslant \frac{\varepsilon}{2}$$

so that $\lambda \in \mathbf{C}\backslash\sigma_a(T)$. Hence $\sigma_a(T)$ is closed.

Next, let λ_0 be a point on the boundary of $\sigma(T)$ and let $\varepsilon > 0$ be given. Then we can find λ in $\sigma(T)$ such that $|\lambda - \lambda_0| < \varepsilon/2$. By Corollary 1.5

$$\|(\lambda I - T)^{-1}\| \geqslant \frac{1}{d(\lambda)} \geqslant \frac{2}{\varepsilon}$$

where $d(\lambda)$ denotes the distance from λ to the spectrum of T. Hence we can choose x in X so that $\|x\| = 1$ and

$$\|(\lambda I - T)^{-1}x\| \geqslant \frac{1}{\varepsilon}.$$

Define $x_0 = \|(\lambda I - T)^{-1}x\|^{-1}(\lambda I - T)^{-1}x$. Then $\|x_0\| = 1$ and

$$\|(\lambda_0 I - T)x_0 - (\lambda I - T)x_0\| < \frac{\varepsilon}{2}.$$

Hence it follows that

$$\begin{aligned}\|(\lambda_0 I - T)x_0\| &\leqslant \|(\lambda_0 I - T)x_0 - (\lambda I - T)x\| + \|(\lambda I - T)x_0\| \\ &< \frac{\varepsilon}{2} + \frac{\|x\|}{\|(\lambda I - T)^{-1}x\|} < \frac{3\varepsilon}{2}.\end{aligned}$$

Since $\varepsilon > 0$ is arbitrary it follows that $\lambda_0 \in \sigma_a(T)$. Now, by Theorem 1.6, $\sigma(T)$ is non-empty. Since a non-empty compact subset of **C** has a non-empty boundary, $\sigma_a(T)$ is non-empty. The proof of (i) and (ii) is now complete.

If $\lambda \in \sigma_p(T)$ there is an x in X with $\|x\| = 1$ and $Tx = \lambda x$. Therefore $\sigma_p(T) \subseteq \sigma_a(T)$ and (iii) is proved.

Finally, suppose that $\lambda \in \sigma_c(T)$ but $\lambda \notin \sigma_a(T)$. Then there is some $\varepsilon > 0$ such that $\|(\lambda I - T)x\| \geqslant \varepsilon\|x\|$ for all x in X. We show that this implies that the range of $\lambda I - T$ is closed. Let

$$y_n = (\lambda I - T)x_n \quad (n = 1, 2, 3, \ldots)$$

where $\{y_n\}$ converges to y in X. Then $\{y_n\}$ is a Cauchy sequence and

$$\begin{aligned}\|y_n - y_m\| &= \|(\lambda I - T)(x_n - x_m)\| \\ &\geqslant \varepsilon\|x_n - x_m\|.\end{aligned}$$

It follows that $\{x_n\}$ is a Cauchy sequence in X. Since X is complete, there is x in X such that $\{x_n\}$ converges to x. The continuity of $\lambda I - T$ implies that $y = (\lambda I - T)x$ and so $y \in (\lambda I - T)X$. Since the range of $\lambda I - T$ is closed and dense in X it must be the whole of X. This contradicts the assumption that $\lambda \in \sigma_c(T)$. Therefore $\sigma_c(T) \subseteq \sigma_a(T)$ and the proof of the theorem is complete.

Later in this chapter we shall give an example to show that the approximate point spectrum is not in general equal to the spectrum.

Next, we introduce and develop the functional calculus for a bounded linear operator on a non-zero complex Banach space.

Definition 1.17. Let $T \in L(X)$. We denote by $\mathscr{F}(T)$ the family of all functions which are analytic on some neighbourhood of $\sigma(T)$. (The neighbourhood need not be connected, and can depend on the particular function in $\mathscr{F}(T)$.)

Definition 1.18. Let $T \in L(X)$, let $f \in \mathscr{F}(T)$, and let U be an open subset of **C** whose boundary B consists of a finite number of rectifiable Jordan curves. We assume throughout that B is oriented so that

$$\int_B (\lambda - \mu)^{-1}\, d\lambda = 2\pi i \quad (\mu \in U),$$

$$\int_B (\lambda - \mu)^{-1}\, d\lambda = 0 \qquad (\mu \notin U \cup B).$$

Suppose that $U \supseteq \sigma(T)$, and that $U \cup B$ is contained in the domain of analyticity of f. Then the operator $f(T)$ is defined by the equation

$$f(T) = \frac{1}{2\pi i} \int_B f(\lambda)\,(\lambda I - T)^{-1}\, d\lambda.$$

The integral exists as a limit of Riemann sums in the norm of $L(X)$. It follows from Theorem 1.4 and Cauchy's theorem that $f(T)$ depends only on the function f and not on the open set U chosen to define this operator.

THEOREM 1.19. *Let $T \in L(X)$. If f, g are in $\mathscr{F}(T)$ and $\alpha, \beta \in \mathbf{C}$, then*

(i) *$\alpha f + \beta g \in \mathscr{F}(T)$ and $(\alpha f + \beta g)(T) = \alpha f(T) + \beta g(T)$;*
(ii) *$f . g \in \mathscr{F}(T)$ and $(f . g)(T) = f(T)\, g(T)$;*
(iii) *if f has power series expansion $f(\lambda) = \sum_{k=0}^{\infty} \alpha_k \lambda^k$, valid in a neighbourhood of $\sigma(T)$, then $f(T) = \sum_{k=0}^{\infty} \alpha_k T^k$;*
(iv) *$f \in \mathscr{F}(T^*)$ and $f(T^*) = (f(T))^*$.*

Proof. Statement (i) is obvious. It is clear that $f . g \in \mathscr{F}(T)$; let U_1 and U_2 be two neighbourhoods of $\sigma(T)$ whose boundaries B_1 and B_2 consist of a finite number of rectifiable Jordan curves, and suppose that $U_1 \cup B_1 \subseteq U_2$. Suppose also that $U_2 \cup B_2$ is contained in a common region of analyticity of f and g. Then

$$f(T)\, g(T) = -\frac{1}{4\pi^2} \left\{ \int_{B_1} f(\lambda)(\lambda I - T)^{-1}\, d\lambda \right\} \left\{ \int_{B_2} g(\mu)(\mu I - T)^{-1}\, d\mu \right\}.$$

A standard argument invoking Fubini's theorem shows that we may rewrite this repeated integral as a double integral to get

$$f(T)\,g(T) = -\frac{1}{4\pi^2}\int_{B_1}\int_{B_2} f(\lambda)\,g(\mu)(\lambda I - T)^{-1}(\mu I - T)^{-1}\,d\mu d\lambda.$$

From the resolvent equation (Proposition 1.11) we obtain

$$f(T)\,g(T) = -\frac{1}{4\pi^2}\int_{B_1}\int_{B_2} \frac{f(\lambda)\,g(\mu)[(\lambda I - T)^{-1} - (\mu I - T)^{-1}]}{\mu - \lambda}\,d\mu d\lambda.$$

It now follows from the Cauchy integral formula that

$$f(T)\,g(T) = -\frac{1}{4\pi^2}\int_{B_1} f(\lambda)(\lambda I - T)^{-1}\left\{\int_{B_2}\frac{g(\mu)}{\mu - \lambda}\,d\mu\right\}d\lambda$$

$$+\frac{1}{4\pi^2}\int_{B_2} g(\mu)(\mu I - T)^{-1}\left\{\int_{B_1}\frac{f(\lambda)}{\mu - \lambda}\,d\lambda\right\}d\mu$$

$$= \frac{1}{2\pi i}\int_{B_1} f(\lambda)\,g(\lambda)(\lambda I - T)^{-1}\,d\lambda = (f\,.g)(T).$$

This proves (ii). To prove (iii), we note that the power series $\sum_{k=0}^{\infty} \alpha_k \lambda^k$ converges uniformly on the circle $C = \{\lambda : |\lambda| = \nu(T) + \varepsilon\}$ where $\varepsilon > 0$ is sufficiently small. Consequently

$$f(T) = \frac{1}{2\pi i}\int_C \left\{\sum_{k=0}^{\infty} \alpha_k \lambda^k\right\}(\lambda I - T)^{-1}\,d\lambda$$

$$= \frac{1}{2\pi i}\sum_{k=0}^{\infty} \alpha_k \int_C \lambda^k (\lambda I - T)^{-1}\,d\lambda$$

$$= \frac{1}{2\pi i}\sum_{k=0}^{\infty} \alpha_k \int_C \lambda^k \left\{\sum_{r=0}^{\infty} T^r \lambda^{-r-1}\right\}d\lambda$$

$$= \sum_{k=0}^{\infty} \alpha_k T^k$$

by Proposition 1.8 and the Cauchy integral formula. Statement (iv) follows immediately from Proposition 1.12.

THEOREM 1.20. (Spectral mapping theorem). *Let* $T \in L(X)$ *and let* $f \in \mathscr{F}(T)$. *Then* $f(\sigma(T)) = \sigma(f(T))$.

Proof. Let $\lambda \in \sigma(T)$. Define a function g with the same domain as f by

$$g(\lambda) = f'(\lambda), \qquad g(\xi) = \frac{f(\lambda) - f(\xi)}{\lambda - \xi} \quad (\xi \neq \lambda).$$

By Theorem 1.19(ii), $f(\lambda)I - f(T) = (\lambda I - T)\,g(T)$. Hence, if $f(\lambda)I - f(T)$ has an inverse A in $L(X)$ then $g(T)A$ would be an inverse for $\lambda I - T$ in $L(X)$. Consequently $f(\lambda) \in \sigma(f(T))$. Conversely, let $\mu \in \sigma(f(T))$ and suppose that $\mu \notin f(\sigma(T))$. Then the function h defined by

$$h(\xi) = (f(\xi) - \mu)^{-1}$$

belongs to $\mathscr{F}(T)$. By Theorem 1.19(ii)

$$h(T)[f(T) - \mu I] = [f(T) - \mu I]\,h(T) = I$$

which contradicts the assumption that $\mu \in \sigma(f(T))$. The proof is complete.

THEOREM 1.21. *Let* $T \in L(X)$, *let* $f \in \mathscr{F}(T)$, *let* $g \in \mathscr{F}(f(T))$ *and* $F = g \circ f$. *Then* $F \in \mathscr{F}(T)$ *and* $F(T) = g(f(T))$.

Proof. The statement $F \in \mathscr{F}(T)$ follows immediately from the spectral mapping theorem. Let U be a neighbourhood of $\sigma(f(T))$ whose boundary B consists of a finite number of rectifiable Jordan curves, and suppose that $U \cup B$ is contained in the domain of analyticity of g. Let V be a neighbourhood of $\sigma(T)$ whose boundary C consists of a finite number of rectifiable Jordan curves, and suppose that $V \cup C$ is contained in the domain of analyticity of f. Suppose, moreover, that $f(V \cap C) \subseteq U$. By Theorem 1.19, if $\lambda \in \rho(f(T))$, the operator

$$A(\lambda) = \frac{1}{2\pi i}\int_C \frac{(\xi I - T)^{-1}}{\lambda - f(\xi)}\,d\xi$$

satisfies the equations $[\lambda I - f(T)]\,A(\lambda) = A(\lambda)[\lambda I - f(T)] = I$. Hence

$A(\lambda) = (\lambda I - f(T))^{-1}$. Consequently,

$$g(f(T)) = \frac{1}{2\pi i}\int_B g(\lambda)(\lambda I - f(T))^{-1}\, d\lambda$$
$$= -\frac{1}{4\pi^2}\int_B g(\lambda)\left(\int_C \frac{(\xi I - T)^{-1}}{\lambda - f(\xi)}\, d\xi\right) d\lambda.$$

A standard argument invoking Fubini's theorem shows that we may rewrite this repeated integral as a double integral and then change the order of integration. We obtain

$$g(f(T)) = -\frac{1}{4\pi^2}\int_B\int_C \frac{g(\lambda)(\xi I - T)^{-1}}{\lambda - f(\xi)}\, d\xi d\lambda$$
$$= \frac{1}{2\pi i}\int_C (\xi I - T)^{-1}\, g(f(\xi))\, d\xi = F(T)$$

by Cauchy's integral formula.

THEOREM 1.22. *Let $T \in L(X)$. For each positive integer n let $f_n \in \mathscr{F}(T)$. Suppose that all the functions f_n are analytic in a fixed neighbourhood V of $\sigma(T)$. Then if $\{f_n\}$ converges uniformly to f on V, $f_n(T)$ converges to $f(T)$ in the norm of $L(X)$.*

Proof. Let U be a neighbourhood of $\sigma(T)$ whose boundary B consists of a finite number of rectifiable Jordan curves and such that $U \cup B \subseteq V$. Then $f_n \to f$ uniformly on B, and consequently

$$\frac{1}{2\pi i}\int_B f_n(\lambda)(\lambda I - T)^{-1}\, d\lambda \to \int_B f(\lambda)(\lambda I - T)^{-1}\, d\lambda$$

in the norm of $L(X)$. This completes the proof.

Next, we develop the properties of the exponential function of a bounded linear operator.

THEOREM 1.23. (i) *Let $A \in L(X)$. Then* $\exp A = \sum_{n=0}^{\infty} \frac{A^n}{n!}$.

(ii) *Let* $A, B \in L(X)$ *and* $AB = BA$. *Then*

$$\exp(A + B) = \exp A . \exp B.$$

(iii) *Let* $A \in L(X)$. *Then* $\exp A . \exp(-A) = I$.
(iv) *Let* $A \in L(X)$. *Then* $\exp A$ *is invertible in* $L(X)$.

Proof. Statement (i) follows immediately from Theorem 1.19(iii). Next, let $X_n, Y_n, Z_n, \xi_n, \eta_n, \zeta_n$ be defined by

$$X_n = I + \sum_{k=1}^{n} \frac{A^k}{k!}, \qquad Y_n = I + \sum_{k=1}^{n} \frac{B^k}{k!},$$

$$Z_n = I + \sum_{k=1}^{\infty} \frac{(A + B)^k}{k!}, \qquad \xi_n = 1 + \sum_{k=1}^{n} \frac{\|A\|^k}{k!},$$

$$\eta_n = 1 + \sum_{k=1}^{n} \frac{\|B\|^k}{k!}, \qquad \zeta_n = 1 + \sum_{k=1}^{n} \frac{1}{k!}(\|A\| + \|B\|)^k.$$

We have

$$X_n Y_n - Z_n = \sum_{j,k=1}^{n} \alpha_{jk} A^j B^k$$

where $\alpha_{jk} \geqslant 0$ for all j, k. Therefore

$$\|X_n Y_n - Z_n\| \leqslant \sum_{j,k=1}^{n} \alpha_{jk} \|A\|^j \|B\|^k = \xi_n \eta_n - \zeta_n.$$

However, $\xi_n \eta_n - \zeta_n \to \exp\|A\| \exp\|B\| - \exp(\|A\| + \|B\|) = 0$ as $n \to \infty$. This proves statement (ii). Statements (iii) and (iv) are then clear.

PROPOSITION 1.24. *Let* $A \in L(X)$. *Then* $\exp A = \lim_{n \to \infty} (I + A/n)^n$.

Proof. Let X_n, Y_n, ξ_n, η_n be defined by

$$X_n = I + \sum_{k=1}^{n} \frac{A^k}{k!}, \qquad Y_n = \left(I + \frac{A}{n}\right)^n,$$

$$\xi_n = 1 + \sum_{k=1}^{n} \frac{\|A\|^k}{k!}, \qquad \eta_n = \left(1 + \frac{\|A\|}{n}\right)^n.$$

We have

$$Y_n = I + \frac{1}{1!}A + \left(1 - \frac{1}{n}\right)\frac{1}{2!}A^2 + \dots$$
$$+ \left(1 - \frac{1}{n}\right)\left(1 - \frac{2}{n}\right)\dots\left(1 - \frac{n-1}{n}\right)\frac{1}{n!}A^n$$

and so

$$X_n - Y_n = \sum_{k=2}^{n} \alpha_k A^k,$$

where $\alpha_k \geqslant 0$ for $k = 2, \dots, n$. Therefore

$$\|X_n - Y_n\| \leqslant \sum_{k=2}^{n} \alpha_k \|A\|^k = \xi_n - \eta_n.$$

However, $\lim_{n\to\infty}(\xi_n - \eta_n) = 0$ since

$$\exp\|A\| = \lim_{n\to\infty}\left(1 + \frac{1}{n}\|A\|\right)^n$$

and this completes the proof.

The next main result (Theorem 1.26) gives an extension of Taylor's theorem to functions of an operator. A preliminary lemma is required.

LEMMA 1.25. *Let $T \in L(X)$. Let C be a set whose minimum distance from $\rho(T)$ is greater than some positive number ε. Then there is a real number K such that*

$$\|(\lambda I - T)^{-n}\| < K\varepsilon^{-n} \ (n = 0, 1, 2, \dots;\ \lambda \in C).$$

Proof. There is an open set U with the following four properties. (i) $\sigma(T) \subseteq U$. (ii) The boundary B of U consists of a finite number of rectifiable Jordan curves. (iii) For every λ in C and every ξ in $U \cup B$ we have $|\lambda - \xi| > \varepsilon$. (iv) For $n = 0, 1, 2, 3, \dots$ we have

$$(\lambda I - T)^{-n} = \frac{1}{2\pi i}\int_B (\lambda - \xi)^{-n}(\xi I - T)^{-1}\, d\xi.$$

Then there is a positive real number K such that for all λ in C

$$\|(\lambda I - T)^{-n}\| = \left\| \frac{1}{2\pi i} \int_B (\lambda - \xi)^{-n} (\xi I - T)^{-1} \, d\xi \right\|$$
$$\leqslant K\varepsilon^{-n} \ (n = 0, 1, 2, \ldots)$$

since the resolvent of T is bounded on the compact set B by Theorem 1.4.

THEOREM 1.26. *Let $S, N \in L(X)$ and let $SN = NS$. Let f be a function analytic in a connected open set including $\sigma(S)$ and every point within a distance from $\sigma(S)$ not greater than some positive number ε. Suppose also that $\sigma(N)$ lies within the circle centre 0 and radius ε. Then f is analytic on a neighbourhood of $\sigma(S + N)$ and*

$$f(S + N) = \sum_{n=0}^{\infty} \frac{f^{(n)}(S) \, N^n}{n!}$$

the series converging in the norm of $L(X)$.

Proof. Let $\delta = \sup\{|\lambda| : \lambda \in \sigma(N)\}$ so that by hypothesis $\delta < \varepsilon$. Choose $\theta < 1$ so that $\delta < \theta\varepsilon < \varepsilon$. Let B be the circle centre 0 and radius $\theta\varepsilon$ described once counterclockwise. Then

$$\|N^n\| = \left\| \frac{1}{2\pi i} \int_B \lambda^n (\lambda I - N)^{-1} \, d\lambda \right\|$$
$$\leqslant K(\theta\varepsilon)^{n+1} \ (n = 0, 1, 2, \ldots).$$

This inequality, together with Lemma 1.25, shows that the series $\sum_{n=0}^{\infty} (\lambda I - S)^{-n-1} N^n$ converges uniformly for λ in any set C whose minimum distance from $\sigma(S)$ is greater than ε. Let $V(\lambda)$ denote the sum of this series. Since S and N commute it is seen, by direct multiplication, that

$$V(\lambda)(\lambda I - S - N) = (\lambda I - S - N) \, V(\lambda) = I.$$

Thus if λ has distance greater than ε from $\sigma(S)$ then $\lambda \in \rho(S + N)$ and

$$(\lambda I - S - N)^{-1} = \sum_{n=0}^{\infty} (\lambda I - S)^{-n-1} N^n.$$

Hence f is analytic on a neighbourhood of $\sigma(S + N)$.

Now let C denote the union of a finite collection of rectifiable Jordan curves which bound a connected open set D containing every point whose distance from $\sigma(S)$ is less than ε and which lie together with D entirely in the domain of analyticity of f. Then

$$f(S+N) = \frac{1}{2\pi i}\int_C f(\lambda)(\lambda I - S - N)^{-1}\,d\lambda$$

$$= \frac{1}{2\pi i}\sum_{n=0}^{\infty}\int_C f(\lambda)(\lambda I - S)^{-n-1}N^n\,d\lambda.$$

On the other hand, we see from the resolvent equation

$$(\lambda_1 I - S)^{-1} - (\lambda_2 I - S)^{-1} = (\lambda_2 - \lambda_1)(\lambda_1 I - S)^{-1}(\lambda_2 I - S)^{-1}$$

that $(d/d\lambda)(\lambda I - S)^{-1} = -(\lambda I - S)^{-2}$ and inductively that

$$(d/d\lambda)^n(\lambda I - S)^{-1} = (-1)^n n!(\lambda I - S)^{-n-1}.$$

Hence

$$\int_C f(\lambda)(\lambda I - S)^{-n-1}\,d\lambda = \frac{(-1)^n}{n!}\int_C f(\lambda)\left(\frac{d}{d\lambda}\right)^n(\lambda I - S)^{-1}\,d\lambda.$$

Integrating by parts n times we find that

$$\int_C f(\lambda)(\lambda I - S)^{-n-1}\,d\lambda = \frac{1}{n!}\int_C\left\{\left(\frac{d}{d\lambda}\right)^n f(\lambda)\right\}(\lambda I - S)^{-1}\,d\lambda$$

so that

$$f(S+N) = \sum_{n=0}^{\infty}\left\{\frac{1}{2\pi i}\int_C f^{(n)}(\lambda)(\lambda I - S)^{-1}\,d\lambda\right\}\frac{N^n}{n!}$$

$$= \sum_{n=0}^{\infty}\frac{f^{(n)}(S)\,N^n}{n!}$$

and the proof is complete.

THEOREM 1.27. *Let $S, N \in L(X)$, where $SN = NS$ and N is quasinilpotent. Then $\sigma(S + N) = \sigma(S)$. If f is analytic on a neighbourhood of $\sigma(S)$ then*

$$f(S + N) = \sum_{n=0}^{\infty} \frac{f^{(n)}(S)\, N^n}{n!}.$$

Proof. By Proposition 1.10 (ii), $\sigma(N) = \{0\}$. Let $\lambda \in \rho(S)$. By Theorem 1.26 the function $(\lambda - z)^{-1}$ is analytic on some neighbourhood of $\sigma(S + N)$ and so $\lambda \in \rho(S + N)$. Hence $\sigma(S + N) \subseteq \sigma(S)$. Similarly,

$$\sigma(S) = \sigma(S + N - N) \subseteq \sigma(S + N).$$

It follows that $\sigma(S + N) = \sigma(S)$. The other statement of the theorem is immediate from Theorem 1.26.

Let Y be a closed subspace of X. Then Y is a complex Banach space under the norm of X. The *annihilator* $Y^{\perp}$ of Y is the closed subspace of X^* defined by

$$Y^{\perp} = \{f \in X^*: f(y) = 0 \text{ for all } y \text{ in } Y\}.$$

For a full discussion on annihilators the reader is referred to [**4**; p. 194]. The main property of $Y^{\perp}$ which we shall use is that $x \in Y$ if and only if $f(x) = 0$ for all f in $Y^{\perp}$.

Now let $T \in L(X)$ and let Y be a closed subspace of X. Y is said to be *invariant* under T if and only if $TY \subseteq Y$. If this is the case we can define an operator $T|Y$ in $L(Y)$ by

$$(T|Y)y = Ty \quad (y \in Y).$$

$T|Y$ is called the *restriction* of T to Y.

It is an open question whether every bounded linear operator on a separable infinite-dimensional complex Banach space E has a proper closed invariant subspace; that is, a closed invariant subspace other than the trivial ones $\{0\}$ and E. In fact the answer to this question is not known for any specific separable infinite-dimensional complex Banach space. The determination of the family of all closed invariant subspaces of a given operator is an important but in general a very difficult problem.

Next we prove some results on relationships between the spectrum of an operator and the spectrum of its restriction to a closed invariant subspace. A preliminary lemma is required.

LEMMA 1.28. *Let T be an invertible element of $L(X)$. Let Y be a closed subspace of X with $TY \subseteq Y$. Then $(T|Y)^{-1}$ exists as a bounded linear operator on Y if and only if $T^{-1}Y \subseteq Y$.*

Proof. If $T^{-1}Y \subseteq Y$, then clearly $T^{-1}|Y$ is bounded, linear and inverse to $T|Y$. Conversely, suppose that $(T|Y)^{-1} \in L(Y)$ and let $y_0 \in Y$. Then there is a unique element y in Y such that $Ty = y_0$. However, $T(T^{-1}y_0) = y_0$. Hence $T^{-1}y_0 = y \in Y$ and so $T^{-1}Y \subseteq Y$.

THEOREM 1.29. *Let $T \in L(X)$ and let Y be a closed subspace of X with $TY \subseteq Y$. Let D_∞ be the unbounded component of $\rho(T)$ and $D_1, D_2, D_3, \ldots$ the bounded components of $\rho(T)$. Then*

$$D_\infty \cap \sigma(T|Y) = \varnothing$$

and for each $n = 1, 2, 3, \ldots$

$$\begin{aligned} \textit{either} \quad & D_n \cap \sigma(T|Y) = D_n \\ \textit{or} \quad & D_n \cap \sigma(T|Y) = \varnothing. \end{aligned}$$

Proof. Suppose that $\lambda_0 \in \sigma(T|Y) \cap D_r$, for some r. Then by Lemma 1.28

$$(\lambda_0 I - T)^{-1} Y \nsubseteq Y.$$

Hence there is x in Y and y in $Y^\perp$ such that $\langle (\lambda_0 I - T)^{-1} x, y \rangle \neq 0$. The function $\lambda \to \langle (\lambda I - T)^{-1} x, y \rangle$ is analytic and not identically zero on D_r. Hence on D_r it is non-zero except on a discrete set. If $r = \infty$ this gives a contradiction since $\sigma(T|Y)$ is compact. In case $r = n$ is finite, D_n is contained in $\sigma(T|Y)$ except possibly for a discrete set. Now, $\sigma(T|Y)$ is closed and so $D_n \subseteq \sigma(T|Y)$. This completes the proof.

THEOREM 1.30. *Let $T \in L(X)$ and let Y be a closed subspace of X with $TY \subseteq Y$. Then*

(i) $\sigma_p(T|Y) \subseteq \sigma_p(T)$;
(ii) $\sigma_a(T|Y) \subseteq \sigma_a(T)$;
(iii) *if* $\lambda \in \sigma(T|Y) \cap \rho(T)$ *then* $\lambda \in \sigma_r(T|Y)$.

Proof. (i) and (ii) are obvious from the definitions of point spectrum and approximate point spectrum. By Theorem 1.16

$$\sigma_p(T \mid Y) \cup \sigma_c(T \mid Y) \subseteq \sigma_a(T \mid Y) \subseteq \sigma_a(T) \subseteq \sigma(T)$$

and so (iii) follows.

We are now in a position to give an example to show that in general the spectrum and approximate point spectrum need not be the same. A preliminary result is required.

PROPOSITION 1.31. *Let U be an invertible element of $L(X)$. Suppose that there is a real number M such that*

$$\|U^n\| \leqslant M < \infty \quad (n \in \mathbf{Z}).$$

Then $\sigma(U) \subseteq \{z : |z| = 1\}$.

Proof. Observe that if $|\lambda| > 1$ then $\sum_{n=0}^{\infty} \|U^n\| \, |\lambda|^{-n-1}$ converges and so $\sum_{n=0}^{\infty} U^n \lambda^{-n-1}$ converges in the norm of $L(X)$. If $S(\lambda)$ denotes the sum of the series then

$$S(\lambda)(\lambda I - U) = (\lambda I - U)\, S(\lambda) = I.$$

Also if $|\lambda| < 1$ then $\sum_{n=0}^{\infty} |\lambda|^n \|U^{-n-1}\|$ converges and so $-\sum_{n=0}^{\infty} \lambda^n U^{-n-1}$ converges in the norm of $L(X)$. If $S(\lambda)$ denotes the sum of the series then

$$S(\lambda)(\lambda I - U) = (\lambda I - U)\, S(\lambda) = I.$$

It follows that $\sigma(U) \subseteq \{z : |z| = 1\}$.

Example 1.32. Let H be a separable complex Hilbert space and let $\{\phi_n : n = 0, \pm 1, \pm 2, \ldots\}$ be an orthonormal basis for H. Define U in $L(H)$ by

$$U\left(\sum_{n=-\infty}^{\infty} c_n \phi_n\right) = \sum_{n=-\infty}^{\infty} c_n \phi_{n+1} \qquad \left(\sum_{n=-\infty}^{\infty} |c_n|^2 < \infty\right).$$

U is an invertible isometric operator and so by Proposition 1.31 we have

$$\sigma(U) \subseteq \{z : |z| = 1\} = \Gamma.$$

Define

$$Y = \text{clm}\{\phi_r : r = 1, 2, 3, \ldots\}.$$

Then $UY \subseteq Y$. However, $U^{-1}\phi_1 = \phi_0 \notin Y$ and so, by Lemma 1.28, $0 \in \sigma(U \mid Y)$. Suppose that $\sigma(U) \neq \Gamma$. Then $\rho(U)$ is connected, so by Theorem 1.29.

$$\sigma(U \mid Y) \subseteq \sigma(U) \subseteq \Gamma.$$

This contradicts $0 \in \sigma(U \mid Y)$. Hence

$$\sigma(U) = \{z : |z| = 1\}.$$

Also by Theorem 1.29

$$\sigma(U \mid Y) = \{z : |z| \leqslant 1\}$$

since $\sigma(U \mid Y)$ is closed. By Theorem 1.16

$$\sigma_a(U) = \{z : |z| = 1\},$$
$$\sigma_a(U \mid Y) \supseteq \{z : |z| = 1\}.$$

However, by Theorem 1.30,

$$\sigma_a(U \mid Y) \subseteq \sigma_a(U) = \{z : |z| = 1\}.$$

Hence $\sigma_a(U \mid Y) = \{z : |z| = 1\}$ and so $\sigma_a(U \mid Y) \neq \sigma(U \mid Y)$.

Let Y be a closed subspace of X. Introduce an equivalence relation on X by

$$x_1 \sim x_2 \Leftrightarrow x_1 - x_2 \in Y.$$

The set of equivalence classes of elements of X corresponding to this equivalence relation is a complex vector space under the operations defined by

$$[x_1]_Y + [x_2]_Y = [x_1 + x_2]_Y$$
$$\alpha[x]_Y = [\alpha x]_Y \qquad (\alpha \in \mathbf{C}).$$

This vector space is called the *quotient space of X modulo Y* and is denoted by X/Y. Define

$$\|[x]_Y\| = \inf\{\|x + y\| : y \in Y\}.$$

This is indeed a norm on X/Y and moreover X/Y is a complex Banach space under this norm. The mapping ϕ defined by

$$\phi(x) = [x]_Y$$

is called the *canonical mapping* of X onto X/Y. ϕ is continuous, linear and $\|\phi\| \leqslant 1$. For a complete discussion and proofs of these facts the reader is referred to [**4**; p. 99–101].

Now let $T \in L(X)$ and let Y be a closed subspace of X invariant under T. The map

$$T_Y[x]_Y = [Tx]_Y$$

is well-defined. Moreover $T_Y \in L(X/Y)$, since it is the composition $\phi \circ T$ of two continuous linear maps, and $\|T_Y\| \leqslant \|T\|$.

We require the following representation theorem for the dual space of X/Y.

PROPOSITION 1.33. *Let Y be a closed subspace of X. Then there is a linear isometry J_1 of $(X/Y)^*$ onto Y which is given by*

$$\langle x, J_1 z\rangle = \langle [x]_Y, z\rangle$$

for all z in $(X/Y)^*$ and all x in X.

For a proof of this result, see Theorem 5.4.5 of [**4**; p. 196].

Let $T \in L(X)$ and let Y be a closed subspace of X invariant under T. The equation

$$\langle Tx, z\rangle = \langle x, T^*z\rangle \qquad (x \in X, z \in X^*)$$

shows that $T^*Y^\perp \subseteq Y^\perp$. In view of this and Proposition 1.33 we may and shall identify T_Y^* and $T^*|Y^\perp$.

Also we require the following representation theorem for the dual space of Y.

PROPOSITION 1.34. *Let Y be a closed subspace of X. Then there is a linear isometry J_2 of $X^*/Y^\perp$ onto Y^* which is given by*

$$\langle y, J_2[z]_{Y^\perp}\rangle = \langle y, z\rangle$$

for all z in X^* and all y in Y.

For a proof of this result, see Theorem 5.4.4 of [**4**; p. 195].

Let $T \in L(X)$ and let Y be a closed subspace of X invariant under T. In view of Proposition 1.34 we may and shall identify $(T|Y)^*$ and $T^*_{Y^\perp}$.

Next we prove the analogue of Theorem 1.29 for operators induced on quotient spaces. A preliminary lemma is required.

LEMMA 1.35. *Let T be an invertible element of $L(X)$. Let Y be a closed subspace of X invariant under T. Then T_Y has a bounded inverse if and only if $T^{-1}Y \subseteq Y$.*

Proof. Suppose that $T^{-1}Y \subseteq Y$. Then $(T^{-1})_Y$ is a bounded linear operator on X/Y and clearly inverse to T_Y. Conversely, suppose that T_Y has a bounded inverse. Let $y \in Y$ and let $[x]$ denote the image of x under the canonical mapping of X onto X/Y. Then, since $T^{-1} \in L(X)$

$$T_Y[T^{-1}y] = [y] = [0]$$

and therefore

$$T_Y^{-1}T_Y[T^{-1}y] = [T^{-1}y] = [0]$$

so that $T^{-1}y \in Y$, proving the lemma.

THEOREM 1.36. *Let $T \in L(X)$ and let Y be a closed subspace of X invariant under T.*

(i) *Let $\lambda \in \rho(T)$. Then $\lambda \in \rho(T_Y)$ if and only if $(\lambda I - T)^{-1}Y \subseteq Y$.*

(ii) $\sigma(T|Y) \cap \rho(T) = \sigma(T_Y) \cap \rho(T)$.

(iii) *If D_∞ denotes the unbounded component of $\rho(T)$ and $D_1, D_2, D_3, \ldots$ the bounded components of $\rho(T)$ then*

$$D_\infty \cap \sigma(T_Y) = \varnothing$$

and for each $n = 1, 2, 3, \ldots$

$$\textit{either} \quad D_n \cap \sigma(T_Y) = D_n$$

$$\textit{or} \quad D_n \cap \sigma(T_Y) = \varnothing.$$

Proof. (i) follows at once from the lemma. (ii) follows from (i) and the corresponding result for restrictions, Lemma 1.28. (iii) follows from (ii) and the corresponding result for restrictions, Theorem 1.29.

Let $E \in L(X)$. E is called a *projection* if and only if $E^2 = E$. If E is a projection there are closed subspaces X_1 and X_2 of X such that

(i) X_1 is the range of E
(ii) X_2 is the null-space of E
(iii) $X = X_1 \oplus X_2$.

Conversely, let X_1 and X_2 be closed subspaces of X such that

$$X = X_1 \oplus X_2.$$

Then there is a projection E in $L(X)$ whose range is X_1 and whose null-space is X_2. Moreover E is uniquely determined by these conditions. For a full discussion and proofs of these facts the reader is referred to [**4**; p. 336–40].

Now let $T, E \in L(X)$ and let E be a projection. Suppose that T leaves invariant the range of E. This is equivalent to

$$ETEx = TEx \quad (x \in X)$$

and hence to $ETE = TE$.

Let $T \in L(X)$. Closed subspaces X_1 and X_2 of X are said to *reduce* T or to be *reducing subspaces for* T if $X = X_1 \oplus X_2$ and X_1, X_2 are invariant under T. Let E be the unique projection whose range is X_1 and whose null-space is X_2. Then the condition that X_1, X_2 reduce T is by the preceding paragraph equivalent to $ETE = TE$ *and*

$$(I - E)\,T(I - E) = T(I - E).$$

These conditions are equivalent to $TE = ET$.

PROPOSITION 1.37. *Let* $T \in L(X)$. *Suppose that the closed subspaces* X_1, X_2 *of* X *reduce* T. *Then* $\sigma(T) = \sigma(T|X_1) \cup \sigma(T|X_2)$.

Proof. Let E be the unique projection whose range space is X_1 and whose null-space is X_2. Then $TE = ET$. Let $\lambda \in \rho(T)$. Then $(\lambda I - T)E = E(\lambda I - T)$ and so

$$E(\lambda I - T)^{-1} = (\lambda I - T)^{-1} E.$$

Hence $(\lambda I - T)^{-1}$ leaves X_1 invariant. Its restriction to this subspace is clearly a bounded linear operator inverse to $(\lambda I - T)|X_1$. Thus $\lambda \in \rho(T|X_1)$. Hence

$$\rho(T) \subseteq \rho(T|X_1), \qquad \sigma(T|X_1) \subseteq \sigma(T).$$

Similarly $\sigma(T|X_2) \subseteq \sigma(T)$. Hence $\sigma(T|X_1) \cup \sigma(T|X_2) \subseteq \sigma(T)$.

Now let $\lambda \in \rho(T|X_1) \cap \rho(T|X_2)$. Then

$$(\lambda I|X_1 - T|X_1)^{-1} \oplus (\lambda I|X_2 - T|X_2)^{-1} \in L(X)$$

and is clearly inverse to $\lambda I - T$. Hence

$$\rho(T|X_1) \cap \rho(T|X_2) \subseteq \rho(T),$$
$$\sigma(T|X_1) \cup \sigma(T|X_2) \supseteq \sigma(T).$$

This completes the proof.

The problem of the existence of proper closed reducing subspaces for a bounded linear operator is much easier than the corresponding question for invariant subspaces. We now give an example of an operator which has no proper closed reducing subspaces.

Example 1.38. Let Γ be the unit circle $\{z : |z| = 1\}$ in $\mathbf{C}$ and let μ be Lebesgue measure on the σ-algebra of Borel subsets of Γ, normalized so that $\mu(\Gamma) = 1$. Define

$$e_n(z) = z^n \quad (z \in \Gamma,\ n \in \mathbf{Z}).$$

The set $\{e_n : n \in \mathbf{Z}\}$ forms an orthonormal basis for the complex Hilbert space $L^2(\mu)$. The space $H^2(\mu)$ is the closed subspace $\text{clm}\{e_n : n = 0, 1, 2, \ldots\}$ of $L^2(\mu)$. The *unilateral shift* operator U on the Hilbert space $H^2(\mu)$ is defined by

$$U\left(\sum_{n=0}^{\infty} c_n e_n\right) = \sum_{n=0}^{\infty} c_n e_{n+1} \qquad \left(\sum_{n=0}^{\infty} |c_n|^2 < \infty\right).$$

Now let A be a bounded linear operator on $H^2(\mu)$ that commutes with U. Let $\phi = Ae_0$. Since for each $n = 0, 1, 2, 3, \ldots$ multiplication by e_n leaves $H^2(\mu)$ invariant it follows that $\phi . e_n \in H^2(\mu)$. Since, moreover,

$$\phi . e_n = e_n . \phi = U^n \phi = U^n A e_0 = A U^n e_0 = A e_n$$

it follows that, for each polynomial p, the product $\phi . p \in H^2(\mu)$ and $\phi . p = Ap$. If $f \in H^2(\mu)$, there is a sequence $\{p_n\}$ of polynomials such that $\{p_n\}$ converges to f in the norm of $H^2\{\mu\}$. It follows that $Ap_n \to Af$ in $H_2(\mu)$. There is no loss of generality in assuming that $p_n \to f(\text{a.e.}\mu)$ and $Ap_n \to Af(\text{a.e.}\mu)$. If this were not true for the sequence $\{p_n\}$ it is true for a suitable subsequence. Since $p_n \to f(\text{a.e.}\mu)$ it follows that $\phi . q_n \to \phi . f(\text{a.e.}\mu)$; since at the same time $\phi . p_n \to Af(\text{a.e.}\mu)$ it follows that $\phi . f = Af(\text{a.e.}\mu)$.

The assertion that U has no proper closed reducing subspaces is equivalent to this: if E is a projection that commutes with U then $E = 0$ or $E = I$. The result proved above shows that if E is a projection with $EU = UE$ then the multiplier ϕ representing E satisfies $\phi^2 = \phi$, since $E^2 e_0 = Ee_0$. Hence ϕ must be the characteristic function of a Borel subset of Γ. Therefore ϕ is real. We complete the proof by showing that this implies that ϕ is constant. Now since $\phi = Ee_0 \in H^2(\mu)$

$$\alpha_n = \int_\Gamma \phi(z)\, e_n(z)\, d\mu = 0 \qquad (n = -1, -2, -3, \ldots).$$

Also for each positive integer n

$$\alpha_{-n} = \int_\Gamma \phi(z)\, e_{-n}(z)\, d\mu = \int_\Gamma \overline{\phi(z)}\, \overline{e_n(z)}\, d\mu = \bar{\alpha}_n$$

and so

$$\alpha_n = \int_\Gamma \phi(z)\, e_n(z)\, d\mu = 0 \quad (n = 1, 2, 3, \ldots).$$

By considering the Fourier expansion of ϕ in $L^2(\mu)$ we see that ϕ is constant.

The argument of Example 1.32 shows that the spectrum of the unilateral shift operator is the closed unit disc, a connected set. We now show that if the spectrum of an operator is disconnected there exist a pair of proper closed subspaces which reduce the operator.

Let $T \in L(X)$ and let τ be an open-and-closed subset of $\sigma(T)$. There is a function f in $\mathscr{F}(T)$ which is identically one on τ and which vanishes on the rest of $\sigma(T)$. We put $E(\tau; T) = f(T)$. If the operator T is understood we may write $E(\tau; T)$ simply as $E(\tau)$. It is clear from Cauchy's theorem that $E(\tau)$ depends only on τ and not on the particular f in $\mathscr{F}(T)$ chosen to define it. $E(\tau)$ is called the *spectral projection corresponding to* τ. If the open-and-closed set τ consists of the single point λ, the symbol $E(\lambda)$ will be used instead of $E(\{\lambda\})$. It will be convenient also to use the symbol $E(\tau)$ for any set τ of complex numbers for which $\tau \cap \sigma(T)$ is an open-and-closed subset of $\sigma(T)$. In this case we put

$$E(\tau) = E(\tau \cap \sigma(T)).$$

Thus $E(\tau) = 0$ if $\tau \cap \sigma(T)$ is void.

THEOREM 1.39. *Let $T \in L(X)$. Let Σ_0 denote the Boolean algebra of open-and-closed subsets of $\sigma(T)$. If $\tau \in \Sigma_0$, $E(\tau)$ is a projection, $TE(\tau) = E(\tau)\,T$ and $\sigma(T \mid E(\tau)\,X) = \tau$. The map $\tau \to E(\tau)$ is an isomorphism of Σ_0 onto a Boolean algebra of projections in $L(X)$. This means that $E(\varnothing) = 0$ and*

$$E(\sigma(T)\backslash\tau) = I - E(\tau) \qquad (\tau \in \Sigma_0),$$

$$R(\tau_1 \cup \tau_2) = E(\tau_1) + E(\tau_2) - E(\tau_1)\,E(\tau_2),$$

$$E(\tau_1 \cap \tau_2) = E(\tau_1)\,E(\tau_2) \qquad (\tau_1, \tau_2 \in \Sigma_0).$$

Proof. That $E(\tau)$ is a projection follows from Theorem 1.19(ii). Also

$$T(\lambda I - T)^{-1} = (\lambda I - T)^{-1}\,T \quad (\lambda \in \rho(T))$$

and so $TE(\tau) = E(\tau)\,T$. Hence $TE(\tau)\,X \subseteq E(\tau)\,X$. Suppose that $\lambda \in \mathbf{C}\backslash\tau$. Define h to be equal to $(\lambda - \mu)^{-1}$ for μ in a neighbourhood of τ not containing λ, and to be identically zero on a neighbourhood of $\sigma(T)\backslash\tau$. We have $h \in \mathscr{F}(T)$ and

$$h(T)(\lambda I - T) = (\lambda I - T)\,h(T) = E(\tau)$$

$$h(T)\,E(\tau) = E(\tau)\,h(T)$$

by Theorem 1.19(ii). It follows that $(\lambda I - T)|E(\tau)\,X$ is invertible in $L(E(\tau)\,X)$ with inverse $h(T)|E(\tau)X$. Hence $\lambda \notin \sigma(T \mid E(\tau)X)$ and so $\sigma(T \mid E(\tau)X) \subseteq \tau$. By

Theorem 1.19(i), $E(\sigma(T)\backslash\tau) = I - E(\tau)$. We can show by a similar argument that $\sigma(T \mid E(\sigma(T)\backslash\tau)X) \subseteq \sigma(T)\backslash\tau$. However, by Proposition 1.37

$$\sigma(T) = \sigma(T \mid E(\tau)X) \cup \sigma(T \mid E(\sigma(T)\backslash\tau)X).$$

It follows that $\sigma(T \mid E(\tau)X) = \tau$.

By Theorem 1.19 the map $\tau \to E(\tau)$ is a homomorphism. To verify that it is an isomorphism it will suffice to show that $E(\tau) = 0$ only when τ is empty. Now if $E(\tau) = 0$ then $E(\tau)X = \{0\}$ and so $\sigma(T \mid E(\tau)X) = \varnothing$. It follows that $\tau = \sigma(T \mid E(\tau)X) = \varnothing$. The proof is complete.

Next we discuss in detail the case of an isolated point of the spectrum. Observe that such a point is an open-and-closed subset of the spectrum.

Let $T \in L(X)$. The map $\lambda \to (\lambda I - T)^{-1}$ is analytic on $\rho(T)$, and an isolated point λ_0 of $\sigma(T)$ is an isolated singular point of the resolvent of T. Hence there is a Laurent expansion of this function in powers of $\lambda - \lambda_0$. We write this in the form

$$(\lambda I - T)^{-1} = \sum_{n=0}^{\infty} (\lambda - \lambda_0)^n A_n + \sum_{n=1}^{\infty} (\lambda - \lambda_0)^{-n} B_n.$$

The coefficients A_n and B_n are members of $L(X)$, and this series representation of the resolvent of T is valid when $0 < |\lambda - \lambda_0| < \delta$ for any δ such that all of $\sigma(T)$ except λ_0 lies on or outside the circle $|\lambda - \lambda_0| = \delta$. These coefficients are given by the standard formulae

$$A_n = \frac{1}{2\pi i} \int_\Gamma (\lambda - \lambda_0)^{-n-1} (\lambda I - T)^{-1}\, d\lambda \tag{1}$$

$$B_n = \frac{1}{2\pi i} \int_\Gamma (\lambda - \lambda_0)^{n-1} (\lambda I - T)^{-1}\, d\lambda \tag{2}$$

where Γ is any circle $|\lambda - \lambda_0| = \rho$ with $0 < \rho < \delta$ described once counter-clockwise.

It turns out that there are several important relationships among these coefficients. The function f_n defined by

$$f_n(\lambda) = \begin{cases} (\lambda - \lambda_0)^{n-1} & \text{if } |\lambda - \lambda_0| \leqslant \rho < \delta \\ 0 & \text{otherwise} \end{cases}$$

is in $\mathscr{F}(T)$ and moreover

$$B_n = f_n(T) \quad (n = 1, 2, 3, \ldots).$$

For each positive integer n we have $(\lambda - \lambda_0) f_n(\lambda) = f_{n+1}(\lambda)$, so by Theorem 1.19(ii)

$$(T - \lambda_0 I) B_n = B_{n+1} \tag{3}$$

and by induction

$$(T - \lambda_0 I)^n B_1 = B_{n+1}. \tag{4}$$

We note in passing that

$$B_1 = E(\lambda_0) \tag{5}$$

the spectral projection corresponding to the open-and-closed subset $\{\lambda_0\}$ of $\sigma(T)$.

Consider for each non-negative integer n the function g_n defined by

$$g_n(\lambda) = \begin{cases} 0 & \text{if } |\lambda - \lambda_0| \leqslant \rho < \delta \\ (\lambda - \lambda_0)^{-n-1} & \text{otherwise.} \end{cases}$$

$g_n \in \mathscr{F}(T)$ and moreover

$$g_n(T) = \frac{1}{2\pi i}\left(\int_{\Gamma_r} - \int_{\Gamma_0}\right)((\lambda - \lambda_0)^{-n-1}(\lambda I - T)^{-1})\, d\lambda$$

where Γ_r is the circle $|\lambda - \lambda_0| = r$ described once counterclockwise, r being chosen to be so large that this circle lies entirely in $\rho(T)$, and Γ_0 is the circle $|\lambda - \lambda_0| = \rho'$ (where $0 < \rho < \rho' < \delta$) described once counterclockwise. Letting $r \to \infty$ and noting that $\|(\lambda I - T)^{-1}\| \to 0$ on Γ_r we obtain

$$A_n = -g_n(T).$$

For each non-negative integer n we have $(\lambda - \lambda_0)\, g_{n+1}(\lambda) = g_n(\lambda)$ and so by Theorem 1.19(ii)

$$(T - \lambda_0 I) A_{n+1} = A_n. \tag{6}$$

Similarly $(\lambda - \lambda_0) g_0(\lambda) + f_1(\lambda) = 1$ and so

$$(T - \lambda_0 I) A_0 = B_1 - I. \tag{7}$$

As in the classical theory of functions, λ_0 is said to be a *pole of the resolvent of T of order m* if and only if $B_m \neq 0$ and $B_n = 0$ when $n > m$. From (3) we see that $B_{n+1} = 0$ if $B_n = 0$. Hence λ_0 is a pole of order m if and only if $B_m \neq 0$ and $B_{m+1} = 0$. If this is the case $B_1, \ldots, B_m$ are all non-zero. The case of a pole is of particular interest and so we state explicitly the following necessary and sufficient condition for an isolated point of the spectrum of T to be a pole.

PROPOSITION 1.40. *Let $T \in L(X)$ and let λ_0 be an isolated point of $\sigma(T)$. Then λ_0 is a pole of order m of the resolvent of T if and only if*

$$(\lambda_0 I - T)^m E(\lambda_0) = 0, \qquad (\lambda_0 I - T)^{m-1} E(\lambda_0) \neq 0.$$

Proof. This follows from the formula $B_1 = E(\lambda_0)$, equation (4) and the preceding discussion.

The next result is deeper. It is the analogue for the infinite-dimensional case of the minimal polynomial theorem in the finite-dimensional case.

THEOREM 1.41. (Minimal equation theorem). *Let $T \in L(X)$ and $f \in \mathscr{F}(T)$. Then $f(T) = 0$ if and only if the following condition is satisfied. For every point λ of $\sigma(T)$*

either (i) *f is identically 0 on a neighbourhood of λ*
or (ii) *λ is a pole of order $v(\lambda)$ of the resolvent of T and f has a zero of order at least $v(\lambda)$ at the point λ;*

moreover there are only a finite number of points satisfying (ii).

Proof. We show first that the condition is sufficient. Let f vanish identically on an open set containing all of $\sigma(T)$ but the poles. Then it is clear from Definition 1.18 that

$$f(T) = \sum_{r=1}^{k} \frac{1}{2\pi i} \int_{C_r} f(\lambda)(\lambda I - T)^{-1}\, d\lambda$$

where C_r is a sufficiently small circle enclosing λ_r and described once counter-clockwise. If f has a zero of order at least $v(\lambda_i)$ at λ_i then, since the resolvent of T has a pole of order $v(\lambda_i)$ at λ_i, it follows that the function $\lambda \to f(\lambda)(\lambda I - T)^{-1}$ is analytic on and inside C_i. By Cauchy's theorem $f(T) = 0$.

Conversely, let $f(T) = 0$. Then, by the spectral mapping theorem, $f(\sigma(T)) = 0$. Let f be analytic on a neighbourhood U of $\sigma(T)$. For each α in $\sigma(T)$ there is an $\varepsilon(\alpha) > 0$ such that the open ball $B(\alpha, \varepsilon(\alpha)) \subseteq U$. Since $\sigma(T)$ is compact, a finite subfamily of these open balls $B(\alpha_1, \varepsilon(\alpha_1)), \ldots, B(\alpha_n, \varepsilon(\alpha_n))$ cover $\sigma(T)$. If some ball $B(\alpha_r, \varepsilon(\alpha_r))$ contains an infinite number of points of $\sigma(T)$, it follows from the identity theorem that f vanishes identically on $B(\alpha_r, \varepsilon(\alpha_r))$. Thus, if U_1 is the union of those balls $B(\alpha_i, \varepsilon(\alpha_i))$ which contain an infinite number of points of $\sigma(T)$ then f vanishes identically on U_1. Hence, U_1 contains all but a finite number of isolated points of $\sigma(T)$, which we suppose to be the points $\lambda_1, \ldots, \lambda_k$. Suppose that f does not vanish identically in any neighbourhood of λ_r. Then, since $f(\sigma(T)) = 0$, f has a zero of finite order n at λ_r. Consequently, the function g_r defined by

$$g_r(\xi) = \frac{(\lambda_r - \xi)^n}{f(\xi)}$$

is analytic in a neighbourhood of λ_r. Let e be a function identically one in a neighbourhood of λ_r and identically zero in a neighbourhood of every other point of $\sigma(T)$. Also let $g = g_r e$. Then

$$(\lambda_r I - T)^n e(T) = f(T)\, g(T) = 0.$$

The Laurent expansion of the resolvent of T in a neighbourhood $\{\xi : 0 < |\xi - \lambda_r| < \varepsilon\}$ of λ_r is given by

$$(\xi I - T)^{-1} = \sum_{n=0}^{\infty} (\xi - \lambda_r)^n A_n + \sum_{n=1}^{\infty} (\xi - \lambda_r)^{-n} B_n$$

where for $m = 1, 2, 3, \ldots$

$$B_m = \frac{1}{2\pi i} \int_\Gamma (\lambda - \lambda_r)^{m-1} (\lambda I - T)^{-1}\, d\lambda$$

and where Γ is a circle centre λ_r, radius less than ε, described once counter-clockwise. Hence

$$B_{m+1} = (T - \lambda_r I)^m e(T) = 0 \quad (m \geqslant n),$$

and therefore λ_r is a pole of order at most n. We see that either f vanishes identically in a neighbourhood of λ_r, or λ_r is a pole of order $\nu(\lambda_r)$ of the resolvent of T and f has a zero of order at least $\nu(\lambda_r)$ at λ_r. This completes the proof.

COROLLARY 1.42. *Let $T \in L(X)$ and $f \in \mathscr{F}(T)$. If $f(T) = 0$ there is a non-trivial polynomial p such that $p(T) = 0$.*

Proof. This follows immediately from the minimal equation theorem.

We now develop the theory of ascent and descent of linear operators on a non-zero complex Banach space.

Let $T \in L(X)$. Define

$$N(T) = \{x \in X; Tx = 0\},$$

$$R(T) = TX.$$

$N(T)$ and $R(T)$ are called respectively the *null-space* and *range* of T.

PROPOSITION 1.43. *Let $T \in L(X)$. Then*

(i) $N(T^n) \subseteq N(T^{n+1}) \quad (n = 0, 1, 2, \ldots)$;
(ii) *if* $N(T^k) = N(T^{k+1})$ *for some positive integer* k *then* $N(T^n) = N(T^k)$ $(n \geqslant k)$.

Proof. (i) This result is trivial.

(ii) Suppose that $N(T^k) = N(T^{k+1})$ and $x \in N(T^{k+2})$. Then $Tx \in N(T^{k+1})$ and so $Tx \in N(T^k)$. Hence $T^{k+1}x = 0$ and so $x \in N(T^{k+1})$. From this and (i) we obtain $N(T^{k+2}) = N(T^{k+1})$. The desired result now follows by induction.

Definition 1.44. Let $T \in L(X)$. Suppose there is a positive integer n such that $N(T^n) = N(T^{n+1})$. The smallest such integer is called the *ascent* of T and is denoted by $\alpha(T)$. If no such integer exists we put $\alpha(T) = \infty$.

PROPOSITION 1.45. *Let $T \in L(X)$. Then*

(i) $R(T^{n+1}) \subseteq R(T^n)$ $(n = 0, 1, 2, \ldots)$;
(ii) *if* $R(T^k) = R(T^{k+1})$ *for some positive integer* k *then* $R(T^n) = R(T^k)$ $(n \geqslant k)$.

Proof. (i) This result is trivial.

(ii) Suppose that $R(T^{k+1}) = R(T^k)$ and $y \in R(T^{k+1})$. Then $y = Tx$, where

$x \in R(T^k) = R(T^{k+1})$. Therefore $y \in R(T^{k+2})$. From this and (i) we obtain $R(T^{k+2}) = R(T^{k+1})$. The desired result now follows by induction.

Definition 1.46. Let $T \in L(X)$. Suppose there is a positive integer n such that $R(T^n) = R(T^{n+1})$. The smallest such integer is called the *descent* of T and is denoted by $\delta(T)$. If no such integer exists we put $\delta(T) = \infty$.

PROPOSITION 1.47. *Let $T \in L(X)$. If $\alpha(T) < \infty$ and $\delta(T) = 0$ then $\alpha(T) = 0$.*

Proof. Suppose that $\alpha(T) > 0$ and $\delta(T) = 0$. Then $R(T) = X$ and T is not one-to-one. Choose $x_1 \neq 0$ such that $Tx_1 = 0$. Define inductively a sequence $\{x_n\}$ in X so that $Tx_{n+1} = x_n$ for each positive integer n. Then

$$T^n x_{n+1} = x_1, \quad T^{n+1} x_{n+1} = 0.$$

Hence $x_{n+1} \in N(T^{n+1}) \backslash N(T^n)$. Therefore $\alpha(T) = \infty$. This contradiction suffices to complete the proof.

Observe that Propositions 1.43, 1.45 remain valid and Definitions 1.44, 1.46 are meaningful in the case where X is a normed complex vector space and T is a linear mapping with domain X and range contained in X. We shall use this fact without further comment in the course of proving the following result.

PROPOSITION 1.48. *Let $T \in L(X)$. If $\alpha(T)$ and $\delta(T)$ are both finite then $\alpha(T) \leqslant \delta(T)$.*

Proof. Let $p = \delta(T)$. Then $R(T^p) = TR(T^p)$. Let T_1 denote the restriction of T to $R(T^p)$. Then $\delta(T_1) = 0$. Clearly, for each positive integer n, T_1^n is the restriction of T^n to $R(T^p)$ and $N(T_1^n) \subseteq N(T^n)$. Hence

$$N(T_1^{n+1}) \backslash N(T_1^n) \subseteq N(T^{n+1}) \backslash N(T^n)$$

and, since $\alpha(T) < \infty$, it follows that $\alpha(T_1) \leqslant \alpha(T)$. By Proposition 1.47 $\alpha(T_1) = 0$ and so T_1 is one-to-one. Let $x \in N(T^{p+1})$ and $y = T^p x$. Then $y \in R(T^p)$ and $T_1 y = Ty = T^{p+1} x = 0$. Hence $y = 0$ and $x \in N(T^p)$. It follows that $\alpha(T) \leqslant p = \delta(T)$.

PROPOSITION 1.49. *Let $T \in L(X)$. If $\alpha(T)$ and $\delta(T)$ are both finite then $\alpha(T) = \delta(T)$.*

Proof. By Proposition 1.48 it is sufficient to show that $\alpha(T) \geqslant \delta(T)$. The case $\delta(T) = 0$ is covered by Proposition 1.47. Thus we may suppose that $\delta(T) = p \geqslant 1$. Hence, there is y in $R(T^{p-1})\backslash R(T^p)$. Therefore $y = T^{p-1}x$ for some x in X. Let $z = Ty = T^p x$. Now $T^p R(T^p) = R(T^{2p}) = R(T^p)$. Hence there is u in $R(T^p)$ such that $T^p u = z$. Let $v = x - u$. Then

$$T^p v = T^p x - T^p u = z - z = 0,$$

$$T^{p-1} v = T^{p-1} x - T^{p-1} u = y - T^{p-1} u.$$

Now, since $u \in R(T^p)$, we have $T^{p-1}u \in R(T^{2p-1}) = R(T^p)$. Also $y \neq T^{p-1}u$ because $y \notin R(T^p)$. Hence $v \in N(T^p)\backslash N(T^{p-1})$ and so $\alpha(T) \geqslant p = \delta(T)$.

Example 1.50. Let H be a separable complex Hilbert space and let $\{\phi_n: n = 1, 2, 3, \ldots\}$ be an orthonormal basis for H. Define U in $L(H)$ by

$$U\left(\sum_{m=1}^{\infty} c_m \phi_m\right) = \sum_{m=1}^{\infty} c_m \phi_{m+1} \qquad \left(\sum_{m=1}^{\infty} |c_m|^2 < \infty\right).$$

Then U is one-to-one and so $\alpha(U) = 0$. Also

$$U^n H = \operatorname{clm}\{\phi_m : m \geqslant n + 1\}$$

for each positive integer n and so $\delta(U) = \infty$.

PROPOSITION 1.51. *Let $T \in L(X)$. Suppose that $\alpha(T)$, $\delta(T)$ are both finite and hence equal. Let $\alpha(T) = \delta(T) = p$. Then*

$$X = R(T^p) \oplus N(T^p).$$

Moreover T_1, the restriction of T to $R(T^p)$, is one-to-one and onto.

Proof. Suppose that $y \in R(T^p) \cap N(T^p)$. Then $y = T^p x$ for some x in X and $T^p y = 0$. It follows that $T^{2p} x = 0$. However, $\alpha(T) = p$ and so $x \in N(T^{2p}) = N(T^p)$. Hence $y = T^p x = 0$ and so $R(T^p) \cap N(T^p) = \{0\}$. Also, $T^p R(T^p) = R(T^p)$. If $x \in X$ there is u in $R(T^p)$ such that $T^p u = T^p x$. Now if $z = x - u$ then $T^p z = 0$. Hence

$$X = R(T^p) \oplus N(T^p).$$

Since $\delta(T) = p$, T maps $R(T^p)$ onto itself. If $y \in R(T^p)$ and $Ty = 0$ then $y \in R(T^p) \cap N(T^p) = \{0\}$. Hence T_1 is one-to-one and onto.

We now discuss in more detail the range and null-space of the spectral projection corresponding to a pole of the resolvent. The reader is referred back to the discussion of the Laurent expansion of the resolvent about an isolated singular point and in particular to equations (1) to (7).

THEOREM 1.52. *Let $T \in L(X)$. Let λ_0 be a pole of the resolvent of T of order m. Let $\tau = \sigma(T)\backslash\{\lambda_0\}$. Then λ_0 is an eigenvalue of T. The ascent and descent of $\lambda_0 I - T$ are both equal to m. Also*

$$E(\lambda_0)\,X = N((\lambda_0 I - T)^m),$$

$$E(\tau)\,X = R((\lambda_0 I - T)^m).$$

Proof. For convenience we denote the null-space and range of $(\lambda_0 I - T)^k$ by N_k and R_k respectively. If $x \in N_n$, where $n \geqslant 1$, we see by (6), induction and (7) that

$$\begin{aligned} 0 &= A_{n-1}(T - \lambda_0 I)^n x = (T - \lambda_0 I)^n A_{n-1} x \\ &= (T - \lambda_0 I)\, A_0 x \qquad = B_1 x - x \end{aligned}$$

so that by (5) we have $x = B_1 x \in E(\lambda_0)\,X$. Thus $N_n \subseteq E(\lambda)\,X$ if $n \geqslant 1$. On the other hand, it follows from (4) that if $x \in E(\lambda_0)\,X$ then $x = B_1 x$ and $(T - \lambda_0 I)^n x = B_{n+1} x$. Since $B_{n+1} = 0$ if $n \geqslant m$, it follows that $E(\lambda_0) X \subseteq N_n$ and hence $N_n = E(\lambda_0)\,X$ if $n \geqslant m$. However, N_{m-1} is a proper subset of N_m because $B_m \neq 0$. The equations $N_{m-1} = N_m = E(\lambda_0)\,X$ imply that $B_m = 0$ in view of the relation $B_m = (T - \lambda_0 I)^{m-1} B_1$. We have now proved that the ascent of $\lambda_0 I - T$ is m and $N_m = E(\lambda_0)\,X$. In particular, since $m > 0$, λ_0 is an eigenvalue of T.

Now let T_1 and T_2 be the restrictions of T to $E(\tau)X$ and $E(\lambda_0)X$ respectively. By Theorem 1.39, $\lambda_0 \in \sigma(T_2)$ but $\lambda_0 \notin \sigma(T_1)$. Hence the descent of $\lambda_0 I_1 - T_1$ is 0 and $R((\lambda_0 I_1 - T_1)^n) = E(\tau)X$ when $n \geqslant 1$. Thus $E(\tau)X \subseteq R_n$. Now, if $n \geqslant m$, the only point common to R_n and N_n is 0. For, if $x \in R_n \cap N_n$ then $(\lambda_0 I - T)^n x = 0$ and there is y in X such that $x = (\lambda_0 I - T)^n y$. Hence $y \in N_{2n} = N_n$ and so $x = 0$. Now suppose that $n \geqslant m$ and $x \in R_n$. Let $x_1 = E(\tau)x$ and $x_2 = E(\lambda_0)x$. Then $x_2 = x - x_1 \in R_n$ because $E(\tau)X \subseteq R_n$. However, $x_2 \in E(\lambda_0)X = N_n$ and so $x_2 = 0$, whence $x = x_1 \in E(\tau)X$. Thus $R_n \subseteq E(\tau)X$ if $n \geqslant m$. We now know that $R_n = E(\tau)X$ if $n \geqslant m$ and therefore that the descent of $\lambda_0 I - T$ is less than or equal to m. Proposition 1.49 then shows that the descent is exactly m, which we know to be the ascent. This completes the proof of the theorem.

The final theorem in this chapter is a partial converse to Theorem 1.52. A preliminary result is required.

PROPOSITION 1.53. *Let $T \in L(X)$ and let τ be an open-and-closed subset of $\sigma(T)$. Suppose that X_1, X_2 are closed subspaces of X that reduce T and such that $\sigma(T|X_1) \subseteq \tau$, $\sigma(T|X_2) \subseteq \mathbf{C}\backslash\tau$. Then*

$$E(\tau)X = X_1 \quad \textit{and} \quad (I - E(\tau))X = X_2.$$

Proof. Let $x \in X_1$. Then

$$E(\tau)x = \frac{1}{2\pi i}\int_B (\lambda I - T)^{-1}\, d\lambda . x$$

where B is a finite family of rectifiable Jordan curves enclosing τ but no other point of $\sigma(T)$. Now if $\lambda \in B$ then $\lambda \in \rho(T|X_1)$ and $\lambda \in \rho(T)$. By Lemma 1.28, $(\lambda I - T)^{-1}X_1 \subseteq X_1$ and $((\lambda I - T)|X_1)^{-1} = (\lambda I - T)^{-1}|X_1$. Hence

$$E(\tau)x = \frac{1}{2\pi i}\int_B ((\lambda I - T)|X_1)^{-1}\, d\lambda . x = x.$$

Let P be the projection whose range is X_1 and whose null-space is X_2. Then $E(\tau)Px = Px \, (x \in X)$ and so $E(\tau)P = P$. Similarly

$$(I - E(\tau))(I - P) = I - P.$$

Therefore $P = E(\tau)$ and the proof is complete.

THEOREM 1.54. *Let $T \in L(X)$ and $\lambda_0 \in \sigma(T)$. Suppose that $\alpha(\lambda_0 I - T), \delta(\lambda_0 I - T)$ are both finite and hence equal. Let*

$$\alpha(\lambda_0 I - T) = \delta(\lambda_0 I - T) = m.$$

Suppose that $(\lambda_0 I - T)^m X$ is closed. Then λ_0 is a pole of the resolvent of T of order m.

Proof. Since $\lambda_0 \in \sigma(T)$ we have $m \geqslant 1$. Let $X_1 = (\lambda_0 I - T)^m X$ and $X_2 = N((\lambda_0 I - T)^m)$. Then, by Proposition 1.51, $X = X_1 \oplus X_2$ and $\lambda_0 \in \rho(T|X_1)$. Now $(\lambda_0 I - T)|X_2$ is nilpotent and so by the spectral mapping theorem $\sigma(T|X_2) = \{\lambda_0\}$. Hence by Proposition 1.37, $\sigma(T|X_1) = \sigma(T)\backslash\{\lambda_0\}$. It follows that $\{\lambda_0\}$ is an open-and-closed subset of $\sigma(T)$ and so we may consider the spectral projection $E(\lambda_0)$. By Proposition 1.53

$$E(\lambda_0)\, X = N((\lambda_0 I - T)^m).$$

Since $\lambda_0 I - T$ has ascent m,

$$(\lambda_0 I - T)^m\, E(\lambda_0) = 0, \; (\lambda_0 I - T)^{m-1}\, E(\lambda_0) \neq 0,$$

and the desired conclusion follows from Proposition 1.40.

Notes and Comments on Part One

The results of Chapter 1 may be regarded as being a unification of two mathematical theories. First of all, the results are a generalization of those in the theory of matrices, and secondly they are an abstraction of results in the theory of integral equations. Consequently it is impossible to give complete and accurate credit in the references to these ideas. For example, the resolvent operator, its functional equation and expansion were used in both of these theories.

We deal first with spectral theory in a finite dimensional space. This is a part of matrix theory. A number λ_0 is called an *eigenvalue* of the operator T if there exists a vector $x_0 \neq 0$ such that $Tx_0 = \lambda_0 x_0$. The terms "proper value", "characteristic value", "secular value", and "latent root" are also used. The last term is due to Sylvester [**2**]. The term "spectrum" is due to Hilbert.

Polynomials of a matrix were used almost from the beginning of the theory, and by 1867 Laguerre [**1**] had considered infinite power series in a matrix in constructing the exponential function of a matrix. Sylvester [**1**], [**2**] constructed arbitrary functions of a matrix with distinct eigenvalues by means of the Lagrange interpolation formula. His method was generalized by Buchheim [**1**] for the case of multiple eigenvalues. The finite dimensional case of the minimal equation theorem (Theorem 1.41) is due to Giorgi [**1**]. Even before Sylvester, Frobenius [**1**; p. 54] and [**2**] had obtained expansions for the resolvent operator in the neighbourhood of a pole.

Frobenius [**3**; p. 11] stated that if f is analytic, then $f(T)$ can be obtained as the sum of the residues of $(\lambda I - T)^{-1} f(\lambda)$ with respect to all the eigenvalues of T. He asserted that this notion had been used in the dissertation of L. Stickelberger (1881). However, Frobenius did not develop a precise calculus. The first to make a clear use of this device was Poincaré [**1**], who employed it where all the roots are distinct. In the case of multiple roots a formula equivalent to that of Definition 1.18 was derived by Fantappié [**1**] on the basis of certain requirements, including the relations of Theorem 1.19, that would be expected for a "reasonable" operational calculus. The formula in Definition 1.18 was used as a definition for $f(T)$ by Giorgi [**1**] at the suggestion of É. Cartan, who was undoubtedly familiar with Poincaré's use.

The book [**8**] is close in spirit to the treatment in Chapter 1 and will be found invaluable. We have used it as our standard reference for the finite-dimensional case throughout the book.

Credit should be given to Hilbert [**1**] and E. H. Moore [**1**], [**2**] for establishing the unification of linear space theory and the theory of integral equations. However, it was F. Riesz who fully revealed and developed it along the lines presented here. The reader of his book (F. Fiesz [**1**], particularly Sections 71 to 82) will find many of these concepts and results expounded with an approach that is "modern". Although he dealt principally with compact operators in l^2, he established that the resolvent set is open, that the resolvent operator is analytic and indicated that, at least in the case of a pole, the Cauchy integral theorem could be employed to obtain a projection operator commuting with the operator.

In the case of bounded normal operators on Hilbert space, many of the results of this section become simpler and can be proved more directly by other methods. See "Notes and Comments on Part Four". Our remarks here deal only with the Banach space case.

The expansion for the resolvent is due to Neumann [**1**] who established it in potential theory. In a more general context it is due to Hilb [**1**].

The fact that a closed linear operator on an arbitrary complex Banach space has non-empty spectrum was proved by Taylor [**1**]. A special case of the formulae for the spectral radius was proved by Beurling [**1**], and the general case by Gelfand [**1**].

In 1923, Wiener [**1**] observed that Cauchy's integral theorem and Taylor's theorem remain valid for analytic functions with values in a complex Banach space. Not much application was made of this fact for about twelve years when a number of research workers independently found it useful. In 1936, Nagumo [**1**] studied Banach algebras from a function-theoretic viewpoint and proved, among other things, some theorems due to Riesz for compact operators. Later, Taylor [**2**] studied certain abstract analytic functions, and Hille [**1**] applied similar methods in the study of semi-groups. In 1941, the famous paper of Gelfand [**1**] appeared which, although it partially overlapped with Nagumo's work, developed the ideal theory of Banach algebras. In addition, Gelfand used the contour integral to obtain idempotents in Banach algebras. Independently, Lorch [**2**] employed the same device and initiated a study of "spectral sets".

Theorem 1.19 is due to Gelfand [**1**]. The spectral mapping theorem and the minimal equation theorem are due to Dunford [**1**], where there is some additional material.

The concepts of residual and continuous spectrum were first introduced by Dunford [**1**]. The term approximate point spectrum was first introduced by Halmos in [7] in a Hilbert space setting. He proved parts (i), (iii), (iv) of Theorem 1.16. The result that every boundary point of the spectrum lies in the approximate point spectrum was first formulated explicitly in this form by D. R. Smart in his Ph.D dissertation (Cambridge University 1958), and independently the result is obtained implicitly in Theorem 5.1D of [**15**; p. 258]. It should be noted that this result is a special case of a result of Rickart [**1**], published in 1947 and which states that every boundary point of the spectrum of an element of a complex Banach algebra with identity is a two-sided topological divisor of zero.

The results on ascent and descent of bounded linear operators are due to Dunford [**1**]. Taylor [**3**] extended these results to the case of closed linear operators. See [**15**], however, for the terminology used.

Scroggs [**1**] was the first to prove Theorems 1.29 and 1.30 on relationships between the spectra of an operator and its restriction with the generality stated. However, Bram [**1**] had earlier obtained Theorem 1.29 in a Hilbert space setting. Dowson [**3**] obtained analogous results for operators induced on quotient spaces. Theorem 1.26 is due to Schwartz [**1**].

In [**7**; p. 41] there is a simple argument, due to W. A. Howard, which establishes that the bilateral shift operator on l^2 has no proper closed reducing subspaces in the Hilbert space sense. We have preferred the argument of [**9**; p. 272–3], for this adapts easily to show that the bilateral shift operator on l^2 has no proper closed reducing subspaces in the Banach space sense. Finally, perhaps we should modify our comments on the invariant subspace problem and state that there is a preprint which may well contain an example of an operator on a separable complex Banach space having no proper closed invariant subspace.

Part 2

RIESZ OPERATORS

2. Compact Operators

We begin this chapter by introducing the classes of finite-rank and compact operators. Various properties of these classes of operators are proved. Next, we give the Hilden-Lomonosov proof of the theorem on the existence of a proper closed invariant subspace for a compact operator. Then, Ringrose's theory of superdiagonal forms for compact operators is presented.

Definition 2.1. An element T of $L(X)$ is called a *finite-rank operator* if its range is finite-dimensional.

Definition 2.2. Let $y \in X$ and $f \in X^*$. Define $y \otimes f$ by

$$(y \otimes f)(x) = f(x)\, y.$$

Then $y \otimes f$ is a bounded linear operator on X, called an *operator of rank one.*

We shall show that every finite-rank operator can be expressed as a finite sum of operators of rank one. A preliminary lemma is required.

LEMMA 2.3. *Let Y be a normed linear space and let $f_1, \ldots, f_n$ be any finite linearly independent set of unit vectors in Y. Then there is a positive real number δ such that, for every choice of the scalars $\alpha_1, \ldots, \alpha_n$ we have*

$$\left\| \sum_{r=1}^{n} \alpha_r f_r \right\| \geqslant \delta \sum_{r=1}^{n} |\alpha_r|.$$

Proof. Let Y_0 consist of all finite linear combinations of $\{f_r : r = 1, \ldots, n\}$. Define

$$\left|\left|\left| \sum_{r=1}^{n} \alpha_r f_r \right|\right|\right| = \sum_{r=1}^{n} |\alpha_r|.$$

Then the norms $|||\ |||$ and $\|\ \|$ on Y_0 are equivalent. This suffices to prove the lemma.

THEOREM 2.4. *Let T be a finite-rank operator on X. Suppose that the unit vectors $g_1, g_2, \ldots, g_n$ constitute a basis for the range of T. Then there are $l_1, l_2, \ldots, l_n$ in X^* such that for every x in X*

$$Tx = l_1(x)\, g_1 + l_2(x)\, g_2 + \ldots + l_n(x)\, g_n$$

i.e.

$$T = \sum_{r=1}^{n} g_r \otimes l_r.$$

Proof. For any x, Tx must be expressible (in a unique manner) as a linear combination of $g_1, g_2, \ldots, g_n$. The linearity of T clearly implies that the coefficients must be complex-valued linear functions of x. It therefore remains only to show that the linear functionals $l_1, l_2, \ldots, l_n$ are bounded. By the preceding lemma, there is $\delta > 0$ such that

$$\|Tx\| \geqslant \delta \sum_{r=1}^{n} |l_k(x)| \qquad (x \in X).$$

For each index k we obtain

$$|l_k(x)| \leqslant \delta^{-1}\|Tx\| \leqslant \delta^{-1}\|T\|\,\|x\| \quad (x \in X)$$

and so the boundedness of l_k is established.

The next result follows immediately.

THEOREM 2.5. *The finite-rank operators on X form a two-sided ideal in $L(X)$.*

Definition 2.6. Let Y be a complex Banach space. A linear mapping T of X into Y is said to be *compact* if for every bounded sequence $\{f_n\}$ in X the corresponding sequence $\{Tf_n\}$ contains a subsequence which is convergent.

THEOREM 2.7. *Let T be a compact linear mapping of X into a complex Banach space Y. Then T is bounded.*

Proof. Let $B = \{x \in X; \|x\| \leqslant 1\}$. Then $\overline{T(B)}$ is compact. Since the map $x \to \|x\|$ of Y into $\mathbf{R}$ is continuous the set $\{\|y\| : y \in \overline{T(B)}\}$ is compact in $\mathbf{R}$. Hence, there is a constant M such that

$$\|y\| \leqslant M < \infty \qquad (y \in \overline{T(B)})$$

and so $\|Tx\| \leqslant M (x \in B)$. Therefore T is bounded.

THEOREM 2.8. (i) *The product of two operators on X is compact if at least one of the factors is compact.*

(ii) *Any finite linear combination of compact operators on X is compact.*

(iii) *Let the sequence $\{S_n\}$ of compact operators on X converge in the norm of $L(X)$ to S. Then S is compact.*

Proof. (i) Let T be compact and let $\{f_n\}$ be any bounded sequence in X. Then there exists a convergent subsequence of the sequence $\{Tf_n\}$. Denoting this subsequence by $\{g_n\}$ and its limit by g, we obtain $\|Sg - Sg_n\| = \|S(g - g_n)\| \leqslant \|S\| \, \|g - g_n\| \to 0$. Thus the sequence $\{Sg_n\}$ is a convergent subsequence of the sequence $\{STf_n\}$ and so ST is compact. On the other hand, if S is assumed to be compact we argue as follows. Given any bounded sequence of vectors $\{f_n\}$ in X, the sequence $\{Tf_n\}$ is also bounded (since T is bounded), and so it is possible to extract from the sequence $\{STf_n\}$ a convergent subsequence. Therefore, ST is compact.

(ii) Since any scalar multiple of a compact operator is obviously compact, it suffices to prove that the sum of two compact operators is compact, for finite induction then gives the desired result. Therefore, suppose that S and T are compact operators in $L(X)$, and let $\{f_n\}$ be any bounded sequence of vectors in X. Then we can select a subsequence $\{g_n\}$ of this sequence such that the sequence $\{Tg_n\}$ converges. From the sequence $\{g_n\}$ we can, in turn, select a subsequence $\{h_n\}$ such that the sequence $\{Sh_n\}$ converges. Since the sequence $\{Th_n\}$ certainly converges, it follows that the sequence $\{(S + T)h_n\}$ also converges. Thus the sequence $\{(S + T)f_n\}$ contains a convergent subsequence and so $S + T$ is compact.

(iii) Let $\{f_n\}$ be any bounded sequence in X. We can extract a subsequence $\{f_{1n}\}$ such that $\lim_{n\to\infty} S_1 f_{1n}$ exists. From the subsequence $\{f_{1n}\}$ we select a subsequence $\{f_{2n}\}$ such that $\lim_{n\to\infty} S_2 f_{2n}$ exists. Repeating this procedure, we obtain for each positive integer k a subsequence $\{f_{kn}\}$ of the original sequence such that $\lim_{n\to\infty} S_r f_{kn}$ exists for $r = 1, 2, \ldots, k$. Now consider the diagonal sequence $\{f_{nn}\}$. This is clearly a subsequence of the original sequence $\{f_n\}$ and, except perhaps for a finite number (depending on k) of initial terms, it is a subsequence of the sequence $\{f_{kn}\}$. Hence $\lim_{n\to\infty} S_k f_{nn}$ exists for each index k. Let $\varepsilon > 0$ be given. We first determine an index K such that $\|S - S_K\| < \varepsilon$, and then an index N such that $\|S_K(f_{mm} - f_{nn})\| < \varepsilon$ whenever m, n both exceed N. Hence

$$\begin{aligned}\|Sf_{nn} - Sf_{mm}\| &= \|(Sf_{nn} - S_K f_{nn}) + (S_K f_{nn} - S_K f_{mm}) + (S_K f_{mm} - Sf_{mm})\| \\ &\leqslant \|(S - S_K) f_{nn}\| + \|S_K f_{nn} - S_K f_{mm}\| + \|(S - S_K) f_{mm}\| \\ &< \|S - S_K\| \, \|f_{nn}\| + \varepsilon + \|S - S_K\| \, \|f_{mm}\| \\ &\leqslant 2\varepsilon \sup \|f_n\| + \varepsilon \quad (m, n > N).\end{aligned}$$

It follows that the sequence $\{Sf_{nn}\}$ is Cauchy. Since X is complete this sequence is convergent and so S is compact.

COROLLARY 2.9. *The compact operators on X form a closed two-sided ideal in $L(X)$.*

We shall denote this ideal by $K(X)$.

THEOREM 2.10. *A finite-rank operator on X is compact.*

Proof. By Theorems 2.4 and 2.8 it is enough to prove the present theorem for an operator of rank one. Let $y \in X$, $f \in X^*$ and let $\{x_n\}$ be any bounded sequence in X. Then $|f(x_n)| \leqslant \|f\| \, \|x_n\|$ and so $\{f(x_n)\}$ is a bounded sequence in $\mathbf{C}$. It has a convergent subsequence, $\{f(x_{n_k})\}$ say. Then $\{f(x_{n_k})y\}$ converges in X. Hence $y \otimes f$ is compact.

Let Y be a complex Banach space. Let $L(Y, X)$ denote the Banach space of bounded linear mappings from Y into X. Let $K(Y, X)$ be the set of compact linear mappings from Y into X and let $F(Y, X)$ denote the set of bounded linear mappings from Y into X with finite-dimensional range. The arguments of this chapter can be used to prove that

$$F(Y, X) \subseteq K(Y, X) \subseteq L(Y, X).$$

X is said to have the *approximation property* if and only if the following condition is satisfied. Let Y be any complex Banach space and $T \in K(Y, X)$. Then there is a sequence $\{T_n\}$ in $F(Y, X)$ such that $\{T_n\}$ converges to T in the norm of $L(Y, X)$.

Let $F(X)$ denote the two-sided ideal of finite-rank operators in $L(X)$. Its norm closure $\bar{F}(X)$ is a closed two-sided ideal in $L(X)$ and by Theorem 2.10 we have $\bar{F}(X) \subseteq K(X)$. For many years it was an unsolved problem whether in general $\bar{F}(X) = K(X)$. In 1973, Enflo [**1**] proved the existence of a separable complex Banach space which fails to have the approximation property. It follows easily from this that in general $\bar{F}(X)$ may be a proper subset of $K(X)$. Davie [**1**] proved that for $2 < p < \infty$ there is a closed subspace Y of l^p such that Y fails to have the approximation property. Using this, F. E. Alexander [**1**] proved the following result.

THEOREM. *Suppose that $2 < p < \infty$. Then there is a closed subspace Y of l^p such that $\bar{F}(Y) \neq K(Y)$.*

In the next chapter it will be seen that compact linear operators are asymptotically quasi-finite-rank in a sense to be made precise there.

Prior to discussing the spectral theory of a compact operator we prove some further basic properties.

THEOREM 2.11. *Let $T \in L(X)$. If T is compact then so is T^*.*

Proof. Let S, S^* be the closed unit balls in X, X^* respectively. Let $\{y_n\}$ be an arbitrary sequence in S^*. Observe that $\overline{T(S)}$ is a compact metric space. Also if $x, z \in \overline{T(S)}$ then

$$|y_n(x) - y_n(z)| \leqslant \|x - z\| \quad (n = 1, 2, \ldots)$$

and so $y_n \in C(\overline{T(S)})$. By the Ascoli-Arzelà theorem some subsequence, $\{y_{n_k}\}$ say, converges in the norm of $C(\overline{T(S)})$. Hence $y_{n_k}(Tx) = (T^* y_{n_k})(x)$ converges uniformly for x in S. It follows that $\{T^* y_{n_k}\}$ converges in the norm of X^* and so T^* is compact.

THEOREM 2.12. *Let T, in $L(X)$, be compact and let Y be a closed subspace of X invariant under T. Then $T|Y$ is compact.*

Proof. This follows immediately from the definition of a compact linear mapping.

THEOREM 2.13. *Let $T \in L(X)$. Then T is compact if and only if T^* is compact.*

Proof. If T^* is compact then by Theorem 2.11 so is T^{**}. It follows from Theorem 2.12 that T, the restriction of T^{**} to X, is compact. Another application of Theorem 2.11 completes the proof.

THEOREM 2.14. *Let T, in $L(X)$, be compact and let Y be a closed subspace of X invariant under T. Then T_Y, the operator induced by T on the quotient space X/Y, is compact.*

Proof. By Theorem 2.11, T^* is compact. Also $Y^\perp$, the annihilator of Y, is a closed subspace of X^* invariant under T^* and so $T^*|Y^\perp$ is compact by Theorem 2.12. As in Proposition 1.33 et seq. we can identify T_Y^* and $T^*|Y^\perp$. Hence by Theorem 2.13, T_Y is compact.

We now develop the spectral theory of a compact operator.

LEMMA 2.15. *Let Y be a proper closed subspace of X. There is an x in X with $\|x\| = 1$ and*

$$d(x, Y) = \inf\{\|x - y\| : y \in Y\} > \tfrac{1}{2}.$$

Proof. By the Hahn-Banach theorem, there is an f in X^* such that $f \neq 0$ and $f(y) = 0$ $(y \in Y)$. We may assume without loss of generality that $\|f\| = 1$. Hence there is an x in X such that $\|x\| = 1$ and $|f(x)| > \frac{1}{2}$. Then if $y \in Y$

$$\tfrac{1}{2} < |f(x)| = |f(x - y)| \leqslant \|f\|\,\|x - y\| = \|x - y\|.$$

LEMMA 2.16. *Let T, in $L(X)$, be compact and let $\lambda \neq 0$ be an eigenvalue of T. Then the eigenspace $N_\lambda = \{x \in X : Tx = \lambda x\}$ is finite-dimensional.*

Proof. Suppose the lemma is false. Recall that each finite-dimensional subspace of X is closed. We can choose x_1 in N_λ with $\|x_1\| = 1$. By induction and the preceding lemma we can construct an infinite sequence $\{x_n\}$ in N_λ with the following properties: $\|x_n\| = 1$ and the distance of x_n from the linear span of $\{x_r : r = 1, 2, \ldots, n - 1\}$ is greater than $\frac{1}{2}$. Then the sequence $\{Tx_n\} = \{\lambda x_n\}$ can have no convergent subsequence since

$$\|\lambda x_n - \lambda x_m\| > \tfrac{1}{2}|\lambda| \quad (m \neq n).$$

This contradicts the compactness of T.

LEMMA 2.17. *Let T, in $L(X)$, be compact and let $k > 0$. The set of eigenvalues of T with modulus greater than k is finite. Thus the eigenvalues of T form a countable set with 0 as the only possible cluster point.*

Proof. Suppose there is an infinite sequence $\{\lambda_n\}$ of distinct eigenvalues of T with $|\lambda_n| > k > 0$. Define $N(\lambda_n) = \{x \in X : Tx = \lambda_n x\}$ and

$$M_n = N(\lambda_1) + \ldots + N(\lambda_n).$$

Then by the previous lemma each subspace on the right is finite-dimensional. Hence each M_n is finite-dimensional. Also, for $n \geqslant 2$, M_{n-1} is a proper subspace of M_n. To see this suppose that $x_r \in N(\lambda_r)$ and $\|x_r\| = 1$ $(r = 1, \ldots, n)$. Suppose also that it has been shown that $x_1, \ldots, x_{n-1}$ are linearly independent. If $x_n = \sum_{r=1}^{n-1} \alpha_r x_r$ then

$$0 = (T - \lambda_n I)\, x_n = \alpha_1(\lambda_1 - \lambda_n)\, x_1 + \ldots + \alpha_{n-1}(\lambda_{n-1} - \lambda_n)\, x_{n-1}$$

and since $\lambda_r - \lambda_n \neq 0$ for $r \neq n$ we have $\alpha_r = 0 (r = 1, \ldots, n-1)$ and $x_n = 0$, a contradiction. It follows by induction that, for $n \geqslant 2$, M_{n-1} is a proper subspace of M_n. By Lemma 2.15 we can choose x_n in M_n with $\|x_n\| = 1$ and $d(x_n, M_{n-1}) > \frac{1}{2} (n \geqslant 2)$. Now $\{Tx_r\}_{n=2}^{\infty} = \{\lambda_n x_n\}_{n=2}^{\infty}$ can have no convergent subsequence, for if $n > m$,

$$\|\lambda_n x_n - \lambda_m x_m\| = |\lambda_n|\, \|x_n - \lambda_m \lambda_n^{-1} x_m\| \geqslant \tfrac{1}{2}|\lambda_n| > \tfrac{1}{2}k.$$

This contradicts the compactness of T and the proof is complete.

LEMMA 2.18. *Let T, in $L(X)$, be compact. Then each non-zero point of $\sigma(T)$ is an eigenvalue of T.*

Proof. If $\lambda \neq 0$ is in the boundary of $\sigma(T)$ then by Theorem 1.16 there is a sequence $\{x_n\}$ in X with $\|x_n\| = 1$ such that $(T - \lambda I)\, x_n \to 0$ as $n \to \infty$. Since T is compact, there is a subsequence $\{x_{n_k}\}$ such that $\{Tx_{n_k}\}$ converges to z, say. Since $x_{n_k} - \lambda^{-1} Tx_{n_k} \to 0$ and $\lambda^{-1} Tx_{n_k} \to \lambda^{-1} z$ we obtain $x_{n_k} \to \lambda^{-1} z$. Hence $Tx_{n_k} \to \lambda^{-1} Tz$ and so $\lambda^{-1} Tz = z$. Therefore $Tz = \lambda z$. Since $\|x_{n_k}\| = 1$ for each k we have $\|z\| = 1$. Thus λ is an eigenvalue of T. This, in conjunction with the previous lemma, shows that the boundary of $\sigma(T)$ is a countable set whose only possible cluster point is 0. This set is compact and has connected complement. Hence $\sigma(T)$ has void interior and the proof is complete.

LEMMA 2.19. *Let T, in $L(X)$, be compact. There is a positive real number M with the following property: for each y in $(I - T)X$ there is an x in X such that $y = (I - T)x$ and $\|x\| \leqslant M\|y\|$.*

Proof. Suppose that the lemma is false. Then for each positive integer n there is a point y_n in $(I - T)X$ such that if $x_n \in X$ and $(I - T)x_n = y_n$, then $\|x_n\| > n\|y_n\|$. Obviously $y_n \neq 0$ (because $0 = (I - T)0$). Choose $w_n \in X$ such that $(I - T)w_n = y_n$. Then $w_n \notin N(I - T)$. Let $d_n = d(w_n, N(I - T))$. Then by Lemmas 2.16 and 2.18, the subspace $N(I - T)$ is finite-dimensional and therefore closed. Hence $d_n > 0$. Thus there exists v_n in $N(I - T)$ such that $d_n \leqslant \|w_n - v_n\| < 2d_n$. Let $z_n = \|w_n - v_n\|^{-1}(w_n - v_n)$. Then $\|z_n\| = 1$ and so the sequence $\{Tz_n\}$ has a convergent subsequence, $\{Tz_{n_k}\}$ say. From the equation

$$(I - T)(\|w_n - v_n\| z_n) = (I - T)(w_n - v_n) = (I - T)w_n = y_n$$

we conclude that $\|w_n - v_n\|\,\|z_n\| > n\|y_n\|$. Therefore for $n = 1, 2, 3, \ldots$

$$\|(I - T)z_n\| = \|y_n\|\,\|w_n - v_n\|^{-1} < \frac{1}{n}$$

and consequently $\lim_{n\to\infty} (I - T)z_n = 0$. Since we have $z_{n_k} = (I - T)z_{n_k} + Tz_{n_k}$ it follows that $\{z_{n_k}\}$ converges. Let $z = \lim_{k\to\infty} z_{n_k}$. Then

$$(I - T)z = \lim_{k\to\infty} (I - T)z_{n_k} = 0$$

and so $z \in N(I - T)$. However, for $n = 1, 2, 3, \ldots$ we have

$$\begin{aligned}\|z_n - z\| &= \big\|\,\|w_n - v_n\|^{-1}(w_n - v_n) - z\big\| \\ &= \|w_n - (v_n + \|w_n - v_n\| z)\|\,\|w_n - v_n\|^{-1} \\ &\geqslant d_n\|w_n - v_n\|^{-1} \\ &\geqslant \tfrac{1}{2}\end{aligned}$$

because $v_n + \|W_n - v_n\| z \in N(I - T)$. This contradicts the fact that $\lim_{k\to\infty} z_{n_k} = z$ and the proof is complete.

LEMMA 2.20. *Let T, in $L(X)$, be compact. Then $(I - T)X$ is closed.*

Proof. Let $\{y_n\}$ be a sequence in $(I - T)X$ with $\lim_{n\to\infty} y_n = y$ and let M be as in the previous lemma. Choose x_n in X with $(I - T)x_n = y_n$ and $\|x_n\| \leqslant M\|y_n\|$ for $n = 1, 2, 3, \ldots$ Since the sequence $\{y_n\}$ converges it is bounded, and hence $\{x_n\}$ is also bounded. Thus $\{Tx_n\}$ has a convergent subsequence, $\{Tx_{n_k}\}$ say. Since $x_n = y_n + Tx_n$, the sequence $\{x_{n_k}\}$ also converges. Let $x = \lim_{k\to\infty} x_{n_k}$. Then $(I - T)x = \lim_{k\to\infty}(I - T)x_{n_k} = y$, which shows that $y \in (I - T)X$. Hence this subspace is closed.

We are now in a position to prove the main theorem describing the spectral theory of a compact operator.

THEOREM 2.21. *Let T, in $L(X)$, be compact.*

(i) *$\sigma(T)$ is countable and has no cluster point except possibly 0. Every non-zero number in $\sigma(T)$ is an eigenvalue of T and moreover a pole of the resolvent of T.*

Let λ be a non-zero point in $\sigma(T)$, and let $v(\lambda)$ be the order of the pole at λ.

(ii) *For each positive integer n, $(\lambda I - T)^n X$ is closed. Also*

$$(\lambda I - T)^{m+1} X = (\lambda I - T)^m X \quad (m \geqslant v(\lambda))$$

and $v(\lambda)$ is the smallest positive integer with this property.

(iii) *For each positive integer n, $N((\lambda I - T)^n)$ is finite-dimensional. Also*

$$N((\lambda I - T)^m) = N((\lambda I - T)^{m+1}) \quad (m \geqslant v(\lambda))$$

and $v(\lambda)$ is the smallest positive integer with this property.

(iv) *The spectral projection $E(\lambda)$ has a non-zero finite-dimensional range given by*

$$E(\lambda)X = N((\lambda I - T)^{v(\lambda)}).$$

The null-space of $E(\lambda)$ is $(\lambda I - T)^{v(\lambda)} X$.

(v) *If $d(\lambda)$ is the dimension of $E(\lambda)X$ then $1 \leqslant v(\lambda) \leqslant d(\lambda)$.*

Note. The integers $v(\lambda)$ and $d(\lambda)$ are called respectively the *index* and the *algebraic multiplicity* of the eigenvalue λ.

Proof. (i) The first statement follows from Lemmas 2.17 and 2.18. Hence if $\lambda \neq 0$ is in $\sigma(T)$ then λ is an isolated point of $\sigma(T)$. Let T_λ be the restriction of T to $E(\lambda)X$. By Theorem 1.22 we have $0 \neq \lambda$ and $\sigma(T_\lambda) = \{\lambda\}$. Hence T_λ has a bounded inverse. Thus if Ω is the closed unit ball in $E(\lambda)X$ then $T_\lambda^{-1}\Omega$ is bounded and since T_λ is compact $\Omega = T_\lambda T_\lambda^{-1}\Omega$ is compact. This shows that $E(\lambda)X$ is finite-dimensional. Hence there is a smallest positive integer $v(\lambda)$ such that $(T_\lambda - \lambda I_\lambda)^{v(\lambda)} = 0$. (It follows from elementary linear algebra that $v(\lambda) \leqslant d(\lambda)$ and this proves (v).) By Proposition 1.40, λ is a pole of the resolvent of T of order $v(\lambda)$.

(ii) The first statement follows from Lemma 2.20, induction, and the fact that the restriction of a compact operator to a closed invariant subspace is compact. The second statement follows from Theorem 1.52.

(iii) The first statement follows from Lemma 2.16, induction, and the fact that the restriction of a compact operator to a closed invariant subspace is compact. The second statement follows from Theorem 1.52.

(iv) This follows from Theorem 1.52.

We now discuss the invariant subspace problem for compact operators. Let Y be a complex Banach space of dimension $\geqslant 2$ and let T be a compact operator on Y. If T has a non-zero eigenvalue, λ say, then, by Lemma 2.17, λ is an isolated point of $\sigma(T)$. Consider the spectral projection

$$E(\lambda) = \frac{1}{2\pi i}\int_\Gamma (\zeta I - T)^{-1}\,d\zeta$$

where Γ is a circle described once counterclockwise enclosing λ but no other point of $\sigma(T)$. If $A \in L(Y)$ and $AT = TA$ then

$$A(\zeta I - T)^{-1} = (\zeta I - T)^{-1}A \quad (\zeta \in \rho(T))$$

and so by norm continuity $AE(\lambda) = E(\lambda)A$. Also if $x \neq 0$ satisfies $Tx = \lambda x$ then the one-dimensional subspace of Y generated by x is a proper closed invariant subspace for T. It is a remarkable and non-trivial fact that even when T has no non-zero eigenvalue, T has a proper closed invariant subspace. The following neat proof of this result is based on ideas of V. I. Lomonosov and H. M. Hilden.

THEOREM 2.22. *Let Y be an infinite-dimensional complex Banach space. Let*

$T \neq 0$ be a compact operator on Y. Then there is a proper closed subspace of Y invariant under $\mathscr{A} = \{A \in L(Y): AT = TA\}$, i.e. T has a hyperinvariant subspace.

Proof. Suppose the theorem is false. Then for each $x \neq 0$,

$$\mathscr{A}x = \{Ax: A \in \mathscr{A}\}$$

is dense in Y, i.e. $\overline{\{\mathscr{A}x\}} = Y$. Also T is quasinilpotent; for otherwise T has a non-zero eigenvalue λ and $E(\lambda)\, Y$, being finite-dimensional, is a hyperinvariant subspace for T. We may choose a ball $B = \{x \in Y: \|x - x_0\| < \delta\}$ such that $0 \notin \overline{TB}$ and $\overline{TB}$ is compact. Hence if $y \in \overline{TB}$ there is A_y in $\mathscr{A}$ and an open neighbourhood N_y of y with $A_y N_y \subseteq B$. Therefore by compactness there exists a finite subset $A_{y_1}, \ldots, A_{y_k}$ of $\mathscr{A}$, such that for each y in $\overline{TB}$ some A_{y_r} maps y into B. For brevity, write $A_{y_r} = A_r$. For each positive integer n we may find suitable indices $i_1, \ldots, i_n$ such that

$$\prod_{m=1}^{n} A_{i_m} T^n x_0 = A_{i_1} T A_{i_2} T \ldots A_{i_n} T x_0 \in B.$$

Now, if $M = \max\{\|A_p\|: 1 \leqslant p \leqslant k\} < \infty$ then

$$\left\| \prod_{m=1}^{n} A_{i_m} T^n x_0 \right\| \leqslant M^n \|T^n\| \, \|x_0\|.$$

Since T is quasinilpotent, $M\|T^n\|^{1/n} \|x_0\|^{1/n} \to 0$ as $n \to \infty$ and so we obtain

$$\left\| \prod_{m=1}^{n} A_{i_m} T^n x_0 \right\|^{1/n} \to 0 \quad \text{as} \quad n \to \infty. \tag{1}$$

Since $0 \notin \overline{TB}$ then also $0 \notin B$ and so for some $\rho > 0$ we have

$$\rho < \left\| \sum_{m=1}^{n} A_{i_m} T^n x_0 \right\|,$$

$$\rho^{1/n} < \left\| \prod_{m=1}^{n} A_{i_m} T^n x_0 \right\|^{1/n}. \tag{2}$$

Since $\lim_{n \to \infty} \rho^{1/n} = 1$, (1) and (2) give a contradiction. This proves the theorem.

THEOREM 2.23. *Let Y be a complex Banach space of dimension $\geqslant 2$. Let T be a* *compact operator on Y. There is a proper closed subspace of Y invariant under T.*

Proof. If $T = 0$ the result is trivially true. If Y is finite-dimensional $\sigma(T)$ consists of a finite set of eigenvalues. Let λ be one such. Then there is an $x \neq 0$ satisfying $Tx = \lambda x$. Hence the one-dimensional subspace generated by x is a proper closed invariant subspace for T. Hence we may assume that $T \neq 0$ and Y is infinite-dimensional. The required result then follows from the previous theorem.

Theorem 2.23 was first proved with this generality by Aronszajn and Smith [**1**] in 1954 by a different method. This result will be referred to henceforth as the theorem of Aronszajn and Smith. It is of crucial importance in Ringrose's theory of super-diagonal forms for compact operators, which we now present.

A well-known theorem (see for example [**8**; p. 144]) asserts that every $n \times n$ matrix with complex entries may be reduced by unitary transformation to superdiagonal form. This result, together with some related theory concerning eigenvalues, may be re-formulated in the following way [**8**; p. 107, p. 144]. If T is a linear operator on an n-dimensional complex inner-product space X, then there exist subspaces $L_0, L_1, \ldots, L_n$ of X such that

(i) $\{0\} = L_0 \subset L_1 \subset L_2 \subset \ldots \subset L_n = X$;

(ii) L_m is m-dimensional;

(iii) $TL_m \subseteq L_m$ $(m = 0, 1, \ldots, n)$;

(iv) if we choose $e_m \in L_m \backslash L_{m-1}$ $(m = 1, \ldots, n)$ then the eigenvalues of T (counted according to their algebraic multiplicity) may be specified as those numbers $\lambda_1, \ldots, \lambda_n$ such that

$$Te_m - \lambda_m e_m \in L_{m-1} \ (m = 1, \ldots, n);$$

(v) T is nilpotent if and only if $TL_m \subseteq L_{m-1}$ $(m = 1, \ldots, n)$.

It is therefore natural to consider such a nest $\{L_m\}$ of subspaces as defining a super-diagonal form for the operator T.

We consider the extension of this concept of 'super-diagonal form' to a compact operator on a non-zero complex Banach space. It is not in general possible in this case to form nests of invariant subspaces with the simple structure exhibited in the finite-dimensional case. To illustrate this point we refer to the compact operator K on the space $L^2[0, 1]$ defined by the equation

$$(Kf)(x) = \int_0^x f(y)\, dy \quad (f \in L^2; 0 \leqslant x \leqslant 1).$$

It was proved by Donoghue [1] that the only closed invariant subspaces for this operator are the subspaces $E_c (0 \leqslant c \leqslant 1)$ defined by

$$E_c = \{f \in L^2; f(x) = 0 \quad \text{a.e. on} \quad (0, c)\}.$$

It follows that if L_1 and L_2 are distinct invariant subspaces, then either $L_1 \subset L_2$ or $L_2 \subset L_1$, and the quotient space of the larger by the smaller is infinite-dimensional. The subspaces E_c form a continuous nest of invariant subspaces in a sense to be specified below.

Throughout the remainder of this chapter, T denotes a compact operator on a non-zero complex Banach space X. The term *subspace* will be used to describe a closed linear subset of X.

A family $\mathscr{F}$ of subspaces of X, which is totally ordered by the inclusion relation, will be termed a *nest* of subspaces. If in addition each subspace in $\mathscr{F}$ is invariant under T we shall describe $\mathscr{F}$ as an *invariant nest*. A trivial example of an invariant nest is the family consisting of the two subspaces $\{0\}$ and X. Non-trivial invariant nests may be constructed by means of the theorem of Aronszajn and Smith.

We shall use the symbol $\subseteq$ to denote the inclusion relation, and reserve $\subset$ for proper inclusion. The norm closure of a subset S of X will be denoted by $\mathrm{cl}(S)$. Given a nest $\mathscr{F}$ of subspaces of X, and $M \in \mathscr{F}$, we define

$$M_- = \mathrm{cl}[\cup\{L: L \in \mathscr{F}, L \subset M\}].$$

If there is no L in $\mathscr{F}$ such that $L \subset M$, we define $M_- = \{0\}$. It is clear that M_- is a subspace of X, and that it will be an invariant subspace if $\mathscr{F}$ is an invariant nest. Also $M_- \subseteq M$. It should be emphasized that the definition of M_- depends on the particular nest $\mathscr{F}$ under consideration and not merely on the subspace M. We shall say that $\mathscr{F}$ is continuous at M if $M = M_-$.

A nest $\mathscr{F}$ will be termed *simple*

(i) if $\{0\} \in \mathscr{F}, X \in \mathscr{F}$;

(ii) if $\mathscr{F}_0$ is any subfamily of $\mathscr{F}$, then the subspaces $\cap\{L: L \in \mathscr{F}_0\}$ and $\mathrm{cl}[\cup\{L: L \in \mathscr{F}_0\}]$ are in $\mathscr{F}$:

(iii) if $M \in \mathscr{F}$, then the quotient space M/M_- is at most one-dimensional.

We observe that condition (ii) implies that $M_- \in \mathscr{F}$ whenever $M \in \mathscr{F}$. Next we discuss the construction and properties of simple invariant nests.

The class $\mathscr{N}$ of all nests of subspaces of X may be partially ordered by inclusion; if $\mathscr{F}_1, \mathscr{F}_2 \in \mathscr{N}$, we say that $\mathscr{F}_1 < \mathscr{F}_2$ if every subspace in the family $\mathscr{F}_1$ is also a member of $\mathscr{F}_2$. It is easily seen that, in this way, $\mathscr{N}$ is inductively ordered; for if $\mathscr{N}_0 \subseteq \mathscr{N}$ and $\mathscr{N}_0$ is totally ordered by the relation $<$, then

$$\mathscr{F}_0 = \cup\{\mathscr{F} : \mathscr{F} \in \mathscr{N}_0\}$$

is the least upper bound of $\mathscr{N}_0$ in $\mathscr{N}$. We may now deduce from Zorn's lemma the existence of at least one maximal nest of subspaces.

LEMMA 2.24. *Let $\mathscr{F}$ be a nest of subspaces of X. Then $\mathscr{F}$ is maximal if and only if $\mathscr{F}$ is simple.*

Proof. Suppose that $\mathscr{F}$ is maximal. Then clearly $\{0\}, X \in \mathscr{F}$ since otherwise $\mathscr{F}$ could be enlarged by the addition of these subspaces, contrary to the assumption that $\mathscr{F}$ is maximal. Secondly, let $\mathscr{F}_0$ be a subfamily of $\mathscr{F}$, and consider

$$M_0 = \cap\{L : L \in \mathscr{F}_0\}.$$

It is evident that M_0 is a closed subspace of X. Let $M \in \mathscr{F}$. Since $\mathscr{F}$ is totally ordered by inclusion we have *either* (a) $M \subseteq L (L \in \mathscr{F}_0)$, and $M \subseteq M_0$, *or* (b) $L \subset M$ for some L in $\mathscr{F}_0$, and $M_0 \subset M$. It follows that the family obtained by adding M_0 to $\mathscr{F}$ remains totally ordered by inclusion and is therefore a nest. Since $\mathscr{F}$ is maximal we deduce that $M_0 \in \mathscr{F}$. A similar argument shows that

$$\mathrm{cl}[\cup\{L : L \in \mathscr{F}_0\}]$$

is a member of $\mathscr{F}$. Hence properties (i) and (ii) of simple nests have been established.

Finally we have to show that, if $M \in \mathscr{F}$, then the quotient space M/M_- is at most one-dimensional. Suppose that, for some M in $\mathscr{F}$, this is not the case. Then we may find a subspace L of X such that $M_- \subset L \subset M$. Given any subspace N in $\mathscr{F}$, we have *either* (a) $M \subseteq N$, and $L \subset N$, *or* (b) $N \subset M$, and

$$N \subseteq \text{cl}[\cup\{K \in \mathscr{F} : K \subset M\}] = M_- \subset L.$$

It follows that $L \notin \mathscr{F}$, and that the family obtained by adding L to $\mathscr{F}$ is a nest. This contradicts the assumption that $\mathscr{F}$ is maximal. Hence M/M_- is at most one-dimensional for every M in $\mathscr{F}$ and so $\mathscr{F}$ is a simple nest.

Suppose conversely that $\mathscr{F}$ is a simple nest, but that $\mathscr{F}$ is not maximal. Then we may choose a subspace L of X such that $L \notin \mathscr{F}$ and the family obtained by adding L to $\mathscr{F}$ remains totally ordered by inclusion. We shall obtain a contradiction. Let

$$M = \cap\{N : N \in \mathscr{F}, L \subseteq N\} \tag{3}$$

$$M' = \text{cl}[\cup\{N : N \in \mathscr{F}, N \subset L\}]. \tag{4}$$

By virtue of property (ii) of simple nests we have $M, M' \in \mathscr{F}$, and it is clear that $M' \subseteq L \subseteq M$. Since $L \notin \mathscr{F}$ we in fact have

$$M' \subset L \subset M. \tag{5}$$

We shall now show that $M' = M_-$. It is apparent that $M' \subseteq M_-$. Suppose now that $N \in \mathscr{F}$ and $N \subset M$. It follows from (3) that $L \nsubseteq N$. Hence $N \subset L$, and by (4), $N \subseteq M'$. Since M_- is the smallest subspace containing all such N, we have $M_- \subseteq M'$. Hence $M_- = M'$. We may now deduce from (5) that the quotient space M/M_- has dimension greater than one, contrary to hypothesis. This is the required contradiction and the proof of the lemma is complete.

THEOREM 2.25. *Let T be a compact operator on X. There exists a simple nest $\mathscr{F}$, each of whose members is a subspace invariant under T.*

Proof. Let $\mathscr{N}_i$ denote the class of all invariant nests. Then $\mathscr{N}_i$ is not empty since it contains the trivial nest consisting of the subspaces $\{0\}$, X. The class $\mathscr{N}_i$ may be partially ordered by inclusion, and an argument based on Zorn's lemma proves the existence of at least one maximal element. Let $\mathscr{F}$ be a maximal member of $\mathscr{N}_i$. By the type of reasoning used in proving Lemma 2.24 we may show that $\mathscr{F}$ has properties (i) and (ii) of simple nests.

It remains to verify that, given any M in $\mathscr{F}$, the quotient space M/M_- is at most one-dimensional. Suppose that, for some M in $\mathscr{F}$, this is not so. When

$x \in M$ we denote by $[x]$ the coset $x + M_-$. Under the usual norm $\| \|_M$, defined by

$$\|[x]\|_M = \inf\{\|x - y\| : y \in M_-\}$$

M/M_- is a complex Banach space. Since M and M_- are invariant under T, we may define a linear operator T_M on M/M_- by the equation

$$T_M[x] = [Tx] \quad (x \in M).$$

By Theorems 2.12 and 2.14, T_M is compact. Since M/M_- has dimension greater than one, the theorem of Aronszajn and Smith implies the existence of a proper subspace L_M of M/M_- which is invariant under T_M. If we now set

$$L = \{x \in M : [x] \in L_M\}$$

then L is a subspace of X (being the inverse image under the continuous linear map $x \to [x]$ of the subspace L_M) such that $M_- \subset L \subset M$. We may now verify, by the method used at the corresponding stage in the proof of Lemma 2.24, that $L \notin \mathscr{F}$, but that the family $\mathscr{F}_1$ consisting of L and the members of $\mathscr{F}$ is totally ordered by inclusion. Since L_M is invariant under T_M, L is invariant under T. Thus $\mathscr{F}_1$ is an invariant nest, and is a proper enlargement of the maximal invariant nest $\mathscr{F}$. This gives a contradiction. Hence for each M in $\mathscr{F}$, M/M_- is at most one-dimensional and so $\mathscr{F}$ is a simple nest.

Throughout the remainder of this chapter we shall use the symbols T and $\mathscr{F}$ with the meanings attributed to them in the statement of Theorem 2.25. If $M \in \mathscr{F}$, then either $M = M_-$ or M/M_- has dimension one. In the latter case let $z_M \in M \backslash M_-$. Then since M is invariant under T we have $Tz_M \in M$, and hence Tz_M can be expressed (uniquely) in the form

$$Tz_M = \alpha_M z_M + y_M \tag{6}$$

where α_M is a scalar and $y_M \in M_-$. It is easily verified that α_M does not depend on the particular choice of z_M. When $M = M_-$ we define $\alpha_M = 0$. In this way we associate with each M in $\mathscr{F}$ a scalar α_M which we call the *diagonal coefficient* of T at M. (In the finite-dimensional case the elements z_M form a basis of X, and with respect to this basis T has super-diagonal matrix with diagonal elements α_M.)

Let $\alpha \in \mathbf{C}$. We define the *diagonal multiplicity* of α to be the number (possibly infinite) of distinct subspaces M in $\mathscr{F}$ for which $\alpha_M = \alpha$.

LEMMA 2.26. *Let $\varepsilon > 0$ be given, and let $\mathscr{F}_0$ be the family consisting of those subspaces M in $\mathscr{F}$ for which $|\alpha_M| \geqslant \varepsilon$. Then $\mathscr{F}_0$ has only a finite number of members.*

Proof. Suppose that $\mathscr{F}_0$ has infinitely many members. We shall use the symbols z_M and y_M as in equation (6). By Lemma 2.15, we may assume that the element z_M in $M \backslash M_-$ has been chosen in such a way that

$$\|z_M\| = 1 \tag{7}$$

$$\|z_M + y\| \geqslant \tfrac{1}{2} \quad (y \in M_-). \tag{8}$$

Now suppose that L, M are distinct members of $\mathscr{F}_0$. Then either $L \subset M$ or $M \subset L$. We may assume that $L \subset M$, and hence that $L \subseteq M_-$. It follows that $z_L \in M_-$ and hence that $Tz_L \in M_-$. Thus

$$\begin{aligned} Tz_M - Tz_L &= \alpha_M z_M + (y_M - Tz_L) \\ &= \alpha_M (z_M + y) \end{aligned}$$

where $y \in M_-$. From (8) we deduce that

$$\begin{aligned} \|Tz_M - Tz_L\| &= |\alpha_M| \, \|z_M + y\| \\ &\geqslant \tfrac{1}{2}\varepsilon \quad (L, M \in \mathscr{F}_0; L \neq M). \end{aligned}$$

It follows that the infinite family $\{Tz_M : M \in \mathscr{F}_0\}$ contains no convergent subsequence. In view of (7) this contradicts the assumption that T is a compact operator.

COROLLARY 2.27. *Every non-zero scalar has finite diagonal multiplicity.*

Proof. If $\alpha \neq 0$, we may choose ε so that $0 < \varepsilon < |\alpha|$. The preceding lemma then implies that α has finite diagonal multiplicity.

LEMMA 2.28. *Let $M \in \mathscr{F}$ and let $\delta > 0$ be given. Then there exists a subspace L*

in $\mathscr{F}$ such that $L \subset M$ and

$$\|[Tx]\|_L \leqslant \delta\|x\| \quad (x \in M_-),$$

where $[y]$ denotes the coset $y + L(y \in X)$ and $\| \ \|_L$ is the usual norm in the quotient space X/L.

Note. The interest of the lemma lies in the case in which $M = M_-$. When $M \neq M_-$, the result is trivial since we may take $L = M_-$.

Proof. Suppose that the lemma is false. Denote by $\mathscr{F}_0$ the class of all L in $\mathscr{F}$ such that $L \subset M$. Since we are going to vary L, we shall not use the notation $[y]$ for cosets, but throughout the proof will write $y + L$. If $L \in \mathscr{F}_0$, the set

$$S_L = \{x : x \in M_-, \|x\| = 1, \|Tx + L\|_L > \delta\}$$

is not empty. Since $S_L \subseteq S_N$ if $N \subseteq L$, the family $\{S_L : L \in \mathscr{F}_0\}$ forms a filter base on the unit sphere U in X. Hence the family $\{T(S_L) : L \in \mathscr{F}_0\}$ forms a filter base on the compact set $\mathrm{cl}[T(U)]$. It follows that there is a point x_0 common to the norm closures of the sets $T(S_L)(L \in \mathscr{F}_0)$. Since $\|y + L\|_L > \delta(y \in T(S_L))$, we have

$$\|x_0 + L\|_L \geqslant \delta \quad (L \in \mathscr{F}_0). \tag{9}$$

Furthermore we have $S_L \subseteq M_-$, $T(S_L) \subseteq M_-$, and hence

$$x_0 \in M_- = \mathrm{cl}[\cup\{L ; L \in \mathscr{F}_0\}].$$

Thus for some L in $\mathscr{F}_0$, we may choose an element y in L such that $\|x_0 - y\| < \delta$. This contradicts (9) and the lemma is proved.

LEMMA 2.29. *Let ρ be a non-zero eigenvalue of T, and x a corresponding eigenvector. Let*

$$M = \cap\{L : L \in \mathscr{F}, x \in L\}.$$

Then $M \in \mathscr{F}$ and $\rho = \alpha_M$.

Proof. The property (ii) of simple nests immediately implies that $M \in \mathscr{F}$. In proving that $\rho = \alpha_M$ we shall consider separately the two cases in which (respectively) $M = M_-$, and $M \neq M_-$.

(a) Suppose that $M = M_-$. Choose δ so that

$$0 < \delta < \tfrac{1}{2}\rho, \tag{10}$$

and let L be chosen to satisfy the conclusions of Lemma 2.28. Since $L \subset M$ and $L \in \mathscr{F}$, it is an immediate consequence of the definition of M that $x \notin L$. We may choose y in M in such a way that

$$y - x \in L, \tag{11}$$

$$\|y\| < 2\|[y]\|_L = 2\|[x]\|_L. \tag{12}$$

Since L is invariant under T, we deduce from (11) that $Ty - Tx \in L$, and that

$$\begin{aligned} Ty - \rho y = Tx - \rho x &+ (Ty - Tx - \rho y + \rho x) \\ &= Ty - Tx - \rho(y - x) \\ &\in L. \end{aligned}$$

Hence $[Ty] = \rho[y]$, and

$$\begin{aligned} \|[Ty]\|_L &= |\rho| \, \|[y]\|_L \\ &> \tfrac{1}{2}|\rho| \, \|y\| \\ &> \delta\|y\|. \end{aligned}$$

Here we have used (12) and (10). Since $y \in M = M_-$ this contradicts the assumption that L satisfies the conclusions of Lemma 2.28. Hence case (a) cannot occur.

(b) We may now suppose that $M \neq M_-$. Then $x \in M$, but $x \notin M_-$, since M is, by definition, the smallest member of $\mathscr{F}$ containing x. Let $z_M \in M \backslash M_-$ and let y_M in M_- be chosen so that $Tz_M = \alpha_M z_M + y_M$. We may put $x = \beta z_M + y$, where $y \in M_-$ and $\beta \neq 0$. Then

$$\begin{aligned} 0 = Tx - \rho x &= T(\beta z_M + y) - \rho(\beta z_M + y) \\ &= \beta(\alpha_M z_M + y_M) + Ty - \rho(\beta z_M + y) \\ &= \beta(\alpha_M - \rho)\, z_M + \beta y_M + Ty - \rho y. \end{aligned}$$

Now y, y_M and (since M_- is invariant under T) Ty are all elements of M_- but $z_M \notin M_-$. Hence $\beta(\alpha_M - \rho) = 0$, and since $\beta \neq 0$, it follows that $\alpha_M = \rho$.

Lemma 2.29 asserts that a non-zero eigenvalue of T is a diagonal coefficient of T. We now prove a result in the opposite direction.

LEMMA 2.30. *Let $M \in \mathscr{F}$ and suppose that $\alpha_M \neq 0$. Then α_M is an eigenvalue of T.*

Proof. It is sufficient to show that α_M is an eigenvalue of the operator T_M obtained by restricting T to the space M. Since $\alpha_M \neq 0$ we have $M \neq M_-$. Now T_M is compact. From equation (6) it follows that the range of the operator $T_M - \alpha_M I_M$ is contained in M_-, and is therefore not the whole space M. It follows from Theorem 2.21(i) that α_M is an eigenvalue of T_M and hence of T.

LEMMA 2.31. *Let ρ be a non-zero eigenvalue of T. Then the diagonal multiplicity of ρ is equal to its algebraic multiplicity as an eigenvalue of T.*

Proof. Let d denote the diagonal multiplicity, m the algebraic multiplicity, and v the index of ρ relative to T. Then by Theorem 2.21

(a) v is the least integer such that whenever for some x in X we have $(T - \rho I)^{v+1} x = 0$ then also $(T - \rho I)^v x = 0$;

(b) v is the least integer such that

$$(T - \rho I)^{v+1} X = (T - \rho I)^v X;$$

(c) the null-space of $(T - \rho I)^v$ has dimension m.

Let S be the compact operator defined by

$$S - \lambda I = (T - \rho I)^v$$

where $\lambda = -(-\rho)^v$. Then λ is an eigenvalue of S which has index one and algebraic multiplicity m. Since S is a polynomial in T, each subspace M in $\mathscr{F}$ is invariant under S. We may therefore consider the diagonal coefficients of S with respect to the nest $\mathscr{F}$.

Let $M \in \mathscr{F}$ and let α_M, σ_M denote the diagonal coefficients at M of T, S respectively. If $M = M_-$, we have $\alpha_M = \sigma_M = 0$. If $M \neq M_-$ then with the usual notation we may deduce from (6) that

$$(T - \rho I)z_M = (\alpha_M - \rho)z_M + y_M.$$

It easily follows that, for $n = 1, 2, \ldots$, we have

$$(T - \rho I)^n z_M = (\alpha_M - \rho)^n z_M + y^{(n)},$$

where $y^{(n)} \in M_-$. In particular, by taking $n = v$, we obtain

$$Sz_M = \lambda z_M + (\alpha_M - \rho)^v z_M + y^{(v)}.$$

Thus $\sigma_M = \lambda + (\alpha_M - \rho)^v$. We deduce that $\sigma_M = \lambda$ if and only if $\alpha_M = \rho$. Hence the diagonal multiplicity of λ relative to S is d. It is now sufficient to prove the lemma under the additional hypothesis that ρ has index one relative to T, since in the general case we may reduce to this situation by replacing T, ρ by S, λ respectively.

Suppose therefore that ρ has index one relative to T, and let N be the null-space of the operator $T - \rho I$. Given x in N, define

$$M(x) = \cap\{L : L \in \mathscr{F}, x \in L\}.$$

From Lemma 2.29 and its proof we deduce that $M(x) \in \mathscr{F}$, $x \in M(x) \backslash M_-(x)$ and $\alpha_{M(x)} = \rho (x \in N, x \neq 0)$. The remainder of the proof is divided into three stages.

First, we show conversely that, if $M \in \mathscr{F}$ and $\alpha_M = \rho$, then $M = M(x)$ for some non-zero x in N. For this purpose, let T_M denote the restriction of T to M, and let W_M, N_M be the range and null-space (respectively) of the operator $T_M - \rho I_M$. Then T_M is compact, and it is immediate from the definition of index in terms of null-spaces that ρ has index one relative to T_M. Hence by Theorem 2.21(iv) we have $W_M \oplus N_M = M$. Since, as in the proof of Lemma 2.30, $W_M \subseteq M_-$, it follows that N_M meets $M \backslash M_-$. If $x \in N_M \cap (M \backslash M_-)$, it is easily verified that $x \in N$, $x \neq 0$, and $M(x) = M$.

Secondly, let $M_1 \subset M_2 \subset \ldots \subset M_d$ be the distinct members of the nest $\mathscr{F}$ at which T has diagonal coefficient ρ. We may choose non-zero vectors $x_1, \ldots, x_d$ in N such that $M_r = M(x_r) (r = 1, \ldots, d)$. For each $r = 1, \ldots, d$, x_r is not a linear combination of $x_1, \ldots, x_{r-1}$: for this would imply that $x_r \in M(x_{r-1}) \subseteq M_-(x_r)$, which is not so. Hence $x_1, \ldots, x_d$ are linearly independent elements of N, and since $\dim N = m$ we have $m \geqslant d$.

Thirdly, suppose that $m > d$. Let $\phi_1, \ldots, \phi_d$ be bounded linear functionals on X such that $\phi_r(x_r) \neq 0$, but $\phi_r(x) = 0 (x \in M_-(x_r))$. Then if $x \in M(x_r)$ and $\phi_r(x) = 0$, we have $x \in M_-(x_r)$. Now since $\dim N > d$, we may choose a non-zero vector x in N such that $\phi_r(x) = 0 \, (r = 1, \ldots, d)$. Then $\alpha_{M(x)} = \rho$, and therefore $M(x) = M(x_r)$ for some r. Thus $x \in M(x_r)$, $\phi_r(x) = 0$, and we have $x \in M_-(x_r) = M_-(x)$. However, this is impossible. Hence $m \leqslant d$. Since the reverse inequality has already been established, we have $m = d$, and the lemma is proved.

We now state a theorem which summarizes the principal results obtained in the preceding lemmas.

THEOREM 2.32. *Let T be a compact operator on X, and let $\mathscr{F}$ be a simple nest of subspaces of X, each of which is invariant under T. Then*

(i) *a non-zero scalar ρ is an eigenvalue of T if and only if it is a diagonal coefficient of T;*

(ii) *the diagonal multiplicity of ρ is equal to its algebraic multiplicity as an eigenvalue of T;*

(iii) *the operator T is quasinilpotent if and only if $\alpha_M = 0 (M \in \mathscr{F})$; or equivalently if and only if $TM \subseteq M_- (M \in \mathscr{F})$.*

Proof. The only statement not already proved is (iii). Now the spectrum of T consists of zero (except possibly in the finite-dimensional case) and the non-zero eigenvalues of T. From (i) it follows that the spectrum of T reduces to the single point zero if and only if $\alpha_M = 0$ $(M \in \mathscr{F})$.

COROLLARY 2.33. *If there is a continuous nest of subspaces of X, each of which is invariant under T, then T is quasinilpotent.*

Proof. If $M \in \mathscr{F}$, then $M = M_-$ and $\alpha_M = 0$. The result follows from part (iii) of the preceding theorem.

3. Riesz Operators

In this chapter we study a class of operators, called the Riesz operators, which have a spectral theory like that of a compact operator. Numerous characterizations of the Riesz operators are given. It is shown that in general neither the sum nor the product of two Riesz operators are Riesz operators. It is also shown that the class of Riesz operators is not closed in the operator norm. A sufficient condition is given for the limit of a sequence of Riesz operators to be a Riesz operator. The possibility of decomposing a Riesz operator into the sum of a compact operator and a quasinilpotent operator is discussed. West's proof that this can be done in Hilbert space is given.

Our first result properly belongs to the last chapter. It will be used frequently in the present chapter.

PROPOSITION 3.1. *Let $E \in L(X)$ and $E^2 = E$. Then E is compact if and only if EX is finite-dimensional*

Proof. If EX is finite-dimensional then, by Theorem 2.10, E is compact. Conversely if E is compact then the restriction of E to EX, namely the identity operator on EX, is also compact. Hence the closed unit ball in EX is compact and so EX is finite-dimensional.

Next we give some definitions.

Definition 3.2. Let $T \in L(X)$. Define

$$\kappa(T) = \inf\{\|T - C\| : C \in K(X)\},$$

where $K(X)$ denotes the set of compact operators on X. T is said to be *asymptotically quasi-compact* if $\{\kappa(T^n)\}^{1/n} \to 0$ as $n \to \infty$.

Observe that, since $K(X)$ is a closed two-sided ideal in the Banach algebra

$L(X)$, the quotient algebra $L(X)/K(X)$ is also a Banach algebra. The asymptotically quasi-compact operators on X are precisely those operators whose images under the canonical mapping of $L(X)$ into $L(X)/K(X)$ are quasinilpotent elements.

Definition 3.3. Let $T \in L(X)$. Define

$$\phi(T) = \inf\{\|T - C\| : C \in F(X)\},$$

where $F(X)$ denotes the set of finite-rank operators on X. T is said to be *asymptotically quasi-finite-rank* if $\{\phi(T^n)\}^{1/n} \to 0$ as $n \to \infty$.

Definition 3.4. Let $T \in L(X)$. T is said to be a *Riesz operator* if it has the following three properties.

(i) For every $\lambda \neq 0$ and each positive integer n, the set of solutions of the equation $(\lambda I - T)^n x = 0$ forms a finite-dimensional subspace of X, which is independent of n provided that n is sufficiently large.

(ii) For every $\lambda \neq 0$ and each positive integer n, the range of $(\lambda I - T)^n$ is a closed subspace of X which is independent of n provided that n is sufficiently large.

(iii) The eigenvalues of T have at most one cluster point 0.

It will be shown that the classes of asymptotically quasi-compact operators, asymptotically quasi-finite-rank operators, and Riesz operators coincide. We begin by proving some elementary properties of asymptotically quasi-compact operators.

THEOREM 3.5. *Let A, B, in $L(X)$, be asymptotically quasi-compact operators with $AB = BA$. Then AB and $A + B$ are asymptotically quasi-compact. Let $\lambda \in \mathbf{C}$. Then λA is also asymptotically quasi-compact.*

Proof. Let $[A], [B]$ denote respectively the images of A and B under the canonical mapping of $L(X)$ into the quotient algebra $L(X)/K(X)$, and let $r(\cdot)$ denote the spectral radius in the second algebra. By hypothesis $r([A]) = r([B]) = 0$ and $[A][B] = [B][A]$. Hence

$$0 \leqslant r([A][B]) \leqslant r([A])r([B]),$$

$$0 \leqslant r([A] + [B]) \leqslant r([A]) + r([B]).$$

(See for example Corollary 3 of [**3**, p. 19].) It follows that

$$r([A] + [B]) = 0 = r([A][B]).$$

Hence AB and $A + B$ are asymptotically quasi-compact. The last statement of the theorem is trivial.

The sum and product of two asymptotically quasi-compact operators need not be asymptotically quasi-compact, as the following example shows.

Example 3.6. Let $X = l^2$. Define

$$\begin{aligned} x &= \{x_1, x_2, x_3, x_4, \ldots\}, \\ Sx &= \{0, x_1, 0, x_3, \ldots\}, \\ Tx &= \{x_2, 0, x_4, 0, \ldots\}, \\ STx &= \{0, x_2, 0, x_4, \ldots\}, \\ TSx &= \{x_1, 0, x_3, 0, \ldots\}, \\ (S + T)x &= \{x_2, x_1, x_4, x_3, \ldots\}. \end{aligned}$$

Observe that $S, T \in L(X)$ and $S^2 = T^2 = 0$. Hence S and T are asymptotically quasi-compact. However, ST is a projection with an infinite-dimensional range. By Proposition 3.1, $ST \notin K(X)$. Also $(ST)^n = ST$ and $K(X)$ is a closed subspace of $L(X)$. Hence

$$\{\kappa((ST)^n)\}^{1/n} \to 1 \quad \text{as} \quad n \to \infty$$

and so ST is not asymptotically quasi-compact. Furthermore, $(S \subseteq T)^2 = I \notin K(X)$, and again it follows that $S + T$ is not asymptotically quasi-compact. Note also that neither S nor T is compact, for otherwise ST would be compact and so have finite-dimensional range.

LEMMA 3.7. *Let $T \in L(X)$. If T is asymptotically quasi-compact then so is T^*.*

Proof. Let T be asymptotically quasi-compact. Then by Theorem 2.13

$$\{\kappa((T^*)^n)\}^{1/n} \leqslant \{\kappa(T^n)\}^{1/n} \qquad (n \in \mathbf{N})$$

and so T^* is asymptotically quasi-compact.

Next, we prove that an asymptotically quasi-compact operator is a Riesz operator. Some preliminary results are required.

Let T, in $L(X)$, be asymptotically quasi-compact. Then, given any positive real number Λ, we can choose a positive integer q so that

$$\{\kappa(T^q)\}^{1/q} < \frac{1}{6\Lambda}, \kappa(T^q) < \frac{1}{(6\Lambda)^q}.$$

Hence we can find a compact operator V such that

$$\|T^q - V\| < \frac{1}{(6\Lambda)^q}.$$

We set $U = T^q - V$. Also for λ in $\mathbf{C}$ and n a positive integer we define

$$N_n(\lambda) = \{x \in X : (\lambda I - T)^n x = 0\}.$$

LEMMA 3.8. *Let T, in $L(X)$, be asymptotically quasi-compact, and let $\lambda \neq 0$. For any positive integer n, $N_n(\lambda)$ is a finite-dimensional subspace of X.*

Proof. Let $\mu = \lambda^{-1}$. We prove the lemma for $|\mu| \leqslant \Lambda$. Since Λ is arbitrary this involves no loss of generality. Now, since $\|\mu^q U\| < 6^{-q} \leqslant \frac{1}{6}$, we know that $I - \mu^q U$ has a bounded inverse, given by

$$K = I + \mu^q U + (\mu^q U)^2 + \dots.$$

Thus

$$K(I + \mu T + \dots + \mu^{q-1}T^{q-1})(I - \mu T) = K[I - \mu^q(U + V)] = I - \mu^q K V,$$

and so, if $x \in N_1(\lambda)$, we have $(I - \mu^q KV)x = 0$. However, $\mu^q KV$ is a compact operator, and so by Lemma 2.16 the solutions of this equation form a finite-dimensional subspace of X. Hence $N_1(\lambda)$, being a subspace of this space, is also finite-dimensional. To extend this to general n, observe that by Theorem 3.5

$$(\lambda I - T)^n = \lambda^n I - T_0,$$

where T_0 is asymptotically quasi-compact. The proof is complete.

LEMMA 3.9. *Let T, in $L(X)$, be asymptotically quasi-compact, and let $\lambda \neq 0$. There is a positive integer p such that $N_n(\lambda) = N_p(\lambda)$ for $n > p$.*

Proof. As in Proposition 1.43 it is enough to find a positive integer p such that $N_{p+1}(\lambda) = N_p(\lambda)$. Suppose, on the contrary, that $N_p(\lambda)$ is a proper subspace on $K_{p+1}(\lambda)$ for each p. We shall suppose, as before, that $\mu = \lambda^{-1}$ and $|\mu| \leqslant \Lambda$. Then, for every non-negative integer n, there is an element y_n such that $y_n \in N_{n+1}(\lambda)$, $\|y_n\| = 1$ and $d(y_n, N_n(\lambda)) \geqslant \frac{1}{2}$, by Lemma 2.15. Put $z_n = \mu^q V y_n$. Then

$$\begin{aligned} z_n &= \mu^q(T^q y_n - U y_n) \\ &= y_n - (I + \mu T + \ldots + \mu^{q-1}T^{q-1})(I - \mu T)y_n - \mu^q U y_n. \end{aligned}$$

Since the middle term of this last expression is an element of $N_n(\lambda)$, it follows that if $n > m$,

$$z_n - z_m = y_n - y - \mu^q U y_n + \mu^q U y_m,$$

where $y \in N_n(\lambda)$, and so

$$\begin{aligned} \|z_n - z_m\| &\geqslant \|y_n - y\| - \|\mu^q U y_n\| - \|\mu^q U y_m\| \\ &\geqslant \tfrac{1}{2} - \tfrac{1}{6} - \tfrac{1}{6} = \tfrac{1}{6}. \end{aligned}$$

Thus $\{z_n\}$ can contain no convergent subsequence. This contradicts the fact that V is compact and the proof is complete.

LEMMA 3.10. *Let T, in $L(X)$, be asymptotically quasi-compact and let $\lambda \neq 0$. For each positive integer n, $(\lambda I - T)^n X$ is a closed subspace of X. Moreover, there is a positive integer p such that*

$$(\lambda I - T)^n X = (\lambda I - T)^p X \qquad (n > p).$$

Proof. By Lemma 3.7, T^* is asymptotically quasi-compact. Hence, by Lemma 3.8, for each positive integer n the subspace

$$Y = \{y \in X^* : (\lambda I^* - T^*)^n y = 0\}$$

is finite-dimensional. Moreover, Y is a weak*-closed subspace of X^*. To see this, let $\{x_\alpha\}$ be a net of elements of Y converging in the weak*-topology to y. Then, since all Hausdorff linear topologies agree on a finite-dimensional space, $\{x_\alpha\}$ converges to y in the norm of X^*. (See, for example, Theorem 3.3H of [**15**: p. 127].) Since a finite-dimensional subspace of X^* is norm-closed it follows that Y is weak*-closed. Hence Y is the annihilator of $(\lambda I - T)^n X$ and this is a closed subspace of X. The desired conclusion now follows from Lemmas 3.8 and 3.9.

LEMMA 3.11. *Let T, in $L(X)$, by asymptotically quasi-compact. The eigenvalues of T have at most one cluster point* 0.

Proof. Let Λ be a positive real number. It suffices to prove that there are only a finite number of eigenvalues λ such that $|\lambda^{-1}| \leqslant \Lambda$. Suppose, on the contrary, that $\{\lambda_n\}$ is a sequence of distinct eigenvalues with $|\lambda_n^{-1}| \leqslant \Lambda$. Let x_n be an eigenvector corresponding to λ_n (that is, a non-zero element of $N_1(\lambda_n)$), and let Y_n be the subspace spanned by $x_1, \ldots, x_n$, which are linearly independent. (See the proof of Lemma 2.17.) Then we can find a sequence $\{y_n\}$ such that $y_n \in Y_n$, $\|y_n\| = 1$, and $d(y_n, Y_{n-1}) \geqslant \frac{1}{2}$, by Lemma 2.15. The sequence $\{\lambda_n^{-q} y_n\}$ is bounded. Define

$$z_n = \lambda_n^{-q} V y_n = \lambda_n^{-q} T^q y_n - \lambda_n^{-q} U y_n.$$

Now y_n is a linear combination of $x_1, \ldots, x_n$, and so $(I - \lambda_n^{-q} T^q) y_n$ is a linear combination of $x_1, \ldots, x_{n-1}$ since

$$(I - \lambda_n^{-q} T^q) x_r = \{1 - (\lambda_r/\lambda_n)^q\} x_r.$$

Hence $y_n \in Y_{n-1}$. It follows that for $m < n$

$$z_n - z_m = y_n - y - \lambda_n^{-q} U y_n + \lambda_m^{-q} U y_m,$$

where $y \in Y_{n-1}$, and so

$$\begin{aligned} \|z_n - z_m\| &\geqslant \|y_n - y\| - \|\lambda_n^{-q} U y_n\| - \|\lambda_m^{-q} U y_m\| \\ &\geqslant \tfrac{1}{2} - \tfrac{1}{6} - \tfrac{1}{6} = \tfrac{1}{6}. \end{aligned}$$

Hence $\{z_n\}$ can contain no convergent subsequence, and this contradicts the fact that V is compact.

THEOREM 3.12. *Let $T \in L(X)$. The following statements are equivalent.*

(i) *T is asymptotically quasi-compact.*

(ii) *T is a Riesz operator.*

(iii) *T is asymptotically quasi-finite-rank.*

Proof. That (i) implies (ii) follows from Lemmas 3.8, 3.9, 3.10 and 3.11. Now assume that (ii) holds, and let λ be a non-zero complex number. The map $\lambda I - T$ has finite ascent and finite descent. By Proposition 1.51 these numbers are equal to m say. If $m = 0$ then $\lambda \in \rho(T)$. If $m \neq 0$ then λ is an eigenvalue of T and so is an isolated point of $\sigma(T)$. Since $(\lambda I - T)^m X$ is closed, λ is a pole of the resolvent of T of order m by Theorem 1.54. Let $\varepsilon > 0$ be given. Define

$$V = \sum_{|\lambda| > \varepsilon} E(\lambda)T \quad \text{and} \quad U = T - V$$

where the sum is taken over the finite number of points of $\sigma(T)$ of absolute value greater than ε. Observe that $U^n = T^n - V^n$, since $UV = 0$ by Theorem 1.39. By the spectral mapping theorem

$$\sigma(U) \subseteq \{\lambda : |\lambda| \leqslant \varepsilon\}.$$

Hence $\mu I - U$ is invertible in $L(X)$ if $|\mu| > \varepsilon$ and

$$(\mu I - U)^{-1} = \sum_{n=0}^{\infty} U^n \mu^{-(n+1)}.$$

Put $\lambda = \mu^{-1}$. Then

$$(I - \lambda U)^{-1} = \sum_{n=0}^{\infty} \lambda^n U^n \qquad (|\lambda| < \varepsilon^{-1})$$

and so, as in the proof of Proposition 1.8,

$$\limsup_n \|U^n\|^{1/n} \leqslant \varepsilon.$$

However, $\phi(T^n) \leqslant \|U^n\|$, since $U^n = T^n - V^n$ and V^n is of finite-rank. Hence T is asymptotically quasi-finite-rank and so (ii) implies (iii). Clearly (iii) implies (i) and the proof of the theorem is complete.

Observe that, since $\overline{F}(X)$, the norm closure of $F(X)$, is a closed two-sided ideal in the Banach algebra $L(X)$, the quotient algebra $L(X)/\overline{F}(X)$ is a Banach algebra.

THEOREM 3.13. *Let $S(X)$ be the subset of operators in $L(X)$ whose images under the canonical mapping of $L(X)$ into $L(X)/\overline{F}(X)$ are quasinilpotent elements. Then $S(X)$ is the class of Riesz operators.*

Proof. Every operator in $S(X)$ is asymptotically quasi-compact. If T is asymptotically quasi-finite-rank then $T \in S(X)$. The desired result now follows from the previous theorem.

We now present the main theorem describing the spectral theory of a Riesz operator.

THEOREM 3.14. *Let T, in $L(X)$, be a Riesz operator.*

(i) *$\sigma(T)$ is countable and has no cluster point except possibly 0. Every non-zero number in $\sigma(T)$ is an eigenvalue of T and moreover a pole of the resolvent of T.*

Let λ be a non-zero point in $\sigma(T)$, and let $v(\lambda)$ be the order of the pole at λ.

(ii) *For each positive integer n, $N((\lambda I - T)^n)$ is finite-dimensional. Also*

$$N((\lambda I - T)^m) = N((\lambda I - T)^{m+1}) \quad (m \geqslant v(\lambda))$$

and $v(\lambda)$ is the smallest positive integer with this property.

(iii) *For each positive integer n, $(\lambda I - T)^n X$ is closed. Also*

$$(\lambda I - T)^{m+1} X = (\lambda I - T)^m X \quad (m \geqslant v(\lambda))$$

and $v(\lambda)$ is the smallest positive integer with this property.

(iv) *The spectral projection $E(\lambda)$ has a non-zero finite-dimensional range given by*

$$E(\lambda)X = N((\lambda I - T)^{v(\lambda)}).$$

The null-space of $E(\lambda)$ is $(\lambda I - T)^{v(\lambda)} X$.

(v) *If $d(\lambda)$ is the dimension of $E(\lambda)X$ then $1 \leqslant v(\lambda) \leqslant d(\lambda)$.*

Note. The integers $v(\lambda)$ and $d(\lambda)$ are called respectively the *index* and the *algebraic multiplicity* of the eigenvalue λ.

Proof. In the course of proving Theorem 3.12 it was established that each non-zero point of $\sigma(T)$ was a pole of the resolvent of T. This suffices to establish (i). Statements (ii) and (iii) were also established in the course of proving Theorem 3.12. Statement (iv) follows from Theorem 1.52. The last statement follows by considering the restriction of $\lambda I - T$ to the finite-dimensional space $E(\lambda)X$ and applying elementary linear algebra.

Let Y be a complex Banach space of dimension $\geqslant 2$ and let T, in $L(Y)$, be a Riesz operator. If there is a non-zero point λ in $\sigma(T)$ then λ is an eigenvalue of T. Hence there is a non-zero vector x in Y such that $Tx = \lambda x$. The one-dimensional subspace of Y generated by x is a proper closed invariant subspace for T. It is an unsolved problem whether in general a quasinilpotent operator on Y has a proper closed invariant subspace. Indeed it is not known for any specific separable infinite-dimensional complex Banach spacc Y whether every quasinilpotent operator on Y has a proper closed invariant subspace.

Next, we give an example to show that the set of Riesz operators is not in general closed in the norm topology for operators.

Example 3.15. Let $X = l^2$. For each positive integer m let f_m denote the element of X with 1 in the mth place and 0 elsewhere. Then $\{f_m : m = 1, 2, 3, \ldots\}$ is an orthonormal basis for X. Consider the sequence of scalars

$$\alpha_m = e^{-k}, \quad \text{for} \quad m = 2^k(2l + 1),$$

where $k, l = 0, 1, 2, \ldots$ Define A, in $L(X)$, by the equations

$$Af_m = \alpha_m f_{m+1} \quad (m = 1, 2, 3, \ldots).$$

It is easily verified that the norm of this operator is given by

$$\|A\| = \sup\{|\alpha_m| : m = 1, 2, 3, \ldots\}.$$

Observe that for each positive integer n

$$A^n f_m = \alpha_m \alpha_{m+1} \cdots \alpha_{m+n-1} f_{m+n}$$

and hence

$$\|A^n\| = \sup_m (\alpha_m \alpha_{m+1} \cdots \alpha_{m+n-1}).$$

Furthermore, from the definition of the α_m,

$$\alpha_1\alpha_2 \ldots \alpha_{2^k-1} \geqslant \prod_{r=1}^{k-1} \exp(-r2^{k-r-1}).$$

Therefore

$$(\alpha_1\alpha_2 \ldots \alpha_{2^k-1})^{2^{1-k}} \geqslant \left(\prod_{r=1}^{k-1} \exp[-(r/2^{r+1})]\right)^2$$

and, if we set

$$\sigma = \sum_{r=1}^{\infty} \frac{r}{2^{r+1}}$$

then

$$e^{-2\sigma} \leqslant \lim_{n\to\infty} \|A^n\|^{1/n}.$$

In particular, A is not a quasinilpotent operator. For each positive integer k we define the operator A_k by the equations

$$A_k f_m = \begin{cases} 0, & \text{for} \quad m = 2^k(2l + 1), \\ \alpha_m f_{m+1}, & \text{for} \quad m \neq 2^k(2l + 1). \end{cases}$$

It is easy to see that A_k is nilpotent. However

$$(A - A_k)\, f_m = \begin{cases} e^{-k} f_{m+1}, & \text{for} \quad m = 2^k(2l + 1), \\ 0, & \text{for} \quad m \neq 2^k(2l + 1). \end{cases}$$

Therefore $\|A - A_k\| = e^{-k}$ so that $\{A_k\}$ converges to A in the norm of $L(X)$. Suppose that A is a Riesz operator. Since the spectral radius of A is positive, there is a non-zero point λ in $\sigma(A)$ and, by Theorem 3.14, λ is an eigenvalue of A. Let x be a non-zero vector such that $Ax = \lambda x$. Suppose that the Fourier expansion of x with respect to the orthonormal basis $\{f_m\}$ is given by $x = \sum_{m=1}^{\infty} c_m f_m$. Then

$$Ax = \sum_{m=1}^{\infty} c_m \alpha_m f_{m+1} \quad \text{and} \quad \lambda x = \sum_{m=1}^{\infty} \lambda c_m f_m.$$

Hence

$$\lambda c_1 = 0$$

$$c_m \alpha_m = \lambda c_{m+1} \quad (m = 1, 2, 3, \ldots).$$

Therefore $c_m = 0$ $(m = 1, 2, 3, \ldots)$ contradicting $x \neq 0$. Hence A is not a Riesz operator.

Next we prove a result of West which gives a sufficient condition for a sequence of Riesz operators to converge to a Riesz operator. The proof given here is due to Ruston.

THEOREM 3.16. *For each positive integer n let T_n, in $L(X)$, be a Riesz operator. Suppose that $\{T_n\}$ converges in the norm of $L(X)$ to T. Suppose also that T commutes with each T_n. Then T is a Riesz operator.*

Proof. We use the fact that an operator is a Riesz operator if and only if it is asymptotically quasi-compact. Let $\varepsilon > 0$ be given. Choose p so that $\|T - T_p\| < \varepsilon/3$, and put $U_p = T - T_p$, so that $\|U_p\| < \varepsilon/3$. We shall suppose that $U_p \neq 0$, for otherwise $T = T_p$ and the conclusion follows immediately. We next choose q so that

$$\{\kappa(T_p^n)\}^{1/n} < \varepsilon/3 \quad (n > q).$$

Then, since T_p commutes with U_p, we have

$$T^n = (T_p + U_p)^n = \sum_{r=0}^{q} \binom{n}{r} T_p^r U_p^{n-r} + \sum_{r=q+1}^{n} \binom{n}{r} T_p^r U_p^{n-r} \quad (n \geqslant q)$$

and so

$$\kappa(T^n) \leqslant \sum_{r=0}^{q} \binom{n}{r} \|T_p\|^r \|U_p\|^{n-r} + \sum_{r=q+1}^{n} \binom{n}{r} \kappa(T_p^r) \|U^p\|^{n-r}$$

$$= \|U_p\|^n \sum_{r=0}^{q} \binom{n}{r} (\|T_p\|/\|U_p\|)^r + \sum_{r=q+1}^{n} \binom{n}{r} \kappa(T_p^r) \|U_p\|^{n-r}.$$

However, $\sum_{r=0}^{q} \binom{n}{r}(\|T_p\|/\|U_p\|)^r$ is a polynomial in n, and so there is a positive constant k such that

$$\sum_{r=0}^{q} \binom{n}{r}(\|T_p\|/\|U_p\|)^r < k.2^n$$

when n is sufficiently large. Hence

$$\kappa(T^n) < (\varepsilon/3)^n.k.2^n + \sum_{r=q+1}^{n} \binom{n}{r}(\varepsilon/3)^r(\varepsilon/3)^{n-r}$$

$$\leqslant (k+1)(2\varepsilon/3)^n$$

when n is sufficiently large. However, $(k+1)(2/3)^n \to 0$ as $n \to \infty$. Hence

$$\kappa(T^n) < \varepsilon^n$$

and so

$$\{\kappa(T^n)\}^{1/n} < \varepsilon$$

when n is sufficiently large. However, $(k+1)(2/3)^n \to 0$ as $n \to \infty$. Hence is asymptotically quasi-compact.

We now give three more characterizations of Riesz operators.

THEOREM 3.17. *Let $T \in L(X)$. Then T is a Riesz operator if and only if the following two conditions hold.*

(i) *Every non-zero point of $\sigma(T)$ is a pole of the resolvent of T.*

(ii) *For every non-zero point λ of $\sigma(T)$ the spectral projection $E(\lambda)$ has finite-dimensional range.*

Proof. It was established in Theorem 3.14 that a Riesz operator has properties (i) and (ii). Conversely, suppose that T, in $L(X)$, has properties (i) and (ii). By Theorem 1.52, every non-zero point of $\sigma(T)$ is an eigenvalue of T. Hence the eigenvalues of T have at most one cluster point 0. Also, for every $\lambda \neq 0$, the set of solutions of the equation $(\lambda I - T)^n x = 0$ forms a finite-dimensional subspace of X, which is independent of n provided that n is sufficiently large. For every $\lambda \neq 0$, $(\lambda I - T)^n X$ is a subspace of X which is independent of n provided that n is sufficiently large. If $\lambda \in \rho(T)$ this subspace is certainly closed. To complete the proof we have to show that this remains true if $\lambda \in \sigma(T)$.

Let m be the order of λ as a pole of the resolvent of T. Then λ is a pole of the resolvent of T^* of order m. Let $F(\lambda)$ be the spectral projection corresponding to the open-and-closed subset $\{\lambda\}$ of $\sigma(T^*)$. Then $F(\lambda) = \{E(\lambda)\}^*$. Since $E(\lambda)X$ is finite-dimensional, $E(\lambda)$ is compact by Proposition 3.1. Hence, by Theorem 2.13, $F(\lambda)$ is compact. Again by Proposition 3.1 it follows that $F(\lambda)$ has finite-dimensional range. Let n be a non-negative integer and let f, in X^*, satisfy $f(y) = 0$, for all y in $(\lambda I - T)^n X$. By Proposition 1.45, $f(y) = 0$ for all y in $(\lambda I - T)^m X$. Hence

$$y \in \{f : (\lambda I^* - T^*)^m f = 0\} = F(\lambda) X^*$$

and this subspace is finite-dimensional. Thus the annihilator of $(\lambda I - T)^n X$ is a finite-dimensional subspace of X^* and so is weak*-closed. Hence $(\lambda I - T)^n X$ is closed and the proof is complete.

The next characterization of Riesz operators is due to Dieudonné [**3**; p. 323]. A preliminary definition is required.

Definition 3.18. Let $T \in L(X)$. A *Riesz point* for T is a point λ in $\sigma(T)$ such that

(i) λ is isolated in $\sigma(T)$;

(ii) X is the direct sum of a closed subspace $F(\lambda)X$ and a finite-dimensional subspace $N(\lambda)X$ such that T leaves both these subspaces invariant, the restriction of $\lambda I - T$ to $F(\lambda)X$ is a linear homeomorphism and the restriction of $\lambda I - T$ to $N(\lambda)X$ is nilpotent.

THEOREM 3.19. *Let $T \in L(X)$. Then T is a Riesz operator if and only if every non-zero point in $\sigma(T)$ is a Riesz point for T.*

Proof. If T is a Riesz operator then every non-zero point λ in $\sigma(T)$ is a Riesz point for T. To see this take $N(\lambda)X = E(\lambda)X$ and $F(\lambda)X = (I - E(\lambda))X$. Apply Theorem 3.14 in conjunction with the result that $\lambda \in \rho(T|F(\lambda)X)$, which follows from Theorem 1.39. Conversely, suppose that every non-zero point of $\sigma(T)$ is a Riesz point for T. If E denotes the projection onto $N(\lambda)X$ then $ET = TE$. We have $\{\lambda\} = \sigma(T|EX)$ and $\lambda \in \rho(T|(I - E)X)$. By Proposition 1.53, $E = E(\lambda)$. There is a smallest positive integer m such that $(\lambda I - T)^m E(\lambda) = 0$. Hence $(\lambda I - T)^{m-1} E(\lambda) \neq 0$ and so, by Proposition 1.40, λ is a pole of the resolvent of T. It now follows from Theorem 3.17 that T is a Riesz operator.

West proved the following variant of Theorem 3.19.

THEOREM 3.20. *Let $T \in L(X)$. Then T is a Riesz operator if and only if the following condition is satisfied. For every non-zero point λ of $\sigma(T)$, X is the direct sum of a closed subspace $F(\lambda)X$ and a finite-dimensional subspace $N(\lambda)X$ such that T leaves both these subspaces invariant, the restriction of $\lambda I - T$ to $F(\lambda)X$ is a linear homeomorphism, and the restriction of $\lambda I - T$ to $N(\lambda)X$ is nilpotent.*

Proof. As in Theorem 3.19, if T is a Riesz operator the condition is satisfied. Conversely, suppose the condition holds for every non-zero point λ in $\sigma(T)$. If E denotes the projection onto $N(\lambda)X$ then $ET = TE$. We have $\{\lambda\} = \sigma(T|EX)$ and $\lambda \in \rho(T|F(\lambda)X)$. By Proposition 1.37

$$\sigma(T) = \sigma(T|EX) \cup \sigma(T|F(\lambda)X)$$

and so $\{\lambda\}$ is an open-and-closed subset of $\sigma(T)$. Hence every non-zero point in $\sigma(T)$ is isolated and the desired conclusion follows from the previous theorem.

We now prove the analogues for Riesz operators of Theorems 2.12, 2.13, 2.14.

THEOREM 3.21. *Let T, in $L(X)$, be a Riesz operator and let Y be a closed subspace of X invariant under T. Then $T|Y$ is a Riesz operator.*

Proof. Let $\lambda \in \rho(T)$. We show first that $(\lambda I - T)^{-1} Y \subseteq Y$. We have

$$(\lambda I - T)^{-1} = \sum_{n=0}^{\infty} \lambda^{-n-1} T^n \quad (|\lambda| > \|T\|)$$

and so by norm continuity

$$(\lambda I - T)^{-1} Y \subseteq Y \quad (|\lambda| > \|T\|).$$

Let $y \in Y$, $z \in Y^{\perp}$. Then

$$\langle (\lambda I - T)^{-1} y, z \rangle = 0 \quad (|\lambda| > \|T\|).$$

The left-hand side of this equation is an analytic function of λ on $\rho(T)$. Also $\rho(T)$ is connected and so, by the identity theorem,

$$\langle (\lambda I - T)^{-1} y, z \rangle = 0 \quad (\lambda \in \rho(T), y \in Y, z \in Y^{\perp}).$$

Hence $(\lambda I - T)^{-1} Y \subseteq Y(\lambda \in \rho(T))$. Let λ_0 be a non-zero point of $\sigma(T)$. Then

$$E(\lambda_0) = \frac{1}{2\pi i} \int_\Gamma (\lambda I - T)^{-1} \, d\lambda,$$

where Γ is a circle enclosing λ_0 but excluding the rest of $\sigma(T)$ and described once counterclockwise. By norm continuity $E(\lambda_0) Y \subseteq Y$. Let $\varepsilon > 0$ be given. As in the proof of Theorem 3.12 define

$$V = \sum_{|\lambda| > \varepsilon} E(\lambda) T,$$

where the sum is taken over the finite number of points in $\sigma(T)$ of absolute value greater than ε. Observe that $VY \subseteq Y$. It was shown in the proof of Theorem 3.12 that

$$\limsup_n \| T^n - V^n \|^{1/n} \leqslant \varepsilon.$$

Hence

$$\limsup_n \| (T^n - V^n) | Y \|^{1/n} \leqslant \varepsilon.$$

Therefore $T | Y$ is asymptotically quasi-compact. Hence, by Theorem 3.12, $T | Y$ is a Riesz operator.

THEOREM 3.22. *Let $T \in L(X)$. Then T is a Riesz operator if and only if T^* is a Riesz operator.*

Proof. By Lemma 3.7 and Theorem 3.12, if T is a Riesz operator then so is T^*. Now suppose that T^* is a Riesz operator. Then so is T^{**}. Now, by Theorem 3.21, T, being the restriction of T^{**} to its closed invariant subspace X, is a Riesz operator.

THEOREM 3.23. *Let T, in $L(X)$, be a Riesz operator and let Y be a closed subspace of X invariant under T. Then T_Y, the operator induced by T on the quotient space X/Y, is a Riesz operator.*

Proof. By Theorem 3.22, T^* is a Riesz operator. Also $Y^\perp$, the annihilator of Y, is a closed subspace of X^* invariant under T^* and so $T^* | Y^\perp$ is a Riesz operator by Theorem 3.21. As in Proposition 1.33 et seq. we can identify T_Y^* and $T^* | Y^\perp$. Hence, by Theorem 3.22, T_Y is a Riesz operator.

Let T, in $L(X)$, be a Riesz operator. It follows from Theorem 3.22 that T^* is also a Riesz operator. Let λ be a non-zero point of $\sigma(T) = \sigma(T^*)$. It has already been noted that the order of λ as a pole of the resolvent of T is the same as the order of λ as a pole of the resolvent of T^*. Next, we show that the dimensions of the spaces $E(\lambda; T)X$ and $E(\lambda; T^*)X^*$ are equal. Here, $E(\lambda; T)$ and $E(\lambda; T^*)$ denote respectively the spectral projections corresponding to the open-and-closed subset $\{\lambda\}$ of $\sigma(T) = \sigma(T^*)$. Some preliminary definitions and results are required.

Definition 3.24. A subspace M of a vector space F is said to have *finite codimension* in F if and only if the quotient space F/M has finite dimension. If M has finite codimension the dimension of F/M is called the *codimension* of M in F and is denoted by codim M.

LEMMA 3.25. *A subspace M of a vector space F has finite codimension n in F if and only if there exists an n-dimensional subspace N of F such that $F = M \oplus N$.*

Proof. Suppose that M has finite codimension n and let

$$\{x_1 + M, x_2 + M, \ldots, x_n + M\}$$

be a basis for F/M. It is easy to see that the set $\{x_1, x_2, \ldots, x_n\}$ is linearly independent and that its linear span N is such that $F = M \oplus N$. Suppose conversely that there is an n-dimensional subspace N of F such that $F = M \oplus N$. It is easy to verify that the restriction to N of the canonical mapping $x \to x + M$ is a one-to-one linear mapping of N onto F/M, and therefore by the rank and nullity theorem F/M is n-dimensional. The proof is complete.

If F is a finite-dimensional vector space and M, N are subspaces of F with $F = M \oplus N$ then $\dim F = \dim M + \dim N$. Hence, by Lemma 3.25, we have $\dim F = \dim M + \operatorname{codim} M$. It now follows from the rank and nullity theorem that if S is a linear operator on F

$$\operatorname{codim} R(S) = \dim N(S). \tag{1}$$

Let us now return to the study of Riesz operators.

THEOREM 3.26. *Let T, in $L(X)$, be a Riesz operator. Then $R(I - T)$ has finite codimension in X and*

$$\operatorname{codim} R(I - T) = \dim N(I - T).$$

Proof. If $1 \in \rho(T)$ the result is obvious. Hence, we may assume without loss of generality that $1 \in \sigma(T)$. Then $(I - T)E(1)X \subseteq E(1)X$; also

$$(I - T)(I - E(1))X = (I - E(1))X, \text{ and so}$$

$$R(I - T) = (I - T)X = (I - T)E(1)X \oplus (I - T)(I - E(1))X.$$

$$\therefore (I - T)X = (I - T)E(1)X \oplus (I - E(1))X. \tag{2}$$

However, $E(1)X$ is finite-dimensional by Theorem 3.14(iv) and so

$$N((I - T)|E(1)X) = N(I - T)$$

because $N(I - T) \subseteq E(1)X$. Consequently, by (1) applied to the operator $(I - T)|E(1)X$, we see that the dimension of $N(I - T)$ is equal to the codimension of $(I - T)E(1)X$ in $E(1)X$. By Lemma 3.25, there is a subspace N of $E(1)X$, with dimension equal to the codimension of $(I - T)E(1)X$ in $E(1)X$ such that $E(1)X = (I - T)E(1)X \oplus N$. Thus

$$\begin{aligned} X &= E(1)X \oplus (I - E(1))X \\ &= N \oplus (I - T)E(1)X \oplus (I - E(1))X \\ &= N \oplus (I - T)X \end{aligned}$$

where the last step follows from (2). Lemma 3.25 now shows that $R(I - T)$ has finite codimension in X and

$$\operatorname{codim} R(I - T) = \dim N = \dim N(I - T).$$

THEOREM 3.27. *Let T, in $L(X)$, be a Riesz operator. The null-spaces $N(I - T)$ and $N(I^* - T^*)$ have the same dimension.*

Proof. By Theorem 3.26,

$$\dim N(I - T) = \operatorname{codim} R(I - T).$$

It follows that

$$\begin{aligned}\dim N(I - T) &= \dim X/R(I - T)\\ &= \dim\{X/R(I - T)\}^*\\ &= \dim[R(I - T)^\perp]\\ &= \dim N(I^* - T^*).\end{aligned}$$

THEOREM 3.28. *Let T, in $L(X)$, be a Riesz operator and let λ be a non-zero point of $\sigma(T)$. Then for each positive integer n the null-spaces $N((\lambda I - T)^n)$ and $N((\lambda I^* - T^*)^n)$ have the same dimension. In particular, the ranges of the spectral projections $E(\lambda; T)$ and $E(\lambda; T^*)$ have the same dimension.*

Proof. By Theorem 3.5

$$(\lambda I - T)^n = \lambda^n I - K,$$

where K is a Riesz operator. Therefore

$$N((\lambda I - T)^n) = N(I - \lambda^{-n} K),$$

$$N((\lambda I^* - T^*)^n) = N(I^* - \lambda^{-n} K^*),$$

and $\lambda^{-n}K$ is a Riesz operator. The first statement of the theorem follows at once from Theorem 3.27. The second statement follows if we take n to be the index of the eigenvalue λ.

The next result is part of the folk-lore of the subject of Riesz operators.

THEOREM 3.29. *Let K be a compact operator on X and let Q, in $L(X)$, be quasi-nilpotent. Then $K + Q$ is a Riesz operator.*

Proof. Let n be a positive integer. Then, since the compact operators form a two-sided ideal of $L(X)$,

$$(K + Q)^n = Q^n + C,$$

where C is compact. It follows that

$$\{\kappa((K + Q)^n)\}^{1/n} \leqslant \|Q^n\|^{1/n} \quad (n = 1, 2, 3, \ldots)$$

and so $K + Q$ is asymptotically quasi-compact. The desired conclusion now follows from Theorem 3.12.

West [2] has proved the following partial converse to this theorem. A Riesz operator on a complex Hilbert space can be expressed as the sum of a compact operator and a quasinilpotent. The method of proof of this result is analogous to the process of super-diagonalizing a matrix and then splitting it into the sum of diagonal and nilpotent matrices. As shown in Chapter 2, for compact operators, super-diagonalization depends essentially on the existence of proper closed invariant subspaces. No equivalent result is known for Riesz operators, and without one we cannot expect a complete theory of super-diagonalization for such operators. In the proof we super-diagonalize along the direct sum of the ranges of the spectral projections corresponding to the non-zero eigenvalues, and then show that the remainder is a quasinilpotent operator.

Let H be an infinite dimensional complex Hilbert space and let K be a Riesz operator on H. There are three possibilities:

(i) $\sigma(K) = \{0\}$;
(ii) $\sigma(K)$ is a finite set containing non-zero points;
(iii) $\sigma(K)$ is an infinite set.

West's decomposition theorem is trivial in case (i). In case (ii) define

$$C = \sum_{r=1}^{n} KE(\lambda_r; K)$$

where $\{\lambda_r : r = 1, 2, \ldots, n\}$ are the distinct non-zero eigenvalues of K and $E(\lambda_r; K)$ is the spectral projection corresponding to the open-and-closed subset $\{\lambda_r\}$ of $\sigma(K)$. If $Q = K - C$ then $K = C + Q$ is a decomposition of the required type, since by the spectral mapping theorem $\sigma(Q) = \{0\}$. We now consider case (iii) and write $\sigma_0(K)$ for the non-zero spectrum of K. In this case, $\sigma_0(K)$ is an infinite sequence $\{\lambda_r\}_{r=1}^{\infty}$, which we assume to be enumerated in such a way that

$$|\lambda_1| \geqslant |\lambda_2| \geqslant |\lambda_3| \geqslant \ldots$$

Let $N_r = E(\lambda_r; K)\, H$ and, for k a non-negative integer, let

$$L_k = \bigoplus_{r=1}^{k} N_r. \tag{3}$$

(This direct sum is not necessarily orthogonal.) We shall consider the family of subspaces

$$\mathscr{F}_0 = \{\{0\}, L_1, L_2, \ldots\}.$$

Each member of $\mathscr{F}_0$ is finite-dimensional, and $\mathscr{F}_0$ is totally ordered by inclusion.

A *super-diagonalization process* (S.D.P.) for K is a family $\mathscr{F} = \{M_r\}$ of distinct subspaces M_r, which is totally ordered by inclusion, such that for $r = 0, 1, 2, \ldots$

$$KM_r \subseteq M_r \quad \text{and} \quad \dim M_r = r.$$

The subspaces L_k, defined by equation (3), are all invariant under K and are finite-dimensional. Thus, for each k, we can consider a maximal (finite) S.D.P. in L_k for the operator $K|L_k$. Let $\mathscr{F}_1$ be such a maximal S.D.P. in L_1: we can construct a maximal S.D.P. $\mathscr{F}_2$ in L_2 containing $\mathscr{F}_1$ as a subfamily. Proceeding thus we can construct, for each k, a maximal S.D.P. $\mathscr{F}_k$ in L_k containing $\mathscr{F}_1, \ldots, \mathscr{F}_{k-1}$ as subfamilies.

$$\overline{\mathscr{F}} = \bigcup_{k=1}^{\infty} \{\mathscr{F}_k\}$$

is a S.D.P. containing each $\mathscr{F}_k$, and we shall write $\overline{\mathscr{F}} = \{M_r\}_{r=0}^{\infty}$, where $\dim M_r = r$.

Choose an orthonormal system $\{e_r\}_{r=1}^{\infty}$ such that

$$e_r \in M_r \backslash M_{r-1} \quad (r \geqslant 1).$$

Then for each r

$$Ke_r = \alpha_r e_r + f_{r-1} \tag{4}$$

where $f_{r-1} \in M_{r-1}$ and α_r is independent of the particular choice of e_r. Consideration of $K|M_r$ shows that $\alpha_r \in \sigma(K|M_r)$. Since M_r is finite-dimensional, α_r is an eigenvalue of $K|M_r$. Hence $\alpha_r \in \sigma(K)$. If $\lambda \in \sigma_0(K)$, we define the *diagonal multiplicity* of λ to be the number of distinct subspaces M_r with $\alpha_r = \lambda$. Finite-dimensional arguments show that the diagonal multiplicity of λ is equal to its algebraic multiplicity as an eigenvalue of K (i.e. the dimension of $E(\lambda; K)\,H$).

Every vector in H can be written

$$x = \sum_{r=1}^{\infty} \langle x, e_r \rangle\, e_r + y \tag{5}$$

where $\langle y, e_r \rangle = 0$ for $r = 1, 2, \ldots$. Equations (4) and (5) give

$$Kx = \sum_{r=1}^{\infty} \alpha_r \langle x, e_r \rangle\, e_r + \sum_{r=1}^{\infty} \langle x, e_r \rangle\, f_{r-1} + Ky. \tag{6}$$

The separation of the series is justified by the norm convergence of the first. Define linear maps C and Q of H into itself by

$$Cx = \sum_{r=1}^{\infty} \alpha_r \langle x, e_r \rangle\, e_r \quad (x \in H), \tag{7}$$

$$Q = K - C.$$

THEOREM 3.30. (i) *$C \in K(H)$ and $Q \in L(H)$.*
(ii) *$\sigma(C) = \sigma(K)$. Also the non-zero eigenvalues have the same algebraic multiplicities with respect to both operators.*

Proof. (i) For each n, define C_n by

$$C_n x = \sum_{r=1}^{n} \alpha_r \langle x, e_r \rangle\, e_r \quad (x \in H).$$

C_n is an operator of finite rank on H and

$$(C - C_n)\, x = \sum_{r=n+1}^{\infty} \alpha_r \langle x, e_r \rangle\, e_r.$$

Therefore

$$\|(C - C_n)\, x\|^2 = \sum_{r=n+1}^{\infty} |\alpha_r \langle x, e_r \rangle|^2$$

$$\leqslant (\sup_{r \geqslant n+1} |\alpha_r|^2) \sum_{r=n+1}^{\infty} |\langle x, e_r \rangle|^2$$

$$\leqslant (\sup_{r \geqslant n+1} |\alpha_r|^2) \|x\|^2,$$

by Bessel's inequality. Hence

$$\|C - C_n\| \leqslant \sup_{r \geqslant n+1} |\alpha_r| \quad (n = 1, 2, \ldots)$$

and, since $\lim_{r \to \infty} \alpha_r = 0$, C is the limit in the norm of $L(H)$ of a sequence of operators of finite rank. Therefore C is compact and $Q = K - C \in L(H)$. This proves (i).

(ii) The non-zero eigenvalues of C are the α_r, and these are precisely the non-zero eigenvalues of K repeated according to their algebraic multiplicities. 0 is a cluster point of $\sigma(C) = \sigma(K)$ and the proof is complete.

In order to prove that Q is quasinilpotent we assume, to the contrary, that there exists a μ in $\sigma_0(Q)$. We shall see that this implies that $\mu \in \sigma_0(K)$, which leads to a contradiction. Now $Q = K - C$ and $C \in K(H)$. Hence Q is asymptotically quasi-compact and so is a Riesz operator. Therefore μ is an eigenvalue of Q. We require a preliminary lemma.

LEMMA 3.31. *If* $L = \text{clm}\{e_r : r = 1, 2, \ldots\}$ *then* $E(\mu; Q)\, H \cap L = \{0\}$.

Proof. From equation (4) and the definition of C it follows that if r is a positive integer then $Qe_r = f_{r-1} \in M_{r-1}$. Hence $Q^2 e_r \in M_{r-2}$ and $Q^r e_r = 0$. Let $\tau = \sigma(Q) \backslash \{\mu\}$. Then

$$E(\tau; Q) = \frac{1}{2\pi i} \int_B (\lambda I - Q)^{-1}\, d\lambda,$$

where B is a suitable finite family of rectifiable Jordan curves in $\rho(Q)$. Define

$$Y = \{x : Q^r x = 0\}.$$

Then Y is a closed subspace of H invariant under Q. Since Q is a Riesz operator, $\rho(Q)$ is connected, by Theorem 3.14. Hence, by Theorem 1.29,

$$(\lambda I - Q)^{-1}\, Y \subseteq Y \quad (\lambda \in \rho(Q)).$$

It follows that

$$\begin{aligned} E(\tau; Q)\, e_r &= \frac{1}{2\pi i} \int_B (\lambda I - Q)^{-1}\, d\lambda\, e_r \\ &= \frac{1}{2\pi i} \int_B (\lambda I | Y - Q | Y)^{-1}\, d\lambda\, e_r \\ &= e_r \end{aligned}$$

since $\sigma(Q | Y) = \{0\}$ and $0 \in \tau$. Therefore $L \subseteq E(\tau; Q)\, H$. Since $\mu \notin \tau$ the proof of the lemma is complete.

THEOREM 3.32. *Q is quasinilpotent.*

Proof. Let z be a non-zero vector such that $Qz = \mu z$. By Lemma 3.31

$$z = \sum_{r=1}^{\infty} \langle z, e_r \rangle\, e_r + w,$$

where $\langle w, e_r \rangle = 0$ for each r, and $w \neq 0$.

Rewriting the equation $Qz = \mu z$ and using equations (5), (6) and (7) we obtain

$$\sum_{r=1}^{\infty} \langle z, e_r \rangle\, f_{r-1} + Kw = \mu \sum_{r=1}^{\infty} \langle z, e_r \rangle\, e_r + \mu w.$$

Hence

$$\mu w - Kw = \sum_{r=1}^{\infty} \langle z, e_r \rangle\, f_{r-1} - \mu \sum_{r=1}^{\infty} \langle z, e_r \rangle\, e_r \tag{8}$$

and so $(\mu I - K)\, w \in L$.

Since w is a non-zero vector orthogonal to L, $w \notin L$. We put $L_1 = L \oplus [w]$, where $[w]$ is the one-dimensional space generated by w. Then L_1 is a closed subspace of H. Equation (8) shows that $KL_1 \subseteq L_1$ and

$$(\mu I - K)\, L_1 \subseteq L.$$

Observe that the restriction of K to L_1 is a Riesz operator, by Theorem 3.21.

Hence, by Theorem 3.14, $\rho(K|L_1)$ is connected. By Theorem 1.29,

$$\mu \in \sigma(K|L_1) \subseteq \sigma(K),$$

and so

$$E(\tau; K|L_1)\, L_1 \subseteq (\mu I - K)\, L_1 \subseteq L. \tag{9}$$

Now

$$L_1 = E(\tau; K|L_1)\, L_1 \oplus E(\mu; K|L_1)\, L_1 \tag{10}$$

and so by equations (9) and (10)

$$E(\mu; K|L_1)\, L_1 \cap (L_1 \backslash L) \neq \varnothing. \tag{11}$$

However, the argument of Lemma 3.31 shows that $E(\mu; K|L_1) = (E(\mu; K))|L_1$. Hence,

$$E(\mu; K|L_1)\, L_1 \subseteq E(\mu; K)\, H$$

and so

$$E(\mu; K)\, H \cap (L_1 \backslash L) \neq \varnothing. \tag{12}$$

The choice of the M_r and e_r has ensured that if $\mu \in \sigma_0(K)$ then

$$E(\mu; K)\, H \subseteq L.$$

Combining this with equation (12), we get a contradiction which completes the proof.

Noting that West's decomposition theorem is trivial in the finite-dimensional case, we summarize the work of this section as follows.

THEOREM 3.33. *Let H be a complex Hilbert space and let K, in $L(H)$, be a Riesz operator. Then $K = C + Q$, where C is a compact operator on H and Q is quasinilpotent.*

It is an unsolved problem whether this result remains true if 'Hilbert space' is replaced by 'Banach space'. At first glance the following two characterizations of Riesz operators might seem to throw some light on this problem.

THEOREM 3.34. *Let $T \in L(X)$. Then T is a Riesz operator if and only if the following condition holds. For λ in $\rho(T)$*

$$(\lambda I - T)^{-1} = C(\lambda) + B(\lambda^{-1})$$

where $C(\lambda) \in K(X)$ $(\lambda \in \rho(T))$ and B is an entire function.

Proof. We show that T satisfies the stated condition if and only if T is asymptotically quasi-compact. Suppose the condition holds. Then for $|\lambda| > \rho(T)$, $(\lambda I - T)^{-1}$ and $B(\lambda^{-1})$ have expansions in powers of λ^{-1}; thus so also has $C(\lambda)$. Therefore

$$\sum_{n=0}^{\infty} \lambda^{-n-1} T^n = \sum_{n=0}^{\infty} \lambda^{-n-1} C_n + \sum_{n=0}^{\infty} \lambda^{-n-1} B_n \quad (|\lambda| > \|T\|),$$

where $\|B_n\|^{1/n} \to 0$ as $n \to \infty$ and

$$C_n = \frac{1}{2\pi i} \int_\Gamma \zeta^n C(\zeta)\, d\zeta \quad (n = 0, 1, 2, \ldots),$$

where Γ is a circle centre 0, radius $\delta > \|T\|$, described once counterclockwise. Since the compact operators form a closed two-sided ideal in $L(X)$, and since the integral exists as a norm limit of Riemann sums, then by hypothesis C_n is compact for each n. Also

$$T^n = C_n + B_n \quad (n = 0, 1, 2, \ldots).$$

Hence T is asymptotically quasi-compact.

Conversely, let T be asymptotically quasi-compact. Then, as in the proof of Theorem 3.12, we can find sequences $\{C_n\}$ and $\{B_n\}$ such that

(i) $C_n \in K(X)$, for each n;
(ii) $\|B_n\|^{1/n} \to 0$ as $n \to \infty$;
(iii) $T^n = C_n + B_n \quad (n = 0, 1, 2, \ldots)$;
(iv) $C_n B_n = B_n C_n \quad (n = 0, 1, 2, \ldots)$.

Suppose that $|z| > \|T\|$. Then

$$(zI - T)^{-1} = \sum_{n=0}^{\infty} T^n/z^{n+1}$$

$$= \sum_{n=0}^{\infty} z^{-n-1} C_n + \sum_{n=0}^{\infty} z^{-n-1} B_n.$$

The convergence of these last two series follows from the condition on the sequence $\{B_n\}$. Hence, for $|z| > \|T\|$,

$$(zI - T)^{-1} = C(z) + B(z^{-1}),$$

where $C(z) \in K(X)$ and B is an entire function. Let Y be the annihilator (in$(L(X))^*$) of the closed subspace $K(X)$ of $L(X)$. Then

$$\langle C(z), y \rangle = 0 \quad (|z| > \|T\|, y \in Y)$$

and $C(z)$ is defined and analytic for z in $\rho(T)$. By Theorem 3.14, $\rho(T)$ is connected and so it follows that

$$\langle C(z), y \rangle = 0 \quad (z \in \rho(T), y \in Y).$$

Hence $C(z) \in K(X)$, for all z in $\rho(T)$, and the proof is complete.

An examination of the proof of Theorem 3.34 shows that we have established the following result.

THEOREM 3.35. *Let $T \in L(X)$. Then T is a Riesz operator if and only if the following condition holds. For λ in $\rho(T)$*

$$(\lambda I - T)^{-1} = C(\lambda) + B(\lambda^{-1})$$

where $C(\lambda) \in K(X) (\lambda \in \rho(T))$, B is an entire function and

$$C(\lambda) B(\lambda^{-1}) = B(\lambda^{-1}) C(\lambda).$$

It seems plausible that, by integrating the decomposition of the resolvent given by Theorem 3.35, one might be able to extend West's result to the Banach space case. This is not so, as the following example shows. Even in

Hilbert space it may not be possible to choose a decomposition of a Riesz operator so that the compact and quasinilpotent operators commute.

Example 3.36. Let $X = l^2$ and let $\{e_n\}_{n=1}^{\infty}$ be an orthonormal basis for X. Define operators C and Q on X as follows:

$$Ce_r = r^{-1} e_r \quad (r \in \mathbf{N})$$

and

$$\left.\begin{aligned} Qe_{2r-1} &= e_{2r} \\ Qe_{2r} &= 0 \end{aligned}\right\} \quad (r \in \mathbf{N}).$$

Clearly $C \in K(X)$ and $Q^2 = 0$. Put $K = C + Q$. Then K^2 is compact and so K is a Riesz operator. However, K is not compact, since

$$Ke_{2r-1} = (2r-1)^{-1} e_{2r-1} + e_{2r} \quad (r \in \mathbf{N}).$$

Observe that $\{Ke_{2r-1}\}_{r=1}^{\infty}$ does not contain a convergent subsequence. A simple calculation shows that if $\sum_{n=1}^{\infty} |c_n|^2 < \infty$ then

$$\begin{aligned} K\left(\sum_{n=1}^{\infty} c_n e_n\right) &= K\left(\sum_{n=1}^{\infty} c_{2n} e_{2n} + \sum_{n=1}^{\infty} c_{2n-1} e_{2n-1}\right) \\ &= \sum_{n=1}^{\infty} (2n)^{-1} c_{2n} e_{2n} + \sum_{n=1}^{\infty} (2n-1)^{-1} c_{2n-1} e_{2n-1} \\ &\quad + \sum_{n=1}^{\infty} c_{2n-1} e_{2n}. \end{aligned}$$

Hence the non-zero eigenvalues and the corresponding eigenspaces of K are given by the formulae:

$$\lambda_r = r^{-1} \quad (r \in \mathbf{N}),$$

and

$$\left.\begin{aligned} E_{2r-1} &= [e_{2r-1} + (2r-1)\, 2r e_{2r}] \\ E_{2r} &= [e_{2r}] \end{aligned}\right\} \quad (r \in \mathbf{N}) \qquad (13)$$

where $[x]$ denotes the one-dimensional subspace generated by x.

Suppose now that $K = C_1 + Q_1$ where C_1 is compact, Q_1 is quasinilpotent, and $C_1Q_1 = Q_1C_1$. It follows that $KQ_1 = Q_1K$. Hence, if $x \in E_r$ for some r, then

$$K(Q_1x) = Q_1(Kx) = Q_1(r^{-1}x) = r^{-1}Q_1x$$

and therefore $Q_1x \in E_r$. Thus E_r is invariant under Q_1. Since E_r is a one-dimensional subspace, $\sigma(Q_1|E_r)$ consists of one eigenvalue. Hence, since $Q_1|E_r$ is also quasinilpotent,

$$\sigma(Q_1|E_r) \subseteq \sigma(Q_1) = \{0\},$$

and so $Q_1|E_r$ is the zero operator. Formula (13) shows that

$$Q_1e_r = 0 \quad (r \in \mathbf{N}).$$

Accordingly $Q_1 = 0$ and $K = C_1$ is compact. This gives a contradiction.

Notes and Comments on Part Two

The theory presented in Chapter 2 generalizes the work of Fredholm [**1**] on a certain type of integral equation. Fredholm's original approach was by means of expansions in determinants which, while intricate, gives a detailed representation of the resolvent as the quotient of two entire functions. (See Sections 9 and 10 of Hellinger and Toeplitz [**1**] for further references.) Of several other methods, we mention the one due to Schmidt [**1**] depending on the possibility (in Hilbert space) of approximating the compact operator by operators with finite-dimensional ranges.

Probably the most ingenious and elementary approach is due to F. Riesz [**2**] and is valid for an arbitrary real or complex Banach space. Certain of Riesz's results concerning the adjoint operator were not completely general, but were completed by Hildebrandt [**1**] and Schauder [**1**]. For an account of this method, see Chapter 4 of Riesz and Sz-Nagy [**1**] or Chapter 11 of Zaanen [**1**]. A similar treatment is given in Chapter 10 of Banach [**1**].

The line of argument which we presented in Chapter 2 is based on papers by Bonsall [**1**] and Dunford [**1**].

The following result, called the Fredholm alternative, is a consequence of various theorems in Chapters 2 and 3.

Let T be a compact operator on X and let λ be a fixed non-zero complex number. Then the non-homogeneous equations

$$(N) \qquad (\lambda I - T)x = y$$

$$(N^*) \qquad (\lambda I^* - T^*)\xi = \eta$$

have unique solutions for any y in X or η in X^* if and only if the homogeneous equations

$$(H) \qquad (\lambda I - T)x = 0$$

$$(H^*) \qquad (\lambda I^* - T^*)\xi = 0$$

have only the zero solutions. Furthermore if one of the homogeneous equations has a non-zero solution, then they both have the same finite number of linearly independent solutions. In this case the equations (N) *and* (N^*) *have solutions if and only if* y *and* η *are orthogonal to all the solutions of* (H^*) *and* (H), *respectively. Moreover, the general solution for* (N) *is found by adding a particular solution of* (N) *to the general solution of* (H).

The results of Fredholm pertaining to the representation of the resolvent via the determinantal approach have been discussed by Altman [**1**], Graves [**1**], Lezánski [**1**], [**2**], Michal and Martin [**1**], Ruston [**1**], [**2**], [**3**], [**4**], Sikorski [**1**], [**2**] and Smithies [**1**]. See also Grothendieck [**1**] and Chapter 9 of Zaanen [**1**].

For the purposes of applications it is important to be able to compute the eigenvalues of an operator; this is particularly true for compact self-adjoint operators in Hilbert space. Reference should be made to the work of Aronszajn [**1**], [**2**] and Collatz [**1**] for these questions. Also of considerable importance is information concerning the distribution of eigenvalues. For the case of integral operators we refer the reader to Hille and Tamarkin [**1**] and Chang [**1**]. For related results in abstract spaces see Fan [**1**], [**2**], Horn [**1**], Silberstein [**1**], Visser and Zaanen [**1**], Weinberger [**1**], [**2**] and Weyl [**1**].

In 1954, Aronszajn and Smith [**1**] proved that a compact operator on a complex Banach space of dimension at least two has a proper closed invariant subspace. Their proof used some deep results on metric projections. The neat elementary proof of the Aronszajn and Smith theorem given in Chapter 2 was obtained by Hilden and Lomonosov in 1973. The theory of superdiagonal forms for compact operators presented in Chapter 2 is due to Ringrose [**2**].

For an account of the theory of ideals of compact operators on a Hilbert space, the reader is referred to Chapter 11 of [**5**], Gohberg and Kreĭn [**1**], Ringrose [**4**] and Schatten [**1**]. See also the paper by Garling [**1**].

The theory of Riesz operators was initiated by Ruston [**4**], to whom Theorem 3.12 is due. Example 3.15 is due to Kakutani. (See [**12**; p. 282].) West [**1**], [**2**] continued the study of these operators. Theorem 3.16 is due to West [**1**] although the proof given here is due to Ruston [**5**]. Theorem 3.19 is due to Dieudonné [**3**, p. 323]. Theorem 3.20, giving another characterization of Riesz operators, is due to West [**1**]. Theorems 3.21, 3.22, 3.23 are due to West [**2**]. The deep result Theorem 3.33 is due to West [**2**]. Example 3.36 is due to Gillespie and West [**1**].

Let H be a separable Hilbert space. Calkin [**1**; p 841] has shown that $K(H)$ is the unique proper closed two-sided ideal of $L(H)$. An extension of this result was discovered by Gohberg, Markus and Feldman [1], who obtained the same conclusion for the spaces l^p $(1 \leqslant p < \infty)$ and c_0. This problem has been discussed further by Herman [1]. See also Caradus, Pfaffenberger and Yood [1; p. 81–95]. However, Porta [1] has shown that if $X = L^p[0, 1]$, where $1 < p < \infty$ and $p \neq 2$, then $K(X)$ is not the only proper closed two-sided ideal of $L(X)$.

Kleinecke [1] has proved the following theorem. See also Kleinecke [2].

Let J be the intersection of all maximal one-sided ideals in $L(X)$ containing $\bar{F}(X)$. Then the spectrum of any operator in J is a countable set of isolated eigenvalues of finite multiplicity with no cluster points except possibly 0. Further J contains any other ideal in $L(X)$ whose operators have spectra of the nature just specified.

Certainly the two-sided ideal $K(X) \subseteq J$. In the case $X = L^1[0, 1]$ there is a two-sided ideal of Riesz operators strictly larger than $K(X)$ but contained in J. This is the two-sided ideal of weakly compact operators. The square of such an operator is compact. For a proof that this ideal is strictly larger than $K(X)$ see Zaanen [1; p. 322–3].

There have been several notions of compactness introduced for elements of a Banach algebra. We mention in particular the work of Bonsall [1], J. C. Alexander [1] and Vala [1]. The study of various notions of "Riesz element" of a Banach algebra was initiated by J. C. Alexander [2] and has been continued by Pearlman [1] and Smyth [1].

We conclude with the following recent result of Smyth and West [1].

Let $\mathscr{A}$ be the quotient algebra $L(X)/K(X)$ and let $T \in L(X)$. Then the spectral radius of the image of T under the canonical mapping of $L(X)$ into $\mathscr{A}$ is equal to the infimum of the spectral radii of the elements $\{T + W: W \in K(X)\}$.

For many other interesting results in this general area the reader is referred to Caradus, Pfaffenberger and Yood [1].

Part 3

HERMITIAN OPERATORS

4. Hermitian Operators

The purpose of this chapter is to develop those properties of hermitian operators which are important in the theory of prespectral operators. No attempt is made to give an exhaustive treatment of numerical ranges.

Define

$$D = \{f \in (L(X))^* : f(I) = 1 = \|f\|\}.$$

The Hahn-Banach theorem guarantees that D is non-empty.

Definition 4.1. Given A in $L(X)$ let $V(A) = \{f(A) : f \in D\}$. $V(A)$ is called the *numerical range* of A.

LEMMA 4.2. *Let $A \in L(X)$ and $\alpha, \beta \in \mathbf{C}$. Then $V(\alpha + \beta A) = \alpha + \beta V(A)$.*

Proof. This follows immediately.

Definition 4.3. Let $A \in L(X)$. Suppose that $V(A)$ is real. Then A is called a *hermitian* operator.

THEOREM 4.4. (i) *Let $\alpha_1, \ldots, \alpha_n \in \mathbf{R}$ and let $A_1, \ldots, A_n$ be hermitian operators on X. Then $\sum_{r=1}^{n} \alpha_r A_r$ is a hermitian operator.*

(ii) *For each positive integer n let A_n be a hermitian operator on X. If $\{A_n\}$ converges to A in the norm of $L(X)$ then A is a hermitian operator.*

Proof. This follows immediately.

In order to characterize hermitian operators in terms of operator norms two preliminary results are required.

PROPOSITION 4.5. *For each A in $L(X)$*

$$\sup\{\operatorname{Re}\lambda : \lambda \in V(A)\} = \lim_{\alpha \to 0+} \alpha^{-1}\{\|I + \alpha A\| - 1\}.$$

Proof. Let $\mu = \sup\{\operatorname{Re}\lambda : \lambda \in V(A)\}$. Observe that $\mu \leqslant \|A\|$. Given $\alpha > 0$, $f \in D$, we have

$$f(A) = \alpha^{-1}\{f(I + \alpha A) - 1\}$$

and so

$$\operatorname{Re} f(A) \leqslant \alpha^{-1}\{\|I + \alpha A\| - 1\}.$$

Hence

$$\mu \leqslant \inf_{\alpha > 0} \frac{1}{\alpha}\{\|I + \alpha A\| - 1\}. \tag{1}$$

We assume now that $A \neq 0$, the theorem being obvious if $A = 0$. Let $0 < \alpha < \|A\|^{-1}$ and $T \in L(X)$. We can choose f in $(L(X))^*$ with $f(T) = \|T\|$ and $\|f\| = 1$. Define

$$g(B) = f(BT) \quad (B \in L(X)).$$

Observe that $g \in \|T\| D$ and so $\operatorname{Re} f(AT) \leqslant \mu\|T\|$. Hence

$$\begin{aligned}\|(I - \alpha A)T\| &\geqslant \operatorname{Re} f((I - \alpha A)T) = \|T\| - \operatorname{Re} f(\alpha AT)\\ &\geqslant (1 - \alpha\mu)\|T\|.\end{aligned}$$

Taking $T = I + \alpha A$, we obtain

$$\begin{aligned}\|I + \alpha A\| &\leqslant (1 - \alpha\mu)^{-1}\|I - \alpha^2 A^2\|\\ &\leqslant (1 - \alpha\mu)^{-1}(1 + \alpha^2\|A\|^2).\end{aligned}$$

Therefore

$$\frac{\|I + \alpha A\| - 1}{\alpha} \leqslant \frac{\mu + \alpha\|A\|^2}{1 - \alpha\mu} \quad (0 < \alpha < \|A\|^{-1}).$$

Together with the inequality (1) this completes the proof.

PROPOSITION 4.6. *Let $A \in L(X)$. Then*

$$\sup\{\operatorname{Re}\lambda : \lambda \in V(A)\} = \lim_{\alpha \to 0+} \frac{1}{\alpha}\log\|\exp(\alpha A)\| = \sup_{\alpha > 0}\left\{\frac{1}{\alpha}\log\|\exp(\alpha A)\|\right\}.$$

Proof. We assume that $A \neq 0$, the result being obvious if $A = 0$. Let $\mu = \sup\{\operatorname{Re}\lambda : \lambda \in V(A)\}$ and $\alpha \geqslant 0$. For all sufficiently large n we have $0 \leqslant \alpha/n < \|A\|^{-1}$ and $(1 - \alpha\mu/n) \geqslant 0$. As in Proposition 4.5

$$\left\|\left(I - \frac{\alpha A}{n}\right)T\right\| \geqslant \left(1 - \frac{\alpha\mu}{n}\right)\|T\| \quad (T \in L(X)).$$

By induction we have

$$\left\|\left(I - \frac{\alpha A}{n}\right)^n T\right\| \geqslant \left(1 - \frac{\alpha\mu}{n}\right)^n \|T\| \quad (T \in L(X);\, n = 1, 2, \ldots). \tag{2}$$

Letting $n \to \infty$ in (2) we obtain

$$\|\exp(-\alpha A)\, T\| \geqslant \exp(-\alpha\mu)\|T\|,$$

since $\exp(\alpha A) = \lim_{n \to \infty}\left(I + \dfrac{\alpha A}{n}\right)^n$ in the norm of $L(X)$ for each α in $\mathbf{C}$, by Proposition 1.24. Taking $T = \exp(\alpha A)$, we obtain

$$\|\exp(\alpha A)\| \leqslant \exp(\alpha\mu)$$

and so

$$\sup\{\alpha^{-1}\log\|\exp(\alpha A)\| : \alpha > 0\} \leqslant \mu. \tag{3}$$

On the other hand, we have

$$\|\exp(\alpha A)\| = \|I + \alpha A\| + \lambda(\alpha)$$

where, for some positive real number M,

$$|\lambda(\alpha)| \leqslant M\alpha^2 \quad (0 \leqslant \alpha \leqslant 1). \tag{4}$$

Using the inequality $t \log t \geqslant t - 1$ $(t > 0)$ we obtain, for $\alpha > 0$,

$$\alpha^{-1} \log\|\exp(\alpha A)\| \geqslant \frac{\alpha^{-1}\{\|I + \alpha A\| - 1\} + \alpha^{-1}\lambda(\alpha)}{\|I + \alpha A\| + \lambda(\alpha)}.$$

From (4) and Proposition 4.5 we see that the right-hand side of the inequality converges to μ as $\alpha \to 0+$. Combined with (3) this completes the proof.

THEOREM 4.7. *Let $A \in L(X)$. The following statements are equivalent.*

(i) *A is a hermitian operator.*

(ii) $\lim\limits_{\alpha \to 0} \alpha^{-1}\{\|I + i\alpha A\| - 1\} = 0$, *where α is real.*

i.e. $\|I + i\alpha A\| = 1 + 0(\alpha)$ *as $\alpha \to 0$ through real values.*

(iii) $\|\exp(i\alpha A)\| = 1 \quad (\alpha \in \mathbf{R})$.

(iv) $\|\exp(i\alpha A)\| \leqslant 1 \quad (\alpha \in \mathbf{R})$.

Proof. By Lemma 4.2, A is a hermitian operator if and only if

$$\sup \operatorname{Re} V(iA) = \sup \operatorname{Re} V(-iA) = 0.$$

Hence the equivalence of (i) and (ii) follows at once from Proposition 4.5. Also (iii) implies (i) by Proposition 4.6. If (i) holds, then again by Proposition 4.6

$$0 = \sup_{\alpha \neq 0}\left\{\frac{1}{|\alpha|} \log\|\exp(i\alpha A)\|\right\}. \tag{5}$$

Now

$$1 = \|I\| \leqslant \|\exp(i\alpha A)\| \, \|\exp(-i\alpha A)\| \quad (\alpha \in \mathbf{R}).$$

If, for some real β, we have $\|\exp(i\beta A)\| \neq 1$, the last inequality shows that there is a real number $\gamma \neq 0$ such that $\|\exp(i\gamma A)\| > 1$, contradicting (5). The argument just given also shows that (iii) and (iv) are equivalent. This completes the proof.

THEOREM 4.8. *Let A, in $L(X)$, be a hermitian operator. Then $\sigma(A)$ is real.*

Proof. Let $U = \exp(iA)$. Then by Theorem 4.7

$$\|U^n\| \leqslant 1 \quad (n \in \mathbf{Z})$$

and so, by Proposition 1.31, $\sigma(U) \subseteq \{z : |z| = 1\}$. The desired conclusion now follows from the spectral mapping theorem.

The next important result is usually referred to as Sinclair's theorem. It states that the norm and spectral radius of a hermitian operator are equal. A preliminary lemma is required.

Let $\{c_r\}$ be the sequence of non-negative real numbers for which

$$\text{arc sin } t = \sum_{r=1}^{\infty} c_r t^r \quad (|t| \leqslant 1)$$

and let

$$F_n(z) = \sum_{r=1}^{n} c_r (\sin z)^r \quad (z \in \mathbf{C}).$$

LEMMA 4.9. *Let K be a compact subset of $(-\pi/2, \pi/2)$. Then there is an open subset U of $\mathbf{C}$ such that $K \subseteq U$ and $F_n(z) \to z$ uniformly on U.*

Proof. There is an open subset U of $\mathbf{C}$ with $K \subseteq U$ and $|\sin z| \leqslant 1$ $(z \in U)$. By the Weierstrass M-test, $\sum_{r=1}^{\infty} c_r (\sin z)^r$ converges uniformly on U and so the limit function is analytic on U. We have $\lim_{n \to \infty} F_n(t) = t$ $(t \in K)$ by real analysis and the result follows.

The proof of Sinclair's theorem now follows.

THEOREM 4.10. *Let A, in $L(X)$, be a hermitian operator. Then $\|A\| = v(A)$.*

Proof. By Proposition 1.8, $v(A) \leqslant \|A\|$. Let $v(A) < \pi/2$. Then $\sigma(A) \subseteq (-\pi/2, \pi/2)$. It follows from Lemma 4.9 and the functional calculus that

$$A = \lim_{n \to \infty} F_n(A) = \sum_{r=1}^{\infty} c_r (\sin A)^r,$$

convergence being in the norm of $L(X)$. Since A is hermitian, then by Theorem

4.7 we have

$$\|\sin A\| \leqslant \tfrac{1}{2}\|\exp(iA)\| + \tfrac{1}{2}\|\exp(-iA)\| = 1,$$

and therefore

$$\|A\| \leqslant \sum_{n=1}^{\infty} c_n = \frac{\pi}{2}.$$

The desired result follows from the positive homogeneity of the spectral radius and the norm.

Some elementary properties of hermitian operators follow immediately from Theorem 4.7.

PROPOSITION 4.11. *Let $T \in L(X)$. Then T is hermitian if and only if T^* is hermitian.*

Proof. This follows immediately since

$$\|\exp(i\alpha T)\| = \|\exp(i\alpha T^*)\| \quad (\alpha \in \mathbf{R}).$$

PROPOSITION 4.12. *Let T be a hermitian operator on X and let Y be a closed subspace of X invariant under T.*

(i) *$T|Y$ is a hermitian operator on Y.*

(ii) *If T_Y denotes the operator induced by T on the quotient space X/Y then T_Y is hermitian.*

Proof. (i) Observe that

$$\|\exp i\alpha(T|Y)\| \leqslant \|\exp(i\alpha T)\| = 1 \quad (\alpha \in \mathbf{R})$$

and so the required result follows.

(ii) By Proposition 4.11, T^* is hermitian. Also $Y^\perp$, the annihilator of Y, is a closed subspace of X^* invariant under T^* and so $T^*|Y^\perp$ is hermitian by (i). As in Proposition 1.33 et seq. we can identify T_Y^* and $T^*|Y^\perp$. Hence by Proposition 4.11, T_Y is hermitian.

By Theorem 4.4(i), the sum of two hermitian operators is hermitian. However, powers of hermitian operators need not be hermitian, as the following example, due to M. J. Crabb, shows. See [**1**; p. 57–8].

Example 4.13. Let p be defined on $\mathbf{C}^3$ by

$$p(\alpha, \beta, \gamma) = \sup\{|\lambda^{-1}\alpha + \beta + \lambda\gamma| : \lambda \in \mathbf{C}, |\lambda| = 1\}.$$

Then p is a Banach space norm on $\mathbf{C}^3$. Now $\mathbf{C}^3$ is an algebra under pointwise multiplication. If $a \in \mathbf{C}^3$, let A be the operator defined by

$$Ax = ax \quad (x \in \mathbf{C}^3).$$

Observe that the operator norm of A is given by

$$\|A\| = \sup\{p(xa) : x \in \mathbf{C}^3, p(x) = 1\}.$$

Now let $a = (-1, 0, 1)$. We show that A is hermitian but A^2 is not hermitian. Given t in $\mathbf{R}$ and $x = (\alpha, \beta, \gamma)$ in $\mathbf{C}^3$ we have

$$\begin{aligned} p(x \exp(ita)) &= p(\alpha \exp(-it), \beta, \gamma \exp(it)) \\ &= p(\alpha, \beta, \gamma) \end{aligned}$$

so that $\|\exp(itA)\| = 1$. Therefore A is hermitian by Theorem 4.7. Let $s = -\pi/2$ and $y = (i, 1, i)$. Then

$$\begin{aligned} p(y)\|\exp(isA^2)\| &\geqslant p(y \exp(isa^2)) \\ &= p(1, 1, 1). \end{aligned}$$

Since $p(y) = 5^{1/2}$ and $p(1, 1, 1) = 3$ it follows that $\|\exp(isA^2)\| > 1$. Hence, by Theorem 4.7, A^2 is not hermitian.

We next present the basic results on renorming of Banach spaces which occur in the theory of hermitian operators.

PROPOSITION 4.14. *Let* $\mathbf{S} \subseteq L(X)$ *be a bounded multiplicative abelian semigroup of operators with identity element* I. *There is an equivalent norm* $|||\ |||$ *on* X *such that*

$$|||Tx||| \leqslant |||x||| \qquad (x \in X, T \in \mathbf{S}).$$

Proof. Let $\|\ \|$ denote the given norm on X. Define

$$|||x||| = \sup\{\|Tx\| : T \in \mathbf{S}\} \qquad (x \in X).$$

Then $||| \ |||$ has the required properties.

PROPOSITION 4.15. *Let* $\mathbf{G} \subseteq L(X)$ *be a bounded multiplicative abelian group of operators with identity element* I. *There is an equivalent norm on* X *with respect to which every element of* $\mathbf{G}$ *is isometric.*

Proof. Let $\| \ \|$ denote the given norm on X. Define

$$|||x||| = \sup\{\|Tx\| : T \in \mathbf{G}\} \qquad (x \in X).$$

Let $T \in \mathbf{G}$. Observe that

$$|||Tx||| \leqslant |||x||| \qquad (x \in X)$$

$$|||x||| = |||T^{-1}Tx||| \leqslant |||Tx||| \qquad (x \in X)$$

from which the desired result follows.

Prior to proving our next result we require a preliminary definition.

Definition 4.16. Let $\Lambda \subseteq L(X)$. Then Λ is said to be *hermitian-equivalent* if and only if there is an equivalent norm on X with respect to which every operator in Λ is hermitian.

THEOREM 4.17. *Let* Λ *be a commutative subset of* $L(X)$. *Then* Λ *is hermitian-equivalent if and only if each operator in the closed real linear span of* Λ *is hermitian-equivalent.*

Proof. Let Y be the closed real linear span of Λ and suppose that each T in Y is hermitian-equivalent. Let $\| \ \|$ denote the given norm on X. Then there is an equivalent norm $||| \ |||$ on X such that

$$|||\exp(itT)||| = 1 \qquad (t \in \mathbf{R})$$

where we have also used $||| \ |||$ to denote the norm of an operator with respect to $||| \ |||$. Hence

$$\|\exp(itT)\| \leqslant M < \infty \qquad (t \in \mathbf{R}).$$

For $k = 1, 2, 3, \ldots$ let

$$Y_k = \{T \in Y : \|\exp(itT)\| \leqslant k, \text{ for all real } t\}.$$

By hypothesis, we have $Y = \bigcup_{k=1}^{\infty} Y_k$. We show that each Y_k is closed. Let $T_n \in Y_k$, $T_n \to T$, and let $t \in \mathbf{R}$. Then $\exp(itT_n) \to \exp(itT)$. Hence $\|\exp(itT)\| \leqslant k$ and so $T \in Y$. By the Baire category theorem some Y_k contains an open ball of Y. For any S in Y we may therefore choose U, V in Y_k and $r \geqslant 0$ such that $S = U + rV$. Then, since Λ is commutative, we have

$$\|\exp(itS)\| \leqslant \|\exp(itU)\| \, \|\exp(itrV)\| \leqslant k^2 \qquad (t \in \mathbf{R}).$$

It follows that the set $\{\exp(itS): S \in Y, t \in \mathbf{R}\}$ is a bounded abelian group of operators and so by Proposition 4.15 there is an equivalent norm on X with respect to which each of these operators is an isometry. It follows from Theorem 4.7 that Y and hence also Λ are hermitian-equivalent. The converse implication follows from Theorem 4.4.

THEOREM 4.18. *Suppose $\Lambda_i (i = 1, 2, \ldots, n)$ are finitely many commutative subsets of $L(X)$, each hermitian-equivalent. Then if $\bigcup_{i=1}^{n} \Lambda_i$ is commutative it is hermitian-equivalent.*

Proof. Since Λ_i is hermitian-equivalent each operator in the closed real linear span Y_i of Λ_i is hermitian-equivalent. The argument of the proof of Theorem 4.17 shows that

$$\|\exp(itS_i)\| \leqslant M_i < \infty \qquad (S_i \in Y_i, t \in \mathbf{R}).$$

If $T = \sum_{i=1}^{n} \lambda_i S_i$, where each λ_i is real and $S_i \in Y_i$, then

$$\|\exp(itT)\| \leqslant \prod_{i=1}^{n} M_i < \infty \qquad (t \in \mathbf{R}),$$

and this inequality holds also if T is in the closed real linear span of $\bigcup_{i=1}^{n} \Lambda_i$. As in Theorem 4.17 it follows that $\bigcup_{i=1}^{n} \Lambda_i$ is hermitian-equivalent.

The last main result in this chapter is a commutativity theorem for hermitian operators. In order to prove it we require some preliminary results.

Given T in $L(X)$, define D_T in $L(L(X))$ by

$$D_T(A) = AT - TA \qquad (A \in L(X)).$$

LEMMA 4.19. (i) *Suppose that* $S, T \in L(X)$ *and* $ST = TS$. *Then* $D_S D_T = D_T D_S$.
(ii) *Suppose that* $H \in L(X)$ *and* H *is hermitian. Then* D_H *is hermitian.*
(iii) *Suppose that* $Q \in L(X)$ *and* Q *is quasinilpotent. Then* D_Q *is quasinilpotent.*

Proof. (i)
$$(D_S D_T)(A) = (AT - TA)S - S(AT - TA)$$
$$(D_T D_S)(A) = (AS - SA)T - T(AS - SA) \qquad (A \in L(X)).$$

Since $TS = ST$ it follows that $D_S D_T = D_T D_S$.
(ii) Define H_L and H_R in $L(L(X))$ by

$$H_L(A) = HA \qquad (A \in L(X)),$$
$$H_R(A) = AH \qquad (A \in L(X)).$$

Since H is hermitian then

$$\|\exp i\alpha H_L\| = \|\exp i\alpha H_R\| = 1 \qquad (\alpha \in \mathbf{R})$$

and so H_L, H_R are hermitian. By Theorem 4.4 the difference of two hermitian operators is hermitian. It follows that D_H is hermitian.
(iii) Define Q_L and Q_R in $L(L(X))$ by

$$Q_L(A) = QA \qquad (A \in L(X)),$$
$$Q_R(A) = AQ \qquad (A \in L(X)).$$

Clearly Q_L, Q_R are quasinilpotent and $Q_L Q_R = Q_R Q_L$. Since $D_Q = Q_R - Q_L$ it follows from Corollary 3 of [**3**; p. 19] that D_Q is quasiilpotent.

PROPOSITION 4.20. *Let* $\{H, K, Q\}$ *be a commutative subset of* $L(X)$. *Suppose that* H, K *are hermitian,* Q *is quasinilpotent and*

$$H + iK + Q = 0.$$

Then $H = K = Q = 0$.

Proof. By Theorem 1.27

$$\sigma(H + Q) = \sigma(H) \subseteq \mathbf{R}.$$

Since $H + Q = -iK$, it follows that $\sigma(K) \subseteq i\mathbf{R}$. However, $\sigma(K) \subseteq \mathbf{R}$ since K is hermitian. Thus $\sigma(K) = \{0\}$ and so, by Sinclair's theorem, $K = 0$. Therefore $H = -Q$, showing that H is a quasinilpotent hermitian operator. Hence $H = 0$ and so $Q = 0$ also. This completes the proof.

PROPOSITION 4.21. *Let H, K be hermitian operators on X and let $HK = KH$.*

(i) *Suppose that $(H + iK)x = 0$ for some x in X. Then $Hx = Kx = 0$.*

(ii) *Suppose that $A \in L(X)$ and $(H + iK)A = A(H + iK)$. Then $HA = AH$ and $KA = AK$.*

Proof. (i) Let $Y = \{y \in X : (H + iK)y = 0\}$. Then Y is a closed subspace of X invariant under H and K. By Proposition 4.12, $H|Y$ and $K|Y$ are commuting hermitian operators on Y. Clearly $H|Y + iK|Y = 0$. Therefore, by Proposition 4.20, $H|Y = K|Y = 0$ and so $Hx = Kx = 0$.
(ii) Observe that D_H and D_K are commuting hermitian operators on $L(X)$ by Lemma 4.19. Also $(D_H + iD_K)(A) = 0$. Apply part (i) to complete the proof.

We now prove the aforementioned commutativity theorem.

THEOREM 4.22. *Let $H, K, Q \in L(X)$, where H, K are hermitian, $HK = KH$, Q is quasinilpotent, and $Q(H + iK) = (H + iK)Q$.*

(i) *Suppose that $(H + iK + Q)x = 0$ for some x in X. Then $Hx = Kx = Qx = 0$.*

(ii) *Let $A \in L(X)$. Suppose that $A(H + iK + Q) = (H + iK + Q)A$. Then $AH = HA$, $AK = KA$ and $AQ = QA$.*

Proof. Note first that $QH = HQ$ and $QK = KQ$ by Proposition 4.21(ii).
(i) Let $Y = \{y \in X : (H + iK + Q)y = 0\}$. Then Y is a closed subspace of X invariant under H, K and Q. By Proposition 4.12, $H|Y$ and $K|Y$ are hermitian operators on Y. Also $\|Q|Y\| \leqslant \|Q\|$ and so from the definition $Q|Y$ is quasinilpotent. Now $H|Y + iK|Y + Q|Y = 0$, and so, by Proposition 4.20, $H|Y = K|Y = Q|Y = 0$. Hence $Hx = Kx = Qx = 0$.
(ii) Observe that, by Lemma 4.19, D_H, D_K, D_Q are a commutative family of operators on $L(X)$. Also D_H, D_K are hermitian and D_Q is quasinilpotent. Since $(D_H + iD_K + D_Q)(A) = 0$, the first part of the theorem shows that $HA = AH$, $KA = AK$ and $QA = AQ$. The proof is complete.

We conclude this chapter by stating a theorem on intertwining operators. This may be established by arguments similar to those used in proving Theorem 4.22.

THEOREM 4.23. *Let X_1 and X_2 be complex Banach spaces and let $T_1 \in L(X_1)$, $T_2 \in L(X_2)$. Suppose that for $r = 1, 2$*

$$T_r = H_r + iK_r + Q_r,$$

where H_r and K_r are hermitian, $H_rK_r = K_rH_r$, Q_r is quasinilpotent and

$$Q_r(H_r + iK_r) = (H_r + iK_r)Q_r.$$

Let A be a bounded linear mapping of X_1 into X_2 such that $AT_1 = T_2A$. Then $AH_1 = H_2A$, $AK_1 = K_2A$ and $AQ_1 = Q_2A$.

Notes and Comments on Part Three

For an operator on an inner-product space, the numerical range has a very natural definition which was introduced, for finite-dimensional spaces, by Toeplitz [**1**] in 1918 as follows. If H_0 is an inner-product space and T is an operator on H_0, the *numerical range* of T is defined to be $\{\langle Tx, x\rangle: x \in H_0, \|x\| = 1\}$. The study of numerical range was motivated by the classical theory of quadratic forms. Hausdorff [**1**] showed that the numerical range (in the finite-dimensional case) is convex. Stone [**1**] extended this last result to inner-product spaces of arbitrary dimension. Since then there have been numerous papers on the numerical range of an operator on an inner-product space. For an account of this theory, see Chapter 17 of [**9**].

No concept of numerical range appropriate to general normed vector spaces appeared until 1961 and 1962, when distinct, though related concepts were introduced independently by Lumer [**1**] and Bauer [**1**]. Bauer's paper dealt only with finite-dimensional vector spaces but he gave a definition of numerical range applicable to general normed vector spaces. Lumer's paper was extremely important in the development of the subject. He defined the concept of semi-inner-product on a vector space and proved that every normed vector space $(X, \| \|)$ has at least one semi-inner-product $[,]$ such that $[x, x] = \|x\|^2$ for all x in X. The Toeplitz definition of numerical range then extends in the obvious way. However, it may happen that there are infinitely many semi-inner-products compatible with the norm and consequently infinitely many numerical ranges. Lumer [**1**] proved that the closed convex hull of one such numerical range contains any other such numerical range. Since then several other numerical ranges have been introduced and the statement of the last sentence applies to these numerical ranges. The one which we chose, because it was most convenient for our purpose, made its appearance in Bonsall [**2**] in 1969. An operator is called *hermitian* if its numerical range is real. In view of our earlier remarks, if an operator is hermitian with respect to one numerical range it is hermitian with respect to any other. Theorems 4.4, 4.7, 4.8 are due to Lumer [**1**], [**2**].

In 1971, Sinclair [1] established the equality of the norm and spectral radius of $\alpha + \beta H$, where α, β are complex and H is a hermitian operator. The elementary proof of Theorem 4.10 (the special case $\alpha = 0$, $\beta = 1$) is due to Bonsall and Crabb [1]. Lumer [2] showed that there exists a hermitian operator T on an infinite-dimensional Banach space such that some power of T is not hermitian. The elementary example of a hermitian operator on $\mathbf{C}^3$ whose square is not hermitian is due to M. J. Crabb.

The notion of hermitian equivalence of families of operators is due to Lumer [2], to whom Theorems 4.17, 4.18 are due. The commutativity result Proposition 4.21(ii) was first proved by Berkson, Dowson and Elliott [1] by a different method. Theorems 4.22, 4.23 are due to Dowson, Gillespie and Spain [1].

Vidav [1] proved the following theorem.

Let A be a complex Banach algebra with identity such that

(i) *every element of A can be expressed in the form $u + iv$ where u and v are hermitian;*

(ii) *if h is a hermitian element of A then $h^2 = u + iv$ where u, v are hermitian and $uv = vu$.*

Then there is a bicontinuous algebra isomorphism $a \to T_a$ of A onto a C^*-algebra such that T_h is self-adjoint and $|T_h| = \|h\|$ whenever h is hermitian.

Berkson [2] and Glickfeld [1] showed independently (and by different methods) that the mapping $a \to T_a$ is in fact an isometry. Palmer [1] showed that condition (ii) is unnecessary. A complete account of the Vidav-Berkson-Glickfeld-Palmer theorem is given in [3; p. 205–12].

Sinclair [2] has given an example of a hermitian operator A such that the closed subalgebra generated by I and A in the operator norm is not semi-simple. Dowson [8] has given an example of a separable reflexive Banach space X and a hermitian operator T on X such that the closed subalgebra $\mathscr{A}$ of operators generated in the weak operator topology by I and T has the property that its second commutant is strictly larger than $\mathscr{A}$.

For a compact self-adjoint operator T on a Hilbert space there is the classical spectral decomposition given by $T = \sum_n \lambda_n P_n$, where $\{\lambda_n\}$ is the set of non-zero eigenvalues of T and P_n is the spectral projection corresponding to λ_n. Yu. I. Lyubič [1], [2], [3] and Bollobás [2] have investigated the extent to which there is an analogous result for a compact hermitian operator on a Banach space and have linked the problem to some deep problems in harmonic analysis.

A comprehensive account of numerical ranges on normed vector spaces is given in [1] and [2].

Part 4

PRESPECTRAL OPERATORS

5. Prespectral Operators

In this chapter we introduce the concept of a bounded Boolean algebra of projections. The theory of integration with respect to a spectral measure is developed. The class of prespectral operators is introduced and the fundamental properties of such operators are proved. We discuss in some detail an example of a prespectral operator due to Fixman.

Definition 5.1. A *Boolean algebra* **B** *of projections* on X is a commutative subset of $L(X)$ such that

(i) $E^2 = E \qquad (E \in \mathbf{B})$;
(ii) $0 \in \mathbf{B}$;
(iii) if $E \in \mathbf{B}$ then $I - E \in \mathbf{B}$;
(iv) if $E, F \in \mathbf{B}$ then

$$E \vee F = E + F - EF \in \mathbf{B},$$

$$E \wedge F = EF \in \mathbf{B}.$$

Definition 5.2. A Boolean algebra **B** of projections on X is said to be *bounded* if there is a real number M such that $\|E\| \leqslant M \ (E \in \mathbf{B})$.

The following inequality is basic to the theory of integration with respect to a spectral measure.

PROPOSITION 5.3. *Let* $E_1, \ldots, E_n$ *be a finite family of projections on* X *such that* $E_r E_s = 0$ *if* $1 \leqslant r < s \leqslant n$. *Suppose that* $\alpha_1, \ldots, \alpha_n \in \mathbf{C}$. *Then*

$$\left\| \sum_{r=1}^{n} \alpha_r E_r \right\| \leqslant 4M \sup\{|\alpha_r| : 1 \leqslant r \leqslant n\},$$

where

$$M = \sup\{\|E_r\| : 1 \leqslant r \leqslant n\}.$$

Proof. Observe that it is sufficient to show that if $\alpha_r \in [-1, 1]$ $(r = 1, \ldots, n)$ then $\|\sum_{r=1}^{n} \alpha_r E_r\| \leqslant 2M$. In view of this, let Ω be the subset of points of $\mathbf{R}^n$ all of whose co-ordinates lie in $[-1, 1]$. Consider the function ϕ defined on Ω by

$$\phi(\lambda_1, \ldots, \lambda_n) = \left\| \sum_{r=1}^{n} \lambda_r E_r \right\|.$$

ϕ is continuous on the compact set Ω. Also ϕ is convex since, if $0 < t < 1$ and $\mathbf{u}, \mathbf{v} \in \Omega$, then

$$\begin{aligned}\phi(t\mathbf{u} + (1 - t)\mathbf{v}) &\leqslant \phi(t\mathbf{u}) + \phi((1 - t)\mathbf{v}) \\ &= t\phi(\mathbf{u}) + (1 - t)\phi(\mathbf{v}) \\ &\leqslant \max\{\phi(\mathbf{u}), \phi(\mathbf{v})\}.\end{aligned}$$

It follows that ϕ attains its supremum at an extreme point of Ω, say $(\mu_1, \ldots, \mu_n)$. Thus $\mu_r = \pm 1$ $(r = 1, \ldots, n)$. Group together those terms in the sum $\sum_{r=1}^{n} \mu_r E_r$ for which $\mu_r = 1$. Similarly, group together those terms in the sum $\sum_{r=1}^{n} \mu_r E_r$ for which $\mu_r = -1$. It follows that

$$\left\| \sum_{r=1}^{n} \alpha_r E_r \right\| \leqslant 2M \; (-1 \leqslant \alpha_r \leqslant 1 \text{ for } r = 1, \ldots, n)$$

and the proof is complete.

We now show that if $\mathbf{B}$ is a bounded Boolean algebra of projections on X then $\mathbf{B}$ is hermitian-equivalent.

THEOREM 5.4. *Let* $\mathbf{B}$ *be a bounded Boolean algebra of projections on* X. *There is an equivalent norm on* X *with respect to which every operator in* $\mathbf{B}$ *is hermitian.*

Proof. Let $\mathbf{S}$ be the multiplicative abelian semigroup of operators of the form $\sum_{r=1}^{n} \lambda_r E_r$ where $E_r \in \mathbf{B}, |\lambda_r| \leqslant 1$ $(1 \leqslant r \leqslant n)$ and $E_r E_s = 0$ if $1 \leqslant r < s \leqslant n$. Observe that I is the identity element of $\mathbf{S}$ and $\mathbf{S}$ is bounded by Proposition 5.3. Thus

$$\exp(itE) = (I - E) + e^{it}E \in \mathbf{S} \qquad (E \in \mathbf{B}, t \in \mathbf{R}).$$

Hence, if we define

$$|||x||| = \sup\{\|Ax\| : A \in \mathbf{S}\} \qquad (x \in X)$$

then

$$|||\exp itE||| \leqslant 1 \qquad (E \in \mathbf{B}, t \in \mathbf{R})$$

and so $\mathbf{B}$ is hermitian-equivalent by Theorem 4.7.

Next, we introduce the concept of a spectral measure and develop the theory of integration with respect to a spectral measure.

A family $\Gamma \subseteq X^*$ is called *total* if and only if $y \in X$ and $\langle y, f \rangle = 0$, for all f in Γ, together imply that $y = 0$. Let Σ be a σ-algebra of subsets of an arbitrary set Ω. Suppose that a mapping $E(\cdot)$ from Σ into a Boolean algebra of projections on X satisfies the following conditions:

(i) $E(\delta_1) + E(\delta_2) - E(\delta_1)E(\delta_2) = E(\delta_1 \cup \delta_2)$;
(ii) $E(\delta_1)E(\delta_2) = E(\delta_1 \cap \delta_2) \qquad (\delta_1, \delta_2 \in \Sigma)$;
(iii) $E(\Omega \backslash \delta) = I - E(\delta) \qquad (\delta \in \Sigma)$;
(iv) $E(\Omega) = I$;
(v) there is $M > 0$ such that $\|E(\delta)\| \leqslant M$, for all δ in Σ;
(vi) there is a total linear subspace Γ of X^* such that $\langle E(\cdot)x, y \rangle$ is countably additive on Σ, for each x in X and each y in Γ. Then $E(\cdot)$ is called a spectral measure of class (Σ, Γ).

Observe that (Ω, Σ) is a measurable space. If f is a simple measurable function we can write

$$f(\lambda) = \sum_{r=1}^{n} \alpha_r \chi(\tau_r; \lambda) \qquad (\lambda \in \Omega)$$

where $\{\tau_r : r = 1, \ldots, n\}$ is a finite family of pairwise disjoint measurable subsets of Ω and $\alpha_r \in \mathbf{C}(r = 1, \ldots, n)$. Define

$$\int_\Omega f(\lambda)E(d\lambda) = \sum_{r=1}^{n} \alpha_r E(\tau_r)$$

and observe that this definition is independent of the particular representation chosen for f. Note also that by Proposition 5.3

$$\left\| \int_\Omega f(\lambda)E(d\lambda) \right\| \leqslant 4M \sup_{\lambda \in \Omega} |f(\lambda)|. \tag{1}$$

Let $B(\Omega, \Sigma)$ denote the Banach algebra of bounded measurable functions on Ω under the supremum norm, the algebraic operations being defined pointwise. Let $g \in B(\Omega, \Sigma)$. There is a sequence $\{f_n\}$ of simple measurable functions converging uniformly on Ω to g. Define

$$\int_\Omega g(\lambda)E(d\lambda) = \lim_{n \to \infty} \int_\Omega f_n(\lambda)E(d\lambda).$$

Observe that, by the inequality (1), the sequence of integrals converges in norm to an element of $L(X)$, and this element does not depend on the particular sequence $\{f_n\}$ chosen to approximate g. The map

$$g \to \int_\Omega g(\lambda)E(d\lambda)$$

is a continuous algebra homomorphism of $B(\Omega, \Sigma)$ into $L(X)$. Moreover,

$$\left\| \int_\Omega g(\lambda)E(d\lambda) \right\| \leqslant 4M \sup_{\lambda \in \Omega} |g(\lambda)| \qquad (g \in B(\Omega, \Sigma)).$$

We now come to the definition of a prespectral operator.

Definition 5.5. An operator T, in $L(X)$, is called a *prespectral operator of class* Γ if and only if the following conditions (i) and (ii) are satisfied.

(i) There is a spectral measure $E(\cdot)$ of class (Σ_p, Γ) with values in $L(X)$ such that

$$TE(\delta) = E(\delta)T \qquad (\delta \in \Sigma_p),$$

where Σ_p denotes the σ-algebra of Borel subsets of the complex plane. Note that this condition implies that the closed subspaces $E(\delta)X(\delta \in \Sigma_p)$ are invariant under T.

(ii) $$\sigma(T|E(\delta)X) \subseteq \bar{\delta} \qquad (\delta \in \Sigma_p)$$

i.e. the spectrum of the restriction of T to $E(\delta)X$ is contained in the closure of δ.

The spectral measure $E(\cdot)$ is called a *resolution of the identity of class* Γ for T.

LEMMA 5.6. *Let T, in $L(X)$, be a prespectral operator with a resolution of the identity $E(\cdot)$ of class Γ. Then $E(\sigma(T)) = I$.*

Proof. Let τ be a closed subset of $\sigma(T)$. Observe that, by condition (i) of Definition 5.5,

$$TE(\tau) = E(\tau)T;$$

$$(\lambda I - T)E(\tau) = E(\tau)(\lambda I - T) \qquad (\lambda \in \mathbf{C});$$

$$(\lambda I - T)^{-1}E(\tau) = E(\tau)(\lambda I - T)^{-1} \qquad (\lambda \in \rho(T));$$

$$(\lambda I - T)^{-1}E(\tau)X \subseteq E(\tau)X \qquad (\lambda \in \rho(T)).$$

It follows from Lemma 1.28 that $\sigma(T|E(\tau)X) \subseteq \sigma(T)$. However, by condition (ii) of Definition 5.5, we have also $\sigma(T|E(\tau)X) \subseteq \tau \subseteq \rho(T)$. Hence $\sigma(T|E(\tau)X)$ is void and so, by Theorem 1.6, $E(\tau)X = \{0\}$. There is a countable family $\{\tau_n\}$ of closed subsets of $\rho(T)$ whose union is $\rho(T)$. It follows that, for each n, $\langle E(\tau_n)x, y\rangle = 0$ and so

$$\langle E(\rho(T))x, y\rangle = 0 \quad (x \in X, y \in \Gamma).$$

Therefore, since Γ is total, $E(\rho(T)) = 0$ and so $E(\sigma(T)) = I$.

Note 5.7. Let T, in $L(X)$, be a prespectral operator with a resolution of the identity $E(\cdot)$ of class Γ. Let ρ denote the union of all open subsets v in $\mathbf{C}$ such that $E(v) = 0$. Then ρ can be expressed as the union of countably many such open sets; $\rho = \bigcup_n v_n$ say. Then for each n

$$\langle E(v_n)\,x, y\rangle = 0,$$
$$\langle E(\rho)\,x, y\rangle = 0 \quad (x \in X,\ y \in \Gamma).$$

It follows that $E(\rho) = 0$. The complement of ρ, which we denote by K, is called the *support of* $E(\cdot)$. By Lemma 5.6, $E(\rho(T)) = 0$ and so $K \subseteq \sigma(T)$. However, $E(K) = I$, and so $\sigma(T) = \sigma(T|E(K)\,X) \subseteq K$. Therefore $K = \sigma(T)$, and so $\sigma(T)$ is the intersection of all closed sets σ such that $E(\delta) = I$.

The following result shows how we can construct prespectral operators from an arbitrary spectral measure.

PROPOSITION 5.8. *Let Ω be a set and Σ a σ-algebra of subsets of Ω. Let $E(\cdot)$ be a spectral measure of class (Σ, Γ) with values in $L(X)$ and let $f \in B(\Omega, \Sigma)$. Define*

$$\psi(f) = \int_\Omega f(\lambda)\, E(d\lambda).$$

Then, if $x \in X$, $y \in \Gamma$ and $\mu(\tau) = \langle E(\tau)x, y\rangle \quad (\tau \in \Sigma)$,

$$\langle \psi(f)x, y\rangle = \int_\Omega f(\lambda)\, \mu(d\lambda).$$

Define

$$F(\tau) = E(f^{-1}(\tau)) \quad (\tau \in \Sigma_p).$$

Then $\psi(f)$ is a prespectral operator with a resolution of the identity $F(\cdot)$ of class Γ. Also

$$\psi(f) = \int_{\mathbf{C}} \lambda F(d\lambda).$$

Proof. Observe that the first statement to be proved is trivially true for a simple measurable function, and we can find a sequence of such functions converging uniformly on Ω to f. To prove the second statement, note that $F(\cdot)$ is certainly a Boolean algebra homomorphism of Σ_p into a bounded Boolean algebra of projections on X with $F(\mathbf{C}) = I$. Also if $x \in X$ and $y \in \Gamma$

then $\langle F(\cdot)x, y\rangle$ is countably additive on Σ_p. Hence $F(\cdot)$ is a spectral measure of class (Σ_p, Γ). Clearly $\psi(f)$ commutes with all values of $F(\cdot)$. Let $\tau \in \Sigma_p$ and suppose that $\lambda_0 \in \mathbf{C}\backslash\bar{\tau}$. Define

$$U = \int_\Omega (\lambda_0 - f(\lambda))^{-1} \chi(f^{-1}(\tau); \lambda)\, E(d\lambda).$$

Observe that this operator satisfies the equations

$$(\lambda_0 I - \psi(f))\, U = U(\lambda_0 I - \psi(f)) = E(f^{-1}(\tau)).$$

This shows that

$$\sigma(\psi(f), E(f^{-1}(\tau))\, X) \subseteq \bar{\tau}.$$

Hence $\psi(f)$ is a prespectral operator with a resolution of the identity $F(\cdot)$ of class Γ. Let $\varepsilon > 0$ be given. To prove the third statement we decompose the closure of $f(\Omega)$ into a finite number of disjoint parts τ_r each of diameter at most ε. Let $\lambda_r \in \tau_r$. Then

$$\Big|\sum_r \lambda_r \chi(\tau_r; f(\lambda)) - f(\lambda)\Big| < \varepsilon \quad (\lambda \in \Omega)$$

and so

$$\begin{aligned}
\psi(f) &= \lim_{\epsilon \to 0} \sum_r \lambda_r E(f^{-1}(\tau_r)) \\
&= \lim_{\epsilon \to 0} \int_{\mathbf{C}} \Big(\sum_r \lambda_r \chi(\tau_r; \lambda)\Big) F(d\lambda) \\
&= \int_{\mathbf{C}} \lambda F(d\lambda)
\end{aligned}$$

This completes the proof.

PROPOSITION 5.9. *Let T be a prespectral operator on X with a resolution of the identity $E(\cdot)$ of class Γ. Define*

$$\psi(f) = \int_{\sigma(T)} f(\lambda)\, E(d\lambda) \quad (f \in C(\sigma(T))).$$

Then (i) $\sigma(\psi(f)) = f(\sigma(T)) \quad (f \in C(\sigma(T)))$
and (ii) ψ *is a bicontinuous algebra isomorphism from* $C(\sigma(T))$ *into* $L(X)$.

Proof. By Proposition 5.8, $\psi(f)$ is a prespectral operator with a resolution of the identity $F(\cdot)$ of class Γ given by

$$F(\tau) = E(f^{-1}(\tau)) \quad (\tau \in \Sigma_p).$$

Therefore, by Note 5.7,

$$\begin{aligned}\sigma(\psi(f)) &= \cap\{\delta;\, \delta \text{ is closed and } F(\delta) = I\} \\ &= \cap\{\delta\colon \delta \text{ is closed and } E(f^{-1}(\delta)) = I\}.\end{aligned}$$

However, if δ is closed and $E(f^{-1}(\delta)) = I$, then since $f^{-1}(\delta)$ is a closed subset of $\sigma(T)$,

$$\sigma(T) \subseteq f^{-1}(\delta) \subseteq \sigma(T).$$

Hence

$$\sigma(\psi(f)) = \cap\{\delta\colon \delta \text{ is closed and } f^{-1}(\delta) = \sigma(T)\}.$$

The spectral mapping theorem (i) follows immediately together with the fact that the spectral radius of $\psi(f)$ is $\sup\{|f(\lambda)|\colon \lambda \in \sigma(T)\}$. Hence, using this and the inequality (1) we obtain

$$\sup_{\lambda \in \sigma(T)} |f(\lambda)| \leqslant \left\| \int_{\sigma(T)} f(\lambda)\, E(d\lambda) \right\| \leqslant 4M \sup_{\lambda \in \sigma(T)} |f(\lambda)|,$$

for each f in $C(\sigma(T))$, where M is such that

$$\| E(\tau) \| \leqslant M < \infty \quad (\tau \in \Sigma_p).$$

Since it has already been established that ψ is an algebra homomorphism the proof is complete.

Next, we prove an analogue for prespectral operators of Jordan's classical decomposition theorem for a matrix.

THEOREM 5.10. *Let* T *be a prespectral operator on* X *with a resolution of the* *identity $E(\cdot)$ of class Γ. Define*

$$S = \int_{\sigma(T)} \lambda E(d\lambda), \quad N = T - S.$$

Then S is a prespectral operator with a resolution of the identity $E(\cdot)$ of class Γ and N is quasinilpotent.

Proof. The first statement follows from Proposition 5.8. Observe that if $\delta \in \Sigma_p$ then $TE(\delta) = E(\delta)T$ and $SE(\delta) = E(\delta)S$. Hence $NE(\delta) = E(\delta)N$. Let $\varepsilon > 0$ be given. In order to prove that N is quasinilpotent, by Proposition 1.10 it is enough to show that

$$\sigma(N) \subseteq \{\lambda : |\lambda| \leqslant \varepsilon\} = D_\varepsilon.$$

Now let the spectrum of T be decomposed into the union of a finite family of disjoint Borel sets $\tau_1, \ldots, \tau_k$, each having diameter less than a positive number $\alpha < \varepsilon$ which will be specified presently. If $\lambda \in \rho(N|E(\tau_r)X)$ for $r = 1, \ldots, k$ and if $R_r = [(\lambda I - N)|E(\tau_r)X]^{-1}$ then, on putting

$$R = \sum_{r=1}^{k} R_r E(\tau_r)$$

we obtain

$$\begin{aligned}(\lambda I - N)R &= \sum_{r=1}^{k} (\lambda I - N)\, E(\tau_r)\, R_r E(\tau_r) \\ &= \sum_{r=1}^{k} E(\tau_r) = I\end{aligned}$$

and

$$\begin{aligned}R(\lambda I - N) &= \sum_{r=1}^{k} R(\lambda I - N)\, E(\tau_r) \\ &= \sum_{r=1}^{k} R_r(\lambda I - N)E(\tau_r) \\ &= \sum_{r=1}^{k} E(\tau_r) = I.\end{aligned}$$

Thus $\lambda \in \rho(N)$. Consequently

$$\sigma(N) \subseteq \bigcup_{r=1}^{k} \sigma(N|E(\tau_r)X),$$

so that to complete the proof it suffices to show that $\sigma(N|E(\tau_r)X) \subseteq D_\varepsilon$ for $r = 1, \ldots, k$. To show this we write

$$N|E(\tau_r)X = (T - \lambda_r I)|E(\tau_r)X + (\lambda_r I - S)|E(\tau_r)X,$$

where $\lambda_r \in \sigma(T|E(\tau_r)X)$. Since $\sigma(T|E(\tau_r)X) \subseteq \bar{\tau}_r$ we have

$$\sigma((T - \lambda_r I)|E(\tau_r)X) \subseteq \bar{\tau}_r - \lambda_r \subseteq D_\alpha.$$

Now, if M is such that $\| E(\tau) \| \leqslant M < \infty$ $(\tau \in \Sigma_p)$, then

$$\|(\lambda_r I - S)|E(\tau_r)X\| \leqslant 4M \sup_{\lambda \in \tau_r} |\lambda - \lambda_r| \leqslant 4M\alpha$$

and so $(\lambda_r I - S)E(\tau_r)X$ is small in norm if α is sufficiently small. Now, by Corollary 3 of [**3**; p. 19]

$$\nu(N|E(\tau_r)X) \leqslant \nu((T - \lambda_r I)|E(\tau_r)X) + \nu((\lambda_r I - S)|E(\tau_r)X)$$

and so by choosing α sufficiently small we can arrange that

$$\sigma(N|E(\tau_r)X) \subseteq D_\varepsilon \quad (r = 1, \ldots, k).$$

By the above, this shows that $\sigma(N) \subseteq D_\varepsilon$. Since $\varepsilon > 0$ is arbitrary it follows that $\sigma(N) = \{0\}$ and so N is quasinilpotent. This completes the proof.

Definition 5.11. Let T be a prespectral operator on X with a resolution of the identity $E(\cdot)$ of class Γ. Define

$$S = \int_{\sigma(T)} \lambda E(d\lambda), \; N = T - S.$$

Then $S + N$ is called the *Jordan decomposition* of T corresponding to the resolution of the identity $E(\cdot)$. S is called the *scalar part* and N the *radical part* of the decomposition.

Next, we prove a commutativity theorem for prespectral operators.

THEOREM 5.12. *Let T be a prespectral operator on X with a resolution of the identity $E(\cdot)$ of class Γ. Let $A \in L(X)$ and let $AT = TA$. Then*

$$A \int_{\sigma(T)} f(\lambda)\, E(d\lambda) = \int_{\sigma(T)} f(\lambda)\, E(d\lambda)\, A \quad (f \in C(\sigma(T))).$$

Proof. Define

$$H = \int_{\sigma(T)} \operatorname{Re} \lambda E(d\lambda), \; K = \int_{\sigma(T)} \operatorname{Im} \lambda E(d\lambda),$$
$$Q = T - (H + iK).$$

By Theorem 5.4 there is an equivalent norm with respect to which $\{E(\tau): \tau \in \Sigma_p\}$ are all hermitian. By Theorem 4.4 and the way in which the integral with respect to a spectral measure is defined, H and K are hermitian with respect to the new norm. Further, Q is quasinilpotent by Theorem 5.10, and the operators H, K and Q commute. By Theorem 4.22(ii), $AH = HA$ and $AK = KA$. Therefore

$$A \int_{\sigma(T)} p(\lambda, \bar{\lambda})\, E(d\lambda) = \int_{\sigma(T)} p(\lambda, \bar{\lambda})\, E(d\lambda)\, A$$

for every polynomial p in λ and $\bar{\lambda}$. Thus, by the Stone-Weierstrass theorem and Proposition 5.9,

$$A \int_{\sigma(T)} f(\lambda)\, E(d\lambda) = \int_{\sigma(T)} f(\lambda)\, E(d\lambda)\, A \quad (f \in C(\sigma(T))).$$

Our next result summarizes the main properties of prespectral operators.

THEOREM 5.13. *Let T be a prespectral operator on X with a resolution of the identity $E(\cdot)$ of class Γ.*

(i) *If $F(\cdot)$ is any resolution of the identity for T*

$$\int_{\sigma(T)} f(\lambda)\, E(d\lambda) = \int_{\sigma(T)} f(\lambda)\, F(d\lambda) \quad (f \in C(\sigma(T))).$$

(ii) *T has a unique resolution of the identity of class Γ.*

(iii) *T has a unique Jordan decomposition for resolutions of the identity of all classes.*

Proof. Define

$$R = \int_{\sigma(T)} \operatorname{Re} \lambda \, E(d\lambda), \qquad J = \int_{\sigma(T)} \operatorname{Im} \lambda \, E(d\lambda),$$

$$R_0 = \int_{\sigma(T)} \operatorname{Re} \lambda \, F(d\lambda), \qquad J_0 = \int_{\sigma(T)} \operatorname{Im} \lambda \, F(d\lambda).$$

Then, by Theorem 5.12, $RR_0 = R_0R, RJ_0 = J_0R, JR_0 = R_0J$ and $JJ_0 = J_0J$, since R_0 and J_0 commute with T. As in the proof of Theorem 5.12, each of R, R_0, J, J_0 can be made hermitian by equivalent renorming of X. Since these operators commute, it follows from Theorem 4.18 that after some appropriate equivalent renorming of X they are simultaneously hermitian. We assume that renorming has been carried out. Let $S + N$ and $S_0 + N_0$ be respectively the Jordan decompositions of T with respect to $E(\cdot)$ and $F(\cdot)$. Then $T = S + N = S_0 + N_0$ and $SS_0 = S_0S$. Hence $NN_0 = N_0N$. Consider the equation

$$N_0 - N = (R - R_0) + i(J - J_0).$$

By Theorem 4.4 the difference of two hermitian operators is hermitian. Also, by Corollary 3 of [**3**; p. 19], $N - N_0$, being the sum of two commuting quasinilpotents, is also quasinilpotent. By Proposition 4.20

$$R = R_0, \; J = J_0 \text{ and } N = N_0.$$

The last equation suffices to prove (iii). Now, by the standard properties of the integral with respect to a spectral measure,

$$\int_{\sigma(T)} p(\lambda, \bar{\lambda}) \, E(d\lambda) = \int_{\sigma(T)} p(\lambda, \bar{\lambda}) \, F(d\lambda)$$

for any polynomial p in λ and $\bar{\lambda}$. Therefore by the Stone-Weierstrass theorem

$$\int_{\sigma(T)} f(\lambda) \, E(d\lambda) = \int_{\sigma(T)} f(\lambda) \, F(d\lambda) \quad (f \in C(\sigma(T))).$$

This proves (i). Now suppose that $E(\cdot)$ and $F(\cdot)$ are both of class Γ. Let $x \in X$, $y \in \Gamma$. Define

$$\mu(\tau) = \langle E(\tau)\, x, y \rangle \quad (\tau \in \Sigma_p),$$
$$\nu(\tau) = \langle F(\tau)\, x, y \rangle \quad (\tau \in \Sigma_p).$$

By Proposition 5.8,

$$\int_{\sigma(T)} f(\lambda)\, \mu(d\lambda) = \left\langle \int_{\sigma(T)} f(\lambda)\, E(d\lambda)\, x, y \right\rangle$$

$$\int_{\sigma(T)} f(\lambda)\, \nu(d\lambda) = \left\langle \int_{\sigma(T)} f(\lambda)\, F(d\lambda)\, x, y \right\rangle$$

for all f in $C(\sigma(T))$. Hence

$$\int_{\sigma(T)} f(\lambda)\, \mu(d\lambda) = \int_{\sigma(T)} f(\lambda)\, \nu(d\lambda) \quad (f \in C(\sigma(T))).$$

$\mu(\cdot)$ and $\nu(\cdot)$ are finite countably additive measures with supports contained in $\sigma(T)$. Hence they are regular measures, and by the Riesz representation theorem $\mu = \nu$. It then follows that

$$\langle E(\tau)\, x, y \rangle = \langle F(\tau)\, x, y \rangle \quad (\tau \in \Sigma_p, x \in X, y \in \Gamma).$$

Since Γ is total, conclusion (ii) follows and the proof is complete.

The last statement of the theorem leads us to the following definition.

Definition 5.14. Let S be a prespectral operator on X with resolution of the identity $E(\cdot)$ of class Γ such that

$$S = \int_{\sigma(S)} \lambda E(d\lambda).$$

Then S is called a *scalar-type operator of class* Γ.

Observe that, by Theorem 5.13 (iii), if S is a scalar-type operator of class Γ

and $F(\cdot)$ is another resolution of the identity for S then

$$S = \int_{\sigma(S)} \lambda F(d\lambda).$$

Next, we prove a stronger version of Theorem 5.10.

THEOREM 5.15. (i) *Let T be a prespectral operator on X with resolution of the identity $E(\cdot)$ of class Γ. Define*

$$S = \int_{\sigma(T)} \lambda E(d\lambda), \quad N = T - S.$$

Then S is a scalar-type operator with resolution of the identity $E(\cdot)$ of class Γ, and N is a quasinilpotent commuting with $\{E(\tau): \tau \in \Sigma_p\}$. Moreover $\sigma(T) = \sigma(S)$.
(ii) *Let S be a scalar-type operator on X with resolution of the identity $E(\cdot)$ of class Γ. Let N be a quasinilpotent operator on X commuting with $\{E(\tau): \tau \in \Sigma_p\}$. Then $S + N$ is prespectral with resolution of the identity $E(\cdot)$ of class Γ. Moreover $\sigma(S + N) = \sigma(S)$.*

Proof. The first statement of (i) follows from Theorem 5.10. By Note 5.7, $\sigma(T)$ and $\sigma(S)$ are both equal to the support of the spectral measure $E(\cdot)$. To establish (ii), it is sufficient to show that if $T = S + N$ then

$$\sigma(T|E(\tau)X) \subseteq \bar{\tau} \quad (\tau \in \Sigma_p).$$

Let $\tau \in \Sigma_p$. Observe that S, N and T leave $E(\tau)X$ invariant. Also $N|E(\tau)X$ is quasinilpotent. By Theorem 1.27, we have

$$\sigma(T|E(\tau)X) = \sigma(S|E(\tau)X) \subseteq \bar{\tau} \quad (\tau \in \Sigma_p)$$

and the proof is complete.

Let T be a prespectral operator on X with resolution of the identity $E(\cdot)$ of class Γ. Recall that in Definition 1.18 we defined $f(T)$, for each f in $\mathscr{F}(T)$.

THEOREM 5.16. *Let T be a prespectral operator on X with resolution of the identity $E(\cdot)$ of class Γ. Let $f \in \mathscr{F}(T)$. Then $f(T)$ is a prespectral operator with* *resolution of the identity $F(\cdot)$ of class Γ given by*

$$F(\tau) = E(f^{-1}(\tau)) \quad (\tau \in \Sigma_p).$$

Proof. The formula above clearly yields a spectral measure of class (Σ_p, Γ) all of whose values commute with $f(T)$. Let $\tau \in \Sigma_p$. If $\lambda_0 \in \mathbf{C} \backslash \bar{\tau}$, the function h given by

$$h(\lambda) = (\lambda_0 - f(\lambda))^{-1}$$

is analytic on a neighbourhood of the closure of $f^{-1}(\tau)$. Hence, if C is a suitable finite family of rectifiable Jordan curves surrounding the closure of $f^{-1}(\tau)$ we have, by Definition 1.18 and Proposition 1.37,

$$\left(\frac{1}{2\pi i} \int_C h(\lambda)\, T_0(\lambda)\, d\lambda\right)(\lambda_0 I - f(T))\, E(f^{-1}(\tau)) = E(f^{-1}(\tau)),$$

where

$$T_0(\lambda) = \{(\lambda I - T) | E(f^{-1}(\tau))\, X\}^{-1}.$$

This shows that

$$\sigma(f(T) | F(\tau) X) \subseteq \bar{\tau} \quad (\tau \in \Sigma_p)$$

and the proof is complete.

THEOREM 5.17. *Let T be a prespectral operator on X with resolution of the identity $E(\cdot)$ of class Γ. Define*

$$S = \int_{\sigma(T)} \lambda E(d\lambda), \; N = T - S.$$

Let $f \in \mathscr{F}(T)$. Then

$$f(T) = \sum_{n=0}^{\infty} \frac{N^n}{n!} \int_{\sigma(T)} f^{(n)}(\lambda)\, E(d\lambda),$$

the series converging in the norm of $L(X)$.

Proof. By Theorem 5.15, $\sigma(T) = \sigma(S)$ and so $\mathscr{F}(T) = \mathscr{F}(S)$. The present theorem will follow immediately from Theorem 1.27 as soon as we show that

$$f(S) = \int_{\sigma(S)} f(\lambda)\, E(d\lambda) \quad (f \in \mathscr{F}(S)).$$

Let $x \in X$ and $y \in \Gamma$. Then if $\mu(\tau) = \langle E(\tau)x, y\rangle$ $(\tau \in \Sigma_p)$ we have

$$\left\langle \int_{\sigma(S)} g(\lambda)\, E(d\lambda)\, x, y \right\rangle = \int_{\sigma(S)} g(\lambda)\, \mu(d\lambda) \tag{2}$$

for every g in $C(\sigma(S))$. Observe that, if C is a suitable finite family of rectifiable Jordan curves surrounding $\sigma(S)$ and if $f \in \mathscr{F}(S)$, then by Proposition 5.8

$$\begin{aligned}\langle f(S)x, y\rangle &= \frac{1}{2\pi i}\int_C f(\lambda)\langle(\lambda I - S)^{-1}x, y\rangle\, d\lambda \\ &= \frac{1}{2\pi i}\int_C f(\lambda)\left\{\int_{\sigma(S)} (\lambda - \xi)^{-1}\, \mu(d\xi)\right\} d\lambda.\end{aligned}$$

A standard argument invoking Fubini's theorem shows that we may interchange the order of integration in the double integral to get

$$\langle f(S)x, y\rangle = \int_{\sigma(S)} f(\lambda)\, \mu(d\lambda). \tag{3}$$

Since Γ is total, it follows from (2) and (3) that

$$f(S) = \int_{\sigma(S)} f(\lambda)\, E(d\lambda) \quad (f \in \mathscr{F}(S)).$$

This completes the proof.

Definition 5.18. Let T be a prespectral operator on X with resolution of the identity $E(\cdot)$ of class Γ. Define

$$S = \int_{\sigma(T)} \lambda E(d\lambda), \quad N = T - S.$$

T is said to be of *type m* if and only if

$$f(T) = \sum_{n=0}^{m} \frac{N^n}{n!} \int_{\sigma(T)} f^{(n)}(\lambda)\, E(d\lambda) \quad (f \in \mathscr{F}(T)).$$

PROPOSITION 5.19. *Let T be a prespectral operator on X with resolution of the identity $E(\cdot)$ of class Γ. Define*

$$S = \int_{\sigma(T)} \lambda E(d\lambda), \; N = T - S.$$

(i) *T is of type m if and only if $N^{m+1} = 0$.*
(ii) *T is a scalar-type operator if and only if it is of type* 0.

Proof. If $N^{m+1} = 0$, then clearly the formula of Theorem 5.17 reduces to the formula of Definition 5.18. Conversely, if T is of type m, we see by putting

$$f(\lambda) = \lambda^{m+1}/(m+1)!$$

in these two formulae that

$$0 = N^{m+1} \int_{\sigma(T)} E(d\lambda) = N^{m+1}.$$

This completes the proof of (i). Statement (ii) follows immediately.

PROPOSITION 5.20. *Let T be a prespectral operator on X with resolution of the identity $E(\cdot)$ of class Γ. Define*

$$S = \int_{\sigma(T)} \lambda E(d\lambda), \; N = T - S.$$

Let $f \in \mathscr{F}(T)$. The scalar part of the Jordan decomposition of $f(T)$ is equal to $f(S)$. If T is of type m then $f(T)$ is also of type m.

Proof. By Theorem 5.17 and its proof

$$f(T) = \sum_{n=0}^{\infty} f^{(n)}(S) \frac{N^n}{n!},$$

$$f(S) = \int_{\sigma(T)} f(\lambda)\, E(d\lambda).$$

Define

$$F(\tau) = E(f^{-1}(\tau)) \quad (\tau \in \Sigma_p).$$

By Theorem 5.16 and Proposition 5.8, $F(\cdot)$ is a resolution of the identity of class Γ both for $f(T)$ and $f(S)$. Thus, if we can show that the operator

$$N_1 = \sum_{n=1}^{\infty} f^{(n)}(S) \frac{N^n}{n!}$$

is quasinilpotent, the first statement will be proved. Let $\mathscr{A}$ be the closed commutative subalgebra of $L(X)$ generated by N and

$$\left\{\int_{\sigma(T)} f(\lambda)\, E(d\lambda) \colon f \in C(\sigma(T))\right\}.$$

Observe that the radical of $\mathscr{A}$ is a closed ideal of $\mathscr{A}$. Hence N_1 is quasinilpotent. To prove the second statement, note that $N^{m+1} = 0$ and so

$$N_1 = \sum_{n=1}^{m} f^{(n)}(S) \frac{N^n}{n!}.$$

Hence $N_1^{m+1} = 0$ and the proof is complete.

We now consider a continuous algebra homomorphism of $C(K)$ into $L(X)$, where K is a compact Hausdorff space. We denote by Σ_K the σ-algebra of Borel subsets of K.

THEOREM 5.21. *Let K be a compact Hausdorff space, and let ψ be a continuous algebra homomorphism of $C(K)$ into $L(X)$ with $\psi(1) = I$. Let N, in $L(X)$, be a quasinilpotent commuting with $\psi(f)$ for every f in $C(K)$. Then there is a spectral* *measure $E(\cdot)$ of class (Σ_K, X) with values in $L(X^*)$ such that*

(i) $$\psi(f)^* = \int_K f(\lambda)\, E(d\lambda) \quad (f \in C(K))$$

and

(ii) $$N^* E(\tau) = E(\tau)\, N^* \quad (\tau \in \Sigma_K).$$

Moreover if $S \in \psi(C(K))$, then the adjoint of $T = S + N$ is prespectral of class X, and $S^* + N^*$ is the Jordan decomposition of T^*.

Proof. Let $x \in X$, $y \in X^*$. The map $f \to \langle \psi(f)\, x, y \rangle$ is a bounded linear functional on $C(K)$, and so by the Riesz representation theorem there is a uniquely determined regular Borel measure $\mu(\cdot, x, y)$ such that for all f in $C(K)$

$$\langle \psi(f)\, x, y \rangle = \int_K f(\lambda)\, \mu(d\lambda, x, y). \tag{4}$$

Since $\mu(\tau, x, y)$ is uniquely determined by τ, x and y it is, for each τ in Σ_K, bilinear in x and y. Also, since

$$\begin{aligned} |\mu(\tau, x, y)| &\leqslant \operatorname{var} |\mu(\cdot, x, y)| \\ &= \sup\{|\langle \psi(f)\, x, y \rangle| : \|f\| = 1\} \\ &\leqslant M \|x\|\, \|y\|, \end{aligned}$$

for some constant M independent of x and y, we see that $\mu(\tau, x, y)$ is continuous in x and y. Hence, for fixed τ and y, there is a point $E(\tau)\, y$ of X^* such that

$$\mu(\tau, x, y) = \langle x, E(\tau)\, y \rangle \quad (x \in X).$$

It follows from the bilinearity and boundedness of μ that $E(\tau) \in L(X^*)$. It will now be shown that $E(\cdot)$ is a spectral measure. By putting $f = 1$ in equation (4), we see that $I^* = E(K)$ and, since $E(\cdot)$ is additive, that

$$I^* - E(\delta) = E(K) - E(\delta) = E(K \backslash \delta) \quad (\delta \in \Sigma_K).$$

To prove that $E(\cdot)$ is a spectral measure it will therefore suffice to show that $E(\delta \cap \tau) = E(\delta) E(\tau)$ for every pair δ, τ of Borel subsets of K. For fixed x in X, y in X^* and g in $C(K)$, define

$$\nu(\tau) = \int_\tau g(\lambda)\, \mu(d\lambda, x, y) \quad (\tau \in \Sigma_K).$$

If $f \in C(K)$ then

$$\int_K f(\lambda)\, \nu(d\lambda) = \int_K f(\lambda) g(\lambda)\, \mu(d\lambda, x, y) = \langle \psi(f)\, \psi(g) x, y \rangle$$

$$= \int_K f(\lambda)\, \mu(d\lambda, \psi(g)\, x, y).$$

Since $\nu(\cdot)$ and $\mu(\cdot, \psi(g)\, x, y)$ are regular measures,

$$\nu(\tau) = \int_K g(\lambda)\, \chi(\tau; \lambda)\, \mu(d\lambda, x, y) = \langle \psi(g)\, x, E(\tau)\, y \rangle$$

$$= \int_K g(\lambda)\, \mu(d\lambda, x, E(\tau)\, y) \quad (\tau \in \Sigma_K).$$

Since g is an arbitrary element of $C(K)$,

$$\langle x, E(\tau \cap \delta)\, y \rangle = \int_{\tau \cap \delta} \mu(d\lambda, x, y) = \int_\delta \chi(\tau; \lambda)\, \mu(d\lambda, x, y)$$

$$= \langle x, E(\delta)\, E(\tau)\, y \rangle \quad (\delta, \tau \in \Sigma_K).$$

It follows that for all pairs of Borel subsets δ, τ of K

$$E(\delta)\, E(\tau) = E(\delta \cap \tau) = E(\tau)\, E(\delta)$$

and this completes the proof that $E(\cdot)$ is a spectral measure. The statement (i) now follows immediately from equation (4) and Proposition 5.8.

Let $x \in X$ and $y \in X^*$. Define

$$\mu_1(\tau) = \langle Nx, E(\tau)\, y \rangle \quad \text{and} \quad \mu_2(\tau) = \langle x, E(\tau)\, N^* y \rangle \quad (\tau \in \Sigma_K).$$

Then, by Proposition 5.8,

$$\int_K f(\lambda)\,\mu_1(d\lambda) = \langle Nx, \psi(f)^* y\rangle = \langle x, \psi(f)^* N^* y\rangle$$

$$= \int_K f(\lambda)\,\mu_2(d\lambda) \quad (f \in C(K)).$$

It follows that the Borel measures $\langle Nx, E(\cdot)y\rangle$ and $\langle x, E(\cdot)\, N^*y\rangle$ (regular by construction) are identical and (ii) is immediate. If $S = \psi(f)$, for some f in $C(K)$, then by (i) and Proposition 5.8, S^* is a scalar-type operator with a resolution of the identity of class X whose range is contained in the range of $E(\cdot)$. Hence by (ii), $S^* + N^*$ is the Jordan decomposition for T^*, and the proof is complete.

Next, we show that the adjoint of a prespectral operator is also a prespectral operator.

THEOREM 5.22. *Let T be a prespectral operator on X with resolution of the identity $E(\cdot)$ of class Γ. Then T^* is prespectral on X^* with resolution of the identity $F(\cdot)$ of class X such that*

$$\left(\int_{\sigma(T)} f(\lambda)\,E(d\lambda)\right)^* = \int_{\sigma(T)} f(\lambda)\,F(d\lambda) \quad (f \in C(\sigma(T))).$$

Moreover if

$$S = \int_{\sigma(T)} \lambda E(d\lambda), \; N = T - S$$

then $S^* + N^*$ is the Jordan decomposition of T^*.

Proof. By Proposition 5.9, the map ψ defined by

$$\psi(f) = \int_{\sigma(T)} f(\lambda)\,E(d\lambda)$$

is a continuous algebra homomorphism from $C(\sigma(T))$ into $L(X)$, and N commutes with each $\psi(f)$. Let Σ_0 denote the σ-algebra of Borel subsets of $\sigma(T)$. Hence, by Theorem 5.21, there is a spectral measure $G(\cdot)$ of class (Σ_0, X)

such that

$$\psi(f)^* = \int_{\sigma(T)} f(\lambda)\, G(d\lambda) \quad (f \in C(\sigma(T))),$$

$$N^*G(\tau) = G(\tau)\, N^* \qquad (\tau \in \Sigma_0).$$

By Proposition 5.8, S^* is a scalar-type operator on X^* with resolution of the identity $F(\cdot)$ of class X given by

$$F(\tau) = G(\tau \cap \sigma(T)) \quad (\tau \in \Sigma_p).$$

Hence, for each f in $C(\sigma(T))$,

$$\psi(f)^* = \int_{\sigma(T)} f(\lambda)\, G(d\lambda) = \int_{\sigma(T)} f(\lambda)\, F(d\lambda)$$

and the proof is complete.

We now prove some generalizations of Theorems 5.12 and 5.15. In connection with the first of these, we observe that it will be shown later that the sum of a scalar-type operator and a commuting quasinilpotent need not be prespectral of any class.

THEOREM 5.23. *Let $S, N \in L(X)$. Suppose that S is a scalar-type operator and N is quasinilpotent with $SN = NS$. Suppose that A, in $L(X)$, commutes with $S + N$. Then A commutes with each of S and N. Moreover, if $S + N = S_0 + N_0$, where S_0 is a scalar-type operator on X, N_0 is quasinilpotent and $S_0N_0 = N_0S_0$, then $S = S_0$ and $N = N_0$.*

Proof. Let $E(\cdot)$ be a resolution of the identity for S. Then by Theorem 5.12 and the hypothesis $NS = SN$ we obtain

$$N \int_{\sigma(S)} f(\lambda)\, E(d\lambda) = \int_{\sigma(S)} f(\lambda)\, E(d\lambda)\, N \quad (f \in C(\sigma(S))).$$

By Theorem 5.21, $(S + N)^*$ is prespectral on X^* of class X, with Jordan decomposition $S^* + N^*$. Similarly $S_0^* + N_0^*$ is a Jordan decomposition for $(S_0 + N_0)^* = (S + N)^*$ and so the second statement of the theorem follows from Theorem 5.13 (iii). Since A^* commutes with the prespectral operator

$(S + N)^*$, the first statement of the theorem follows readily from Theorem 5.12.

THEOREM 5.24. *Let S be a scalar-type operator on X. Let N, in $L(X)$, be a quasinilpotent operator with $SN = NS$. Then if $T = S + N$ is prespectral, every resolution of the identity for T is also a resolution of the identity for S. Also $T = S + N$ is the unique Jordan decomposition for T. Moreover, N commutes with every resolution of the identity for T.*

Proof. Let $S_0 + N_0$ be the Jordan decomposition of the prespectral operator T. Then from the definition of Jordan decomposition and Theorem 5.23 we obtain $S = S_0$, $N = N_0$. The other statements of the theorem now follow from Theorem 5.15.

THEOREM 5.25. *Let S be a scalar-type operator on X with resolution of the identity $E(\cdot)$ of class Γ. Let N, in $L(X)$, be quasinilpotent with $SN = NS$. Then $S + N$ is prespectral of class Γ if and only if*

$$NE(\tau) = E(\tau)N \quad (\tau \in \Sigma_p).$$

Proof. The sufficiency of the condition follows from Theorem 5.15. Now let $S + N$ be prespectral with resolution of the identity $F(\cdot)$ of class Γ. By Theorem 5.24, $F(\cdot)$ is a resolution of the identity of class Γ for S and

$$NF(\tau) = F(\tau)N \quad (\tau \in \Sigma_p).$$

By Theorem 5.13 (ii), S has a unique resolution of the identity $E(\cdot)$ of class Γ. Hence $F(\cdot) = E(\cdot)$ and

$$NE(\tau) = E(\tau)N \quad (\tau \in \Sigma_p).$$

The proof is complete.

As an application of these ideas we prove an important theorem due to Bade and Curtis on commutative Banach algebras. Some preliminary notation is required.

Let A be a complex Banach algebra (not necessarily possessing an identity element). Define for each a in A

$$T_a x = ax \quad (x \in A),$$
$$T_e x = x \quad (x \in A).$$

In the following theorem we do not assume that the algebra A has an identity. This result is due to Bade and Curtis [**1**; p. 858].

THEOREM 5.26. *Let A be a commutative Banach algebra with radical R, such that for some compact Hausdorff space Ω, the quotient algebra A/R is isomorphic to $C(\Omega)$. If A is the direct sum of a closed subalgebra B and the radical R, then the closed subalgebra B is uniquely determined.*

Proof. It follows from commutative Banach algebra theory that there is a bicontinuous algebra isomorphism ψ of $C(\Omega)$ into $L(A)$ with $\psi(1) = T_e$. Hence, by Theorem 5.21, if $b \in B$ then T_b^* is a scalar-type operator on A^* of class A. Moreover, if $a \in A$ then $T_a^* = T_b^* + T_r^*$ is a prespectral operator on A^* of class A, since T_r is a quasinilpotent commuting with $\{\psi(f): f \in C(\Omega)\}$. Thus, if $A = B_1 \oplus R$, where B_1 is a closed subalgebra of A different from B, there would be an element with two distinct decompositions as the sum of a scalar-type operator and a commuting quasinilpotent, contradicting Theorem 5.23. This completes the proof.

It will be shown later that a prespectral operator may have distinct resolutions of the identity corresponding to distinct total linear subspaces of the dual space. However, as we now show, the projection corresponding to an open-and-closed subset of the spectrum always coincides with the spectral projection. Also, the ranges of the projections corresponding to a closed set are equal.

PROPOSITION 5.27. *Let T be a prespectral operator on X with resolution of the identity $E(\cdot)$ of class Γ. Let τ be an open-and-closed subset of $\sigma(T)$. Then $E(\tau)$ is equal to the spectral projection corresponding to τ.*

Proof. Observe that

$$\sigma(T \,|\, E(\tau)X) \subseteq \tau \quad \text{and} \quad \sigma\big(T \,|\, (I - E(\tau))X\big) \subseteq \mathbf{C} \backslash \tau.$$

The desired conclusion now follows from Proposition 1.53.

In order to prove that the ranges of the projections corresponding to a closed set are equal, it is necessary to introduce the concept of the single-valued extension property.

Let $T \in L(X)$ and let $x \in X$. An X-valued function f_x, defined and analytic on an open subset $D(f_x)$ of $\mathbf{C}$ such that

$$(\zeta I - T) f_x(\zeta) = x \quad (\zeta \in D(f_x))$$

is called a *pre-imaging function for x and T*. It is easy to see that

$$f_x(\zeta) = (\zeta I - T)^{-1} x \quad (\zeta \in \rho(T) \cap D(f_x)).$$

If for all x in X and all pairs $f_x^{(1)}$, $f_x^{(2)}$ of pre-imaging functions for x and T we have

$$f_x^{(1)}(\zeta) = f_x^{(2)}(\zeta) \quad (\zeta \in D(f_x^{(1)}) \cap D(f_x^{(2)}))$$

then T is said to have the *single-valued extension property*. In this case there is a unique pre-imaging function with maximal domain $\rho(x)$, an open set containing $\rho(T)$. The values of this function are denoted by $\{x(\xi) : \xi \in \rho(x)\}$. Let $\sigma(x) = \mathbf{C} \backslash \rho(x)$. Clearly $\sigma(x) \subseteq \sigma(T)$. $\rho(x)$ is called the *resolvent set* of x and $\sigma(x)$ is called the *spectrum* of x.

PROPOSITION 5.28. *Let $T \in L(X)$. Suppose that $\sigma(T)$ is nowhere dense. Then T has the single-valued extension property.*

Proof. Let $x \in X$. Let $f_x^{(1)}$ and $f_x^{(2)}$ be pre-imaging functions for x and T. Now let $\zeta \in D(f_x^{(1)}) \cap D(f_x^{(2)})$. Since $\rho(T)$ is dense in $\mathbf{C}$ there is a sequence $\{\lambda_n\}$ in $\rho(T)$ converging to ζ. Also, since $D(f_x^{(1)})$ and $D(f_x^{(2)})$ are open, so is their intersection. By taking a subsequence if necessary we may assume that $\lambda_n \in D(f_x^{(1)}) \cap D(f_x^{(2)})$ for each n. Now for each n we have

$$f_x^{(1)}(\lambda_n) = (\lambda_n I - T)^{-1} x = f_x^{(2)}(\lambda_n).$$

Since $f_x^{(1)}$ and $f_x^{(2)}$ are analytic and so continuous on $D(f_x^{(1)}) \cap D(f_x^{(2)})$ it follows that $f_x^{(1)}(\zeta) = f_x^{(2)}(\zeta)$. The proof is complete.

Not every operator has the single-valued extension property as the following example shows.

Example 5.29. Let H be a separable complex Hilbert space and let $\{\phi_n : n \in \mathbf{Z}\}$ be an orthonormal basis for H. Let U, in $L(H)$, be defined by

$$U\phi_n = \phi_{n+1} \quad (n \in \mathbf{Z}).$$

Since U is isometric then, by Proposition 1.31, $\sigma(U) \subseteq \{z : |z| = 1\}$. Let Y be the closed subspace of H generated by $\{\phi_n : n = 1, 2, 3, \ldots\}$. Consider the vector $[\phi_0]$ in the quotient space H/Y. Note that $[(\zeta I - U)^{-1}\phi_0]$ and $[(\zeta I - U)^{-1}(\phi_0 + \phi_1)]$ $(|\zeta| \neq 1)$ are analytic (H/Y)-valued functions such that

$$[(\zeta I - U)(\zeta I - U)^{-1}\phi_0] = [\phi_0] \quad (|\zeta| \neq 1),$$

$$[(\zeta I - U)(\zeta I - U)^{-1}(\phi_0 + \phi_1)] = [\phi_0] \quad (|\zeta| \neq 1).$$

At $\zeta = 0$ these functions take the values $-[\phi_{-1}]$ and $-[\phi_{-1} + \phi_0]$. Since $\phi_0 \notin Y$, U_Y, the operator induced by U on the quotient space H/Y, fails to have the single-valued extension property.

In order to show that every prespectral operator has the single-valued extension property a preliminary lemma is required.

LEMMA 5.30. *Let T be a prespectral operator on X with resolution of the identity $E(\cdot)$ of class Γ. Let τ be a closed subset of $\mathbf{C}$ and let $\xi_0 \in \mathbf{C}\backslash\tau$. If for some x_0 in X we have $(\xi_0 I - T)x_0 = 0$ then $E(\tau)x_0 = 0$ and $E(\{\xi_0\})x_0 = x_0$.*

Proof. Observe that since $\xi_0 \notin \bar{\tau}$ then $\xi_0 \in \rho(T|E(\tau)X)$ and so

$$((\xi_0 I - T)|E(\tau)X)^{-1}(\xi_0 I - T)E(\tau)x = E(\tau)x \quad (x \in X).$$

However, since

$$(\xi_0 I - T)E(\tau)x_0 = E(\tau)(\xi_0 I - T)x_0 = 0$$

we have $E(\tau)x_0 = 0$. Now for each positive integer n let $\tau_n = \{\xi : |\xi - \xi_0| \geqslant n^{-1}\}$, so that by the above $E(\tau_n)x_0 = 0$. Since, for each y in Γ, $\langle E(\cdot)x_0, y\rangle$ is countably additive

$$\langle (I - E(\{\xi_0\}))x_0, y\rangle = \lim_{n\to\infty} \langle E(\tau_n)x_0, y\rangle = 0 \quad (y \in \Gamma),$$

and so $x_0 = E(\{\xi_0\})x_0$. The proof is complete.

THEOREM 5.31. *Let T be a prespectral operator on X with resolution of the identity $E(\cdot)$ of class Γ. Then T has the single-valued extension property.*

Proof. Let $f_x^{(1)}$, $f_x^{(2)}$ be pre-imaging functions for x and T. Define

$$h(\xi) = f_x^{(1)}(\xi) - f_x^{(2)}(\xi) \quad (\xi \in D(f_x^{(1)}) \cap D(f_x^{(2)})).$$

Suppose that for some ξ_0 in $D(f_x^{(1)}) \cap D(f_x^{(2)})$ we have $h(\xi_0) \neq 0$. Then there is an open neighbourhood $N(\xi_0)$ of ξ_0 contained in $D(f_x^{(1)}) \cap D(f_x^{(2)})$ such that

$$h(\xi) \neq 0, \ (\xi I - T)\, h(\xi) = 0 \ (\xi \in N(\xi_0)).$$

Let $\{\xi_n\}$ be a sequence in $N(\xi_0)$ converging to ξ_0 such that $\xi_n \neq \xi_0$, for each positive integer n. Then $h(\xi_n) \to h(\xi_0)$ and Lemma 5.30 shows that

$$0 = E(\{\xi_0\})\, h(\xi_n) \to E(\{\xi_0\})\, h(\xi_0) = h(\xi_0)$$

which gives a contradiction and proves the theorem.

PROPOSITION 5.32. *Let T in $L(X)$ have the single-valued extension property, and let $x \in X$. The spectrum $\sigma(x)$ of x is empty if and only if $x = 0$.*

Proof. The function $\xi \to x(\xi)$ is entire. If $|\zeta| > \|T\|$

$$\langle x(\zeta), y\rangle = \langle(\zeta I - T)^{-1} x, y\rangle \quad (y \in X^*)$$

and the right-hand side tends to 0 as $\zeta \to \infty$. By Liouville's theorem $x(\xi) = 0$ $(\xi \in \mathbf{C})$ and so $x = (\xi I - T)\, x(\xi) = 0$.

THEOREM 5.33. *Let T be a prespectral operator on X with resolution of the identity $E(\cdot)$ of class Γ. Let δ be a closed subset of $\mathbf{C}$. Then*

$$E(\delta)\, X = \{x \in X : \sigma(x) \subseteq \delta\}.$$

Proof. Let $x \in E(\delta)\, X$ so that $E(\delta)\, x = x$. Since $\sigma(T \,|\, E(\delta)\, X) \subseteq \delta$, we see from the relation

$$((\xi I - T)|\, E(\delta) X)^{-1}\, E(\delta) x = ((\xi I - T)|\, E(\delta) X)^{-1}\, x$$

that for ξ in $\mathbf{C}\backslash\delta$ the left-hand side is a pre-imaging function for x and T. Hence $\rho(x) \supseteq \mathbf{C}\backslash\delta$ and so $\sigma(x) \subseteq \delta$.

Conversely, assume that $\sigma(x) \subseteq \delta$. Let τ be a closed subset of $\mathbf{C}\backslash\delta$. Observe that the function

$$\xi \to ((\xi I - T)|E(\tau)X)^{-1} E(\tau)x \quad (\xi \in \mathbf{C}\backslash\tau)$$

is a pre-imaging function for $E(\tau)x$ and T. Moreover

$$\xi \to E(\tau)\, x(\xi) \quad (\xi \in \rho(x))$$

is a pre-imaging function for $E(\tau)x$ and T. Hence $\rho(E(\tau)x) = \mathbf{C}$ and so $\sigma(E(\tau)x)$ is void. By Proposition 5.32, $E(\tau)x = 0$. Let $\{\tau_n\}$ be a sequence of closed sets whose union is $\mathbf{C}\backslash\delta$. Then for each n

$$\langle E(\tau_n)\, x, y\rangle = 0 \qquad (y \in \Gamma).$$

Hence

$$\langle E(\mathbf{C}\backslash\delta)\, x, y\rangle = 0 \qquad (y \in \Gamma)$$

and, since Γ is total, $E(\delta)\, x = x$.

We next discuss in detail an example of a prespectral operator, first introduced by Fixman [**1**] and further developed by Berkson and Dowson [**1**]. A preliminary result is required.

LEMMA 5.34. *Let $T \in L(X)$. Let E, F be projections in $L(X)$ such that $EF = F$ and T, E, F commute. Then*

$$\sigma(T|FX) \subseteq \sigma(T|EX) \subseteq \sigma(T).$$

Proof. Let $\lambda \in \rho(T)$. Now E commutes with T and hence also with $(\lambda I - T)^{-1}$. Therefore $(\lambda I - T)^{-1}$ leaves EX invariant, and its restriction to that subspace is a bounded operator, clearly inverse to $(\lambda I - T)|EX$. Hence $\lambda \in \rho(T|EX)$, and $\sigma(T|EX) \subseteq \sigma(T)$. Similarly $T|EX$ commutes with $F|EX$, and $\sigma(T|FX) \subseteq \sigma(T|EX)$.

Example 5.35. On the subspace of l^∞ consisting of convergent sequences, the map which assigns to each such sequence its limit is a linear functional of norm 1. Throughout this section L denotes a fixed linear functional on l^∞ with $\|L\| = 1$ such that for each convergent sequence $\{\xi_n\}$

$$L(\{\xi_n\}) = \lim_n \{\xi_n\}.$$

Define operators S and A on l^∞ by

$$S\{\xi_n\} = \{\eta_n\}, \text{ where } \eta_n = \xi_n, \text{ if } n = 1 \text{ or } 2,$$
$$= \frac{n-2}{n-1}\xi_n, \quad \text{if} \quad n = 3, 4, 5, \ldots;$$

$$A\{\xi_n\} = \{L(\{\xi_n\}), L(\{\xi_n\}), 0, 0, 0, \ldots\}.$$

Clearly $\|A\| = 1$ and $A^2 = 0$. Also

$$S\{\xi_n\} = \{\xi_n\} - \{\gamma_n\}, \text{ where } \gamma_n = 0 \quad \text{if} \quad n = 1, 2$$
$$= \frac{1}{n-1}\xi_n \quad \text{if} \quad n = 3, 4, 5, \ldots.$$

Since $L(\{\gamma_n\}) = 0$, we have $AS\{\xi_n\} = A\{\xi_n\}$. It is easy to see that $SA\{\xi_n\} = A\{\xi_n\}$, and hence

$$AS = SA.$$

$\sigma(S)$ is the totally disconnected set consisting of 1 and the numbers $(n-2)/(n-1)$ for $n = 3, 4, 5, \ldots$. S is a prespectral operator with resolution of the identity $E(\cdot)$ of class l^1 satisfying

$$\begin{aligned} E(\{1\})\{\xi_k\} &= \{\xi_1, \xi_2, 0, 0, 0, \ldots\}, \\ E\left(\left\{\frac{n-2}{n-1}\right\}\right)\{\xi_k\} &= \{\delta_{kn}\xi_k\} \qquad (n = 3, 4, \ldots). \end{aligned} \tag{5}$$

Define the sequence $\{\lambda_n\}$ by setting

$$\lambda_n = 1 \quad \text{if} \quad n = 1, 2; \; \lambda_n = (n-2)/(n-1) \quad \text{if} \quad n = 3, 4, 5, \ldots.$$

Then it is easy to see from (5) that for τ in Σ_p, $E(\tau)$ is the operator which multiplies the nth term of a sequence by 1 if $\lambda_n \in \tau$ and by 0 if $\lambda_n \notin \tau$. The sequence $\{f_n\}$ of functions on $\sigma(S)$, given by

$$\begin{aligned} f_n(\lambda) &= \lambda, \quad \text{if} \quad \lambda < (n-2)/(n-1), \\ &= 1, \quad \text{if} \quad \lambda \geqslant (n-2)/(n-1), \end{aligned}$$

for $n = 3, 4, 5, \ldots$, converges uniformly to the function identically equal to λ on $\sigma(S)$. One sees directly that

$$\int_{\sigma(S)} f_n(\lambda)\, E(d\lambda) \to S$$

in the norm of $L(l^\infty)$ and hence

$$S = \int_{\sigma(S)} \lambda E(d\lambda).$$

Therefore S is a scalar-type operator on l^∞ of class l^1. Since

$$AE(\{1\})\{1, 1, 1, \ldots\} = \{0, 0, 0, \ldots\}$$

and

$$E(\{1\})\, A\{1, 1, 1, \ldots\} = \{1, 1, 0, \ldots\}$$

we have

(i) *A commutes with S but not with the resolution of the identity of class l^1 for S.*

Next, we define $T_1 = S + A$. Thus T_1 is the sum of S and a nilpotent commuting with S. It is clear from (i) and Theorem 5.25 that T_1 is not prespectral of class l^1. In fact we shall show that

(ii) *T_1 is not prespectral of any class.*

Suppose to the contrary that $G(\cdot)$ is a resolution of the identity of class Γ for T_1. By Theorem 5.24, $G(\cdot)$ is a resolution of the identity of class Γ for S, and A commutes with every value of $G(\cdot)$. Now, by Theorem 5.33, the projections $G(\{1\})$ and $E(\{1\})$ have the same range. Also

$$G(\{1\})\{1, 1, 1, \ldots\} \in E(\{1\})\, l^\infty$$

and

$$AG(\{1\})\{1, 1, 1, \ldots\} = \{0, 0, 0, \ldots\}.$$

However

$$A\{1, 1, 1, \ldots\} = \{1, 1, 0, \ldots\} \in E(\{1\})\, l^\infty = G(\{1\})\, l^\infty$$

and

$$G(\{1\})\, A\{1, 1, 1, \ldots\} = \{1, 1, 0, \ldots\}.$$

This gives a contradiction, and so (ii) is established.

If A were to commute with some resolution of the identity for S then by Theorem 5.15 this would contradict (ii). Thus

(iii) *A does not commute with any resolution of the identity for S.*

Resolutions of the identity other than $E(\cdot)$ can be constructed for S by the method of Fixman. Define

$$F(\tau) = E(\tau) + AE(\tau) - E(\tau)A \qquad (\tau \in \Sigma_p). \tag{6}$$

Using the relations $A^2 = 0$ and $AE(\tau)A = 0\,(\tau \in \Sigma_p)$, it is easily verified that $F(\cdot)$ is a homomorphism from Σ_p into a Boolean algebra of projections on l^∞ with $F(\sigma(S)) = I$. Clearly $\|F(\tau)\| \leqslant 3\ (\tau \in \Sigma_p)$. For each positive integer n let e_n in l^1 be given by $e_n = \{\delta_{nk}\}_{k=1}^\infty$, and let e_n^* be the corresponding linear functional on l^∞. Let Γ_1 be the total linear manifold in $(l^\infty)^*$ generated by $e_1^* - L$, $e_2^* - L$, and $\{e_n^* : n = 3, 4, 5, \ldots\}$. Since for each τ in Σ_p and x in l^∞

$$\langle F(\tau)\, x, e_n^* \rangle = \langle E(\tau)\, x, e_n^* \rangle \qquad (n = 3, 4, 5, \ldots),$$

$$\langle F(\tau)\, x, e_n^* - L \rangle = \chi(\tau; 1) \langle x, e_n^* - L \rangle \qquad (n = 1, 2),$$

it follows that $F(\cdot)$ is Γ_1-countably additive. Since $E(\cdot)$ and A commute with S, elementary algebra shows that $F(\tau)S = SF(\tau)\ (\tau \in \Sigma_p)$. In order to prove that $F(\cdot)$ is a resolution of the identity for S it remains only to show that $\sigma(S \mid F(\tau)\, l^\infty) \subseteq \bar{\tau} (\tau \in \Sigma_p)$. By virtue of Lemma 5.34, it suffices to prove this inclusion when τ is a closed subset of $\sigma(S)$. Again by Lemma 5.34, and the fact that $\sigma(S)$ is totally disconnected, it is sufficient to prove the inclusion for an open-and-closed subset τ of $\sigma(S)$ since each closed subset of $\sigma(S)$ is the intersection of open-and-closed subsets that contain it. It is easy to see from the definition of $F(\cdot)$ that $E(\cdot)$ and $F(\cdot)$ agree on finite subsets of $\sigma(S)\backslash\{1\}$. Since every open-and-closed subset of $\sigma(S)$ is such a set or the complement in $\sigma(S)$ of such a set, $F(\cdot)$ and $E(\cdot)$ agree on open-and-closed subsets of $\sigma(S)$. Therefore

$$\sigma(S|F(\tau)\, l^{\infty}) = \sigma(S|E(\tau)\, l^{\infty}) \subseteq \tau$$

for τ open-and-closed in $\sigma(S)$. In establishing (i) it was shown that A and $E(\{1\})$ do not commute. Hence by (6), $F(\{1\}) \neq E(\{1\})$. Therefore $F(\cdot)$ and $E(\cdot)$ are distinct.

In contrast to the property of A stated in (iii) we now show that

(iv) *there is a nilpotent N commuting with $E(\cdot)$ but not with $F(\{1\})$.*

We define N on l^{∞} by setting

$$N(\{\xi_n\}) = \{\xi_2, 0, 0, 0, \ldots\}.$$

Then $\|N\| = 1$, $N^2 = 0$ and N commutes with $E(\cdot)$. Moreover

$$F(\{1\})\{1, \tfrac{1}{2}, 1, 1, 1, \ldots\} = \{0, -\tfrac{1}{2}, 0, 0, 0, \ldots\};$$
$$NF(\{1\})\{1, \tfrac{1}{2}, 1, 1, 1, \ldots\} = \{-\tfrac{1}{2}, 0, 0, 0, \ldots\}.$$

However,

$$N\{1, \tfrac{1}{2}, 1, 1, 1, \ldots\} = \{\tfrac{1}{2}, 0, 0, 0, 0, \ldots\},$$
$$F(\{1\})N\{1, \tfrac{1}{2}, 1, 1, 1, \ldots\} = \{\tfrac{1}{2}, 0, 0, 0, 0, \ldots\}.$$

Therefore (iv) is demonstrated.

Define $T_2 = S + N$. Since N commutes with $E(\cdot)$, T_2 is prespectral of class l^1 by Theorem 5.15. By Theorem 5.24 every resolution of the identity of T_2 is a resolution of the identity for S. Now T_2 does not commute with $F(\cdot)$, and so S has a resolution of the identity $F(\cdot)$ which is not a resolution of the identity of T_2. Moreover, if in the statement of Theorem 5.25 the words "of class Γ" are deleted in both places, then the theorem fails.

To round off the considerations in (iii) and (iv) we prove the following results.

(v) *If Q is a quasinilpotent operator commuting with every resolution of the identity for S, then $Q = 0$.*

Since for $n = 3, 4, 5, \ldots,$

$$Q \,|\, E\left(\left\{\frac{n-2}{n-1}\right\}\right) l^{\infty}$$

is a quasinilpotent on a one-dimensional space, it is 0. Therefore

$$QE\left(\left\{\frac{n-2}{n-1}\right\}\right) = 0$$

and so

$$E\left(\left\{\frac{n-2}{n-1}\right\}\right) Q = 0 \qquad (n = 3, 4, 5, \ldots). \tag{7}$$

Let $\{\xi_k\} \in l^{\infty}$, and let $Q\{\xi_k\} = \{\eta_k\}$. Then from (7) it follows that

$$0 = E\left(\left\{\frac{n-2}{n-1}\right\}\right) Q\{\xi_k\} = \{\delta_{nk}\eta_k\}_{k=1}^{\infty}.$$

Hence $\eta_n = 0$ for $n \geqslant 3$. If further $\xi_1 = \xi_2 = 0$ then clearly

$$QE(\{1\})\{\xi_k\} = Q\{\xi_1, \xi_2, 0, 0, 0, \ldots\} = 0,$$

and so in this case

$$0 = E(\{1\})\, Q\{\xi_k\} = \{\eta_1, \eta_2, 0, 0, 0, \ldots\},$$

which gives

$$Q\{\xi_k\} = 0 \quad \text{if} \quad \xi_1 = \xi_2 = 0. \tag{8}$$

Now we consider $Q \,|\, E(\{1\}) l^{\infty}$. Representing this operator by the matrix

$$\begin{bmatrix} a & b \\ c & d \end{bmatrix}$$

relative to the basis $\{1, 0, 0, \ldots\}$ and $\{0, 1, 0, 0, \ldots\}$, we observe that for any

$\{\xi_k\}$ in l^∞,

$$Q\{\xi_1, \xi_2, 0, 0, 0, \ldots\} = \{a\xi_1 + b\xi_2, c\xi_1 + d\xi_2, 0, 0, 0, \ldots\}.$$

Hence by (8)

$$\begin{aligned} Q\{\xi_k\} &= \{Q\{\xi_1, \xi_2, 0, 0, 0, \ldots\} + Q\{0, 0, \xi_3, \xi_4, \xi_5, \ldots\} \\ &= Q\{\xi_1, \xi_2, 0, 0, 0, \ldots\} \qquad (9) \\ Q\{\xi_k\} &= \{a\xi_1 + b\xi_2, c\xi_1 + d\xi_2, 0, 0, 0, \ldots\}. \end{aligned}$$

Direct computation with (6) shows that

$$F(\{1\})\{1, 0, 1, 1, 1, \ldots\} = \{0, -1, 0, 0, 0, \ldots\}$$

and we see with the aid of (9) that

$$QF(\{1\})\{1, 0, 1, 1, 1, \ldots\} = \{-b, -d, 0, 0, 0, \ldots\}. \qquad (10)$$

However

$$Q\{1, 0, 1, 1, 1, \ldots\} = \{a, c, 0, 0, \ldots\}.$$

The right-hand member of this last equation belongs to the range of $F(\{1\})$. Therefore

$$F(\{1\})\, Q\{1, 0, 1, 1, 1, \ldots\} = \{a, c, 0, 0, 0, 0, \ldots\}.$$

Since Q commutes with $F(\cdot)$ it follows from this equation and (10) that

$$a + b = 0, \quad c + d = 0. \qquad (11)$$

Define an operator A_1 on l^∞ by

$$A_1\{\xi_k\} = \{L(\{\xi_k\}), 0, 0, 0, \ldots\}.$$

As in the proof of the corresponding results for A we have $A_1^2 = 0$, $A_1 S = SA_1$ and $A_1 E(\{1\}) \neq E(\{1\})A_1$. Denote by Γ_2 the total linear manifold in $(l^\infty)^*$

generated by $e_1^* - L$ and $\{e_n^*; n = 2, 3, 4, \ldots\}$. The set function $H(\cdot)$ defined by

$$H(\tau) = E(\tau) + A_1 E(\tau) - E(\tau) A_1 \qquad (\tau \in \Sigma_p)$$

is a resolution of the identity of class Γ_2 for S. From the definition of $H(\cdot)$ we obtain

$$H(\{1\})\{1, 1, 1, 1, 1, \ldots\} = \{0, 1, 0, 0, 0, \ldots\}.$$

Therefore

$$QH(\{1\})\{1, 1, 1, 1, 1, \ldots\} = \{b, d, 0, 0, 0, \ldots\}. \tag{12}$$

However, by (9) and (11)

$$Q\{1, 1, 1, 1, 1, \ldots\} = \{a + b, c + d, 0, 0, 0, \ldots\} = 0.$$

Using this last fact and the relation $QH(\{1\}) = H(\{1\})Q$, we deduce from (12) that $b = d = 0$. Therefore by (9) and (11), $Q = 0$.

We conclude this chapter with some miscellaneous results.

THEOREM 5.36. *Let T, in $L(X)$, have totally disconnected spectrum. In order that T^* be prespectral of class X it is necessary and sufficient that the set $\{A(\delta)$: δ open-and closed in $\sigma(T)\}$ of spectral projections for T be uniformly bounded in norm. If this is the case, T^* has (unique) resolution of the identity $E(\cdot)$ of class X, where for each open-and-closed subset δ of $\sigma(T)$, $E(\delta) = A(\delta)^*$.*

Proof. If T^* is prespectral with resolution of the identity $E(\cdot)$ of class X then, by Proposition 5.27, $E(\delta) = A(\delta)^*$ for each open-and-closed subset δ of $\sigma(T)$. Hence the condition is necessary.

Conversely suppose that $\|A(\delta)\| \leqslant M < \infty$ (δ open-and-closed in $\sigma(T)$). Let $f \in C(\sigma(T))$, and let $\varepsilon > 0$ be given. Since the topology of the compact set $\sigma(T)$ has a base of open-and-closed subsets, it is easily shown that there exist a partition of $\sigma(T)$ into open-and-closed subsets $\{\delta_r : r = 1, 2, \ldots, n\}$ and points λ_r in δ_r $(r = 1, 2, \ldots, n)$ such that

$$|f(\lambda_r) - f(\lambda)| < \varepsilon \qquad (\lambda \in \delta_r : r = 1, 2, \ldots, n).$$

Hence the algebra $\tilde{A}$ of finite linear combinations of characteristic functions of disjoint open-and-closed subsets of $\sigma(T)$ is dense in $C(\sigma(T))$. Define a map ψ from $\tilde{A}$ into $L(X)$ by

$$\psi\left(\sum_{r=1}^{n} \alpha_r \chi(\tau_r)\right) = \sum_{r=1}^{n} \alpha_r A(\tau_r) \qquad (\tau_r \cap \tau_s = \emptyset \text{ if } r \neq s).$$

It is easy to see that ψ is well-defined. By Proposition 5.3,

$$\|\psi(f)\| \leqslant 4M \sup\{|f(\lambda)| : \lambda \in \sigma(T)\} \qquad (f \in \tilde{A}).$$

Therefore ψ can be extended to a continuous algebra homomorphism from $C(\sigma(T))$ into $L(X)$ with $\psi(1) = I$. Let Σ_0 denote the σ-algebra of Borel subsets of $\sigma(T)$. By Theorem 5.21 there is a spectral measure $E_0(\cdot)$ of class (Σ_0, X) such that

$$\psi(f)^* = \int_{\sigma(T)} f(\lambda) E_0(d\lambda) \qquad (f \in C(\sigma(T))).$$

Define $E(\cdot)$ on Σ_p by

$$E(\delta) = E_0(\delta \cap \sigma(T)) \qquad (\delta \in \Sigma_p).$$

Then $E(\cdot)$ is a spectral measure of class (Σ_p, X). Let τ be an open-and-closed subset of $\sigma(T)$. By the integral formula above $A(\tau)^* = E(\tau)$. Also

$$\langle Tx, E(\tau)y \rangle = \langle Tx, A(\tau)^* y \rangle = \langle x, E(\tau) T^* y \rangle \qquad (x \in X, y \in X^*).$$

The topology of $\sigma(T)$ has a countable base of open-and-closed subsets. It follows that the regular measures $\langle Tx, E(\cdot)y \rangle$ and $\langle x, E(\cdot)T^* y \rangle$ are identical for all x in X, y in X^*. Hence

$$T^* E(\delta) = E(\delta) T^* \qquad (\delta \in \Sigma_p).$$

Finally let $\delta \in \Sigma_p$, and let τ be an open-and-closed subset of $\sigma(T)$ with $\bar{\delta} \cap \sigma(T) \subseteq \tau$. Then, since $A(\tau)^* = E(\tau)$ it follows from Lemma 5.34 that

$$\sigma(T^* | E(\delta) X^*) \subseteq \sigma(T^* | E(\bar{\delta} \cap \sigma(T)) X^*) \subseteq \sigma(T^* | E(\tau) X^*) \subseteq \tau.$$

Now since $\sigma(T)$ is totally disconnected, $\bar{\delta} \cap \sigma(T)$ is equal to the intersection of all open-and-closed subsets τ of $\sigma(T)$ with $\bar{\delta} \cap \sigma(T) \subseteq \tau$. Hence

$$\sigma(T^*|E(\delta)X^*) \subseteq \bar{\delta} \cap \sigma(T) \subseteq \bar{\delta} \qquad (\delta \in \Sigma_p)$$

and the proof is complete.

In order to state our next theorem we need some terminology from the theory of uniform approximation in the complex plane. We summarize the main results of the theory here. For a full discussion of this topic the reader is referred to [**14**; p. 382–390]. The first result we state is due to Lavrentieff [**1**].

THEOREM 5.37. *Let K be a compact subset of* **C** *with void interior and connected complement. Let $f \in C(K)$. There is a sequence $\{p_n\}$ of polynomials such that*

$$\sup_{z \in K} |f(z) - p_n(z)| \to 0 \text{ as } n \to \infty.$$

Definition 5.38. A compact subset K of **C** is called an *R-set* if and only if the rational functions with poles in $\mathbf{C} - K$ are uniformly dense in $C(K)$.

We observe that every R-set is nowhere dense, but that there exist nowhere dense compact subsets of **C** which are not R-sets. If a compact subset of **C** has plane Lebesgue measure zero, or if it is nowhere dense and its complement has a finite number of components, then it is an R-set.

We now return to the theory of prespectral operators.

THEOREM 5.39. *Let $S \in L(X)$, and let $\sigma(S)$ be an R-set. In order that S^* be a scalar-type operator of class X it is necessary and sufficient that there exists a real constant M such that for each rational function g with poles outside $\sigma(S)$*

$$\|g(S)\| \leqslant M \sup\{|g(\lambda)| : \lambda \in \sigma(S)\}.$$

Proof. Necessity is obvious, since if S^* is a scalar-type operator with resolution of the identity $E(\cdot)$ of class X, then

$$g(S)^* = \int_{\sigma(S)} g(\lambda) E(d\lambda).$$

Conversely, suppose the condition is satisfied. It follows that the map $r \to r(S)$, which sends a rational function in $C(\sigma(S))$ into an element of $L(X)$, is well-defined. Since $\sigma(S)$ is an R-set, this map can be extended to a continuous algebra homomorphism from $C(\sigma(S))$ into a subalgebra of $L(X)$. The desired conclusion now follows from Theorem 5.21.

We now prove two results concerning relationships between scalar-type operators and hermitian operators.

THEOREM 5.40. *Let S be a scalar-type operator on X with resolution of the identity $E(\cdot)$ of class Γ. There are operators R and J on X such that*

(i) $S = R + iJ$,

(ii) $RJ = JR$,

(iii) *there is an equivalent norm on X with respect to which the operators $\{R^m J^n : m, n = 0, 1, 2, \ldots\}$ are all hermitian.*

Moreover, the operators R and J are uniquely determined by conditions (i), (ii) *and* (iii).

Proof. Define

$$R = \int_{\sigma(S)} \operatorname{Re} \lambda E(d\lambda), \; J = \int_{\sigma(S)} \operatorname{Im} \lambda E(d\lambda).$$

By Theorem 5.4 there is an equivalent norm on X with respect to which $\{E(\tau) : \tau \in \Sigma_p\}$ are all hermitian. Let $p(x, y)$ be a polynomial with real coefficients in the two real variables $x = \operatorname{Re} \lambda$ and $y = \operatorname{Im} \lambda$. There is a sequence of real-valued simple Borel measurable functions of λ converging uniformly on $\sigma(S)$ to $p(x, y)$. It follows from Theorem 4.4 that $p(R, J)$ is hermitian and so (iii) holds. Suppose that R, J can be replaced by R_0, J_0 respectively so that (i), (ii), (iii) remain true. Since R_0 and J_0 commute with S then, by Theorem 5.12, $\{R, R_0, J, J_0\}$ is a commutative family. Each of $\{R, J\}$ and $\{R_0, J_0\}$ is a hermitian-equivalent family. Hence, by Theorem 4.18, there is an equivalent norm on X with respect to which R, R_0, J, J_0 are simultaneously hermitian. Assume that this renorming has been carried out. Then

$$(R - R_0) + i(J - J_0) = 0$$

and so, by Proposition 4.20, $R = R_0$ and $J = J_0$. This completes the proof of the theorem.

THEOREM 5.41. *Let $R, J \in L(X)$ and $RJ = JR$. Suppose that there is an equiva-*

lent norm $\| \ \|$ on X with respect to which $\{R^m J^n : m, n = 0, 1, 2, \ldots\}$ are all hermitian operators. If $S = R + iJ$, then S^* is a scalar-type operator of class X.

Proof. Let $\mathscr{A}$ be the closed subalgebra of $L(X)$ generated by I, R and J. Define $\Omega = \{x + iy : x \in \sigma(R), y \in \sigma(J)\}$. Observe that Ω is compact. Let $p(x, y)$ be a polynomial in the two real variables x and y. We define

$$\|p\|_\infty = \sup\{|p(x, y)| : x + iy \in \Omega\},$$

$$\psi(p) = p(R, J).$$

When p has real coefficients, $\psi(p)$ is hermitian, and so by Sinclair's theorem the norm and spectral radius of $\psi(p)$ are equal. Let $\mathscr{M}$ denote the set of multiplicative linear functionals on $\mathscr{A}$. It follows that

$$\begin{aligned} \|\psi(p)\| &= \sup\{|\phi[p(R, J)]| : \phi \in \mathscr{M}\} \\ &= \sup\{|p(\phi(R), \phi(J))| : \phi \in \mathscr{M}\} \\ &\leqslant \|p\|_\infty. \end{aligned}$$

If p has complex coefficients we can express p in the form $p_1 + ip_2$, where p_1 and p_2 are polynomials in two real variables with real coefficients. Hence

$$\|\psi(p_1)\| \leqslant \|p_1\|_\infty \leqslant \|p\|_\infty,$$

$$\|\psi(p_2)\| \leqslant \|p\|_\infty,$$

and so

$$\|\psi(p)\| \leqslant 2\|p\|_\infty. \tag{13}$$

It follows from the inequality (13) and the Stone-Weierstrass theorem that ψ can be extended to a continuous, identity preserving algebra homomorphism of $C(\Omega)$ into $L(X)$. Let $f_0(\lambda) = \lambda (\lambda \in \Omega)$. Since

$$\psi(f_0) = R + iJ = S$$

the desired conclusion follows from Theorem 5.21.

We now return to the study of bounded Boolean algebras of projections. The following result is valid for an arbitrary Boolean algebra and is usually called the Stone representation theorem. (See for example [**5**; p. 41].)

THEOREM 5.42. *Every Boolean algebra is isomorphic with the Boolean algebra of open-and-closed subsets of a totally disconnected compact Hausdorff space, called the Stone representation space of the Boolean algebra.*

This leads us to the following result.

PROPOSITION 5.43. *Let* **B** *be a bounded Boolean algebra of projections on* X *and let* $\mathscr{A}$ *be the closed subalgebra of* $L(X)$ *generated by* **B**. *Then, if* Ω *is the Stone representation space of* **B**, *there is a bicontinuous algebra isomorphism* ψ *from* $C(\Omega)$ *onto* $\mathscr{A}$.

Proof. Let $E(\tau)$ be the projection corresponding to the open-and-closed subset τ of Ω under the isomorphism of the Stone representation theorem. Suppose that M satisfies

$$\|E(\tau)\| \leqslant M < \infty \text{ } (\tau \text{ open-and-closed in } \Omega).$$

Let $\tau_1, \ldots, \tau_n$ be a finite family of pairwise disjoint non-empty open-and-closed subsets of Ω. Define

$$\psi\left(\sum_{r=1}^{n} \alpha_r \chi(\tau_r)\right) = \sum_{r=1}^{n} \alpha_r E(\tau_r). \tag{14}$$

Observe that, by Proposition 5.3,

$$\sup\{|\alpha_r| : 1 \leqslant r \leqslant n\} \leqslant \left\| \sum_{r=1}^{n} \alpha_r E(\tau_r)\right\| \leqslant 4M \sup\{|\alpha_r| : 1 \leqslant r \leqslant n\}.$$

Since the topology of Ω has a base of open-and-closed sets then, by the Stone-Weierstrass theorem, equation (14) defines ψ on a dense subset of $C(\Omega)$. We may therefore extend ψ to all of $C(\Omega)$ by continuity and the proof is complete.

Definition 5.44. Let **S** be a commutative family of projections on X. Then $S(\mathbf{S})$ denotes the closure in the norm of $L(X)$ of the real linear span of **S**.

LEMMA 5.45. *Let* **S** *be a commutative family of projections on* X. *If every operator in* $S(\mathbf{S})$ *is hermitian-equivalent then* **S** *is bounded.*

Proof. By Theorem 4.17, **S** is a hermitian-equivalent family. Under the corresponding renorming each projection has norm one by Sinclair's theorem. This suffices to complete the proof.

The next result, due to Lumer, shows that in order to get a spectral theory along the lines discussed in this chapter it is necessary to assume that the values of the spectral measure are bounded.

THEOREM 5.46. *Let* **B** *be a Boolean algebra of projections on* X. *A necessary and sufficient condition for the adjoint of every operator in the closed subalgebra of* $L(X)$ *generated by* **B** *to be a scalar-type operator of class* X *is that* **B** *be bounded.*

Proof. By Proposition 5.43 and Theorem 5.21, if **B** is bounded the adjoint of every operator in the closed subalgebra of $L(X)$ generated by **B** is a scalar-type operator of class X.

Conversely, suppose that $T \in S(\mathbf{B})$ and T^* is a scalar-type operator of class X. By Theorem 5.40, $T^* = R + iJ$, where $RJ = JR$ and $\{R, J\}$ are hermitian-equivalent. Assume that X^* has been equivalently renormed so that R and J are hermitian. Observe that the spectrum of any finite real linear combination of projections in **B** is real, and so by Gelfand theory every operator in $S(\mathbf{B})$ has real spectrum. Hence T^* has real spectrum. Let $\mathscr{A}$ be any maximal commutative subalgebra of $L(X^*)$ containing R and J. Note that $T^* \in \mathscr{A}$ and the spectra of T^*, R, J are the same as operators or as elements of $\mathscr{A}$. Consider the Gelfand representation of $\mathscr{A}$. If $\mathscr{M}$ denotes the maximal ideal space of $\mathscr{A}$ then

$$\hat{T}^*(m) = \hat{R}(m) + i\hat{J}(m) \qquad (m \in \mathscr{M}).$$

Since R and J are hermitian, $\hat{R}(m)$ and $\hat{J}(m)$ are real. Also, since $\sigma(T^*)$ is real, $\hat{T}^*(m)$ is real. Hence $\hat{J}(m) = 0 (m \in \mathscr{M})$ and so J is a quasinilpotent hermitian operator. By Sinclair's theorem $J = 0$ and so T^* is hermitian. Therefore, there is a constant M such that, with respect to the original norm on X^* we have

$$\|\exp(itT^*)\| \leqslant M < \infty \qquad (t \in \mathbf{R}).$$

It follows that

$$\|\exp(itT)\| \leqslant M < \infty \qquad (t \in \mathbf{R})$$

and so, by Proposition 4.15, T is hermitian-equivalent. It now follows from Lemma 5.45 that $\mathbf{B}$ is uniformly bounded and so the proof is complete.

Finally we show that the spectrum and the approximate point spectrum of a prespectral operator coincide. For the definition and properties of the approximate point spectrum the reader is referred to Definition 1.15 and Theorem 1.16.

THEOREM 5.47. *Let T be a prespectral operator on X with resolution of the identity $E(\cdot)$ of class Γ. Then $\sigma_a(T) = \sigma(T)$.*

Proof. Let $G = \sigma(T) \backslash \sigma_a(T)$. G is open, as it is equal to the intersection of $\mathbf{C} \backslash \sigma_a(T)$ and the interior of $\sigma(T)$. If δ is any compact subset of G, then $\sigma_a(T|E(\delta)X) = \varnothing$ since

$$\sigma_a(T|E(\delta)X) \subseteq \sigma(T|E(\delta)X) \cap \sigma_a(T) \subseteq \delta \cap \sigma_a(T) = \varnothing.$$

Hence, by Theorem 1.16, the boundary of the compact set $\sigma(T|E(\delta)X)$ is empty and so $\sigma(T|E(\delta)X) = \varnothing$. Therefore $E(\delta)X = \{0\}$, and so for all x in X and y in Γ we have $\langle E(\delta)x, y \rangle = 0$. Each measure $\langle E(\cdot)x, y \rangle$ is countably additive on Σ_p and so $\langle E(G)x, y \rangle = 0$. Since Γ is total $E(G) = 0$. Hence $E(\rho(T) \cup G) = 0$, by Lemma 5.6. Now, by Note 5.7, $\rho(T)$ is the largest open set on which the spectral measure $E(\cdot)$ vanishes. Therefore $G = \varnothing$ and $\sigma_a(T) = \sigma(T)$.

6. Spectral Operators

In this chapter we discuss spectral operators, a very important subclass of the prespectral operators.

Definition 6.1. Let $T \in L(X)$. Then T is called a *spectral operator* if there is a spectral measure $E(\cdot)$ defined on Σ_p with values in $L(X)$ such that

(i) $E(\cdot)$ is countably additive on Σ_p in the strong operator topology,

(ii) $TE(\tau) = E(\tau)T \qquad (\tau \in \Sigma_p)$,

(iii) $\sigma(T \mid E(\tau)X) \subseteq \bar{\tau} \qquad (\tau \in \Sigma_p)$.

Observe that (i) means that the vector-valued measure $E(\cdot)x$ is countably additive on Σ_p for each x in X. We shall show that an operator on X is spectral if and only if it is prespectral of class X^*. For this purpose we require a slight generalization of a result of Barry [**1**].

Definition 6.2. A net $\{T(\alpha)\}$ of operators on X is said to be *naturally ordered* if $T(\alpha) = T(\alpha)\,T(\beta) = T(\beta)T(\alpha)$ whenever $\alpha < \beta$.

If $Y \subseteq X$, we denote by $wk(Y)$ the closure of Y in the weak topology.

Definition 6.3. Let $\{T(\alpha)\}$ be a net of operators on X and let $x \in X$. Then y_x is called a *weak x-cluster point* of $\{T(\alpha)\}$ if $y_x \in \bigcap_{\alpha} wk\{T(\beta)x\colon \beta \geqslant \alpha\}$; that is if y_x is a weak cluster point of the net of vectors $\{T(\alpha)x\}$.

THEOREM 6.4. *Let $\{T(\alpha)\}$ be a naturally ordered uniformly bounded net of operators on X. Then $\{T(\alpha)\}$ converges in the strong operator topology if and only if $\{T(\alpha)\}$ has a weak x-cluster point for each x in X.*

Proof. The necessity of the condition is clear. In order to prove sufficiency we fix x in X and $T(\alpha_0)$, a member of the net. Let y_x be a weak x-cluster point of

$\{T(\alpha)\}$ and let $\varepsilon > 0$ be given. Let $z \in X^*$. Define

$$N_\varepsilon(y_x) = \{y \in X : |\langle y - y_x, T(\alpha_0)^* z\rangle| < \varepsilon\}.$$

By hypothesis, there is $\beta \geqslant \alpha_0$ such that

$$|\langle T(\alpha_0)x - T(\alpha_0)y_x, z\rangle| = |\langle T(\beta)x - y_x, T(\alpha_0)^* z\rangle| < \varepsilon.$$

Hence $T(\alpha_0)x = T(\alpha_0)y_x$. Now, by Theorem V.3.13 of [**5**; p. 422], a convex subset of X is weakly closed if and only if it is norm-closed. Since $y_x \in wk\{T(\alpha)x\}$, it follows that there are complex numbers $c_k^{(n)}$ and a sequence $\{A_n\}$ of operators on X of the form

$$A_n = \sum_{k=1}^{n} c_k^{(n)} T(\alpha_k)$$

such that $y_x = \lim_{n\to\infty} A_n x$ in the norm of X. Clearly, for each n, $\lim_\alpha T(\alpha) A_n = A_n$ in the strong operator topology. Furthermore, for each n,

$$\begin{aligned}\| T(\alpha)x - T(\alpha)A_n x \| &= \| T(\alpha)y_x - T(\alpha)A_n x \| \\ &\leqslant M \| y_x - A_n x \|,\end{aligned}$$

where $M = \sup_\alpha \| T(\alpha) \|$. Hence

$$\lim_{n\to\infty} T(\alpha)A_n x = T(\alpha)x,$$

in the norm of X, uniformly in α. By the E. H. Moore theorem on the interchange of limits (see I.7.6 of [**5**; p. 28–9])

$$\begin{aligned} y_x &= \lim_{n\to\infty} \lim_\alpha T(\alpha)A_n x \\ &= \lim_\alpha \lim_{n\to\infty} T(\alpha)A_n x \\ &= \lim_\alpha T(\alpha)x \qquad (x \in X)\end{aligned}$$

in the norm of X. Put $Tx = y_x (x \in X)$. It is easily verified that $T \in L(X)$ and the proof is complete.

THEOREM 6.5. *Let $T \in L(X)$. Then T is a spectral operator if and only if it is prespectral of class X^*.*

Proof. Clearly, if T is spectral then it is prespectral of class X^*. Now let T be prespectral with resolution of the identity $E(\cdot)$ of class X^*. Let $x \in X$ and let $\{\tau_n\}$ be a sequence of sets in Σ_p with union τ. Observe that $E(\tau)x$ is a weak x-cluster point for the sequence

$$\left\{E\left(\bigcup_{k=1}^{n} \tau_k\right)x : n = 1, 2, 3, \ldots\right\}$$

and so, by Theorem 6.4,

$$\lim_{n \to \infty} E\left(\bigcup_{k=1}^{n} \tau_k\right)x = E(\tau)x \qquad (x \in X)$$

in the norm of X. This completes the proof.

THEOREM 6.6. *Let T be a spectral operator on X and let $E(\cdot)$ be the resolution of the identity of class X^* for T. Let $A \in L(X)$ and $AT = TA$. Then*

$$AE(\tau) = E(\tau)A \qquad (\tau \in \Sigma_p).$$

Proof. Define

$$\psi(f) = \int_{\sigma(T)} f(\lambda)\, E(d\lambda) \qquad (f \in C(\sigma(T))).$$

Then, by Theorem 5.12, $A\psi(f) = \psi(f)A\ (f \in C(\sigma(T)))$. Let $x \in X$ and $y \in X^*$. Define

$$\mu_1(\tau) = \langle E(\tau)Ax, y\rangle, \quad \mu_2(\tau) = \langle E(\tau)x, A^*y\rangle \qquad (\tau \in \Sigma_p).$$

μ_1 and μ_2 are finite measures with supports contained in $\sigma(T)$ and so are regular. Also, by Proposition 5.8 and the above

$$\int_{\sigma(T)} f(\lambda)\, \mu_1(d\lambda) = \int_{\sigma(T)} f(\lambda)\, \mu_2(d\lambda) \qquad (f \in C(\sigma(T))).$$

It follows from the Riesz representation theorem that $\mu_1 = \mu_2$. Hence for all x in X, y in X^* and τ in Σ_p

$$\langle E(\tau)Ax, y\rangle = \langle AE(\tau)x, y\rangle.$$

The desired conclusion follows at once.

This result is known as the *commutativity theorem for a spectral operator.* Two other important properties of spectral operators now follow.

THEOREM 6.7. *Let T be a spectral operator on X. Then T has a unique resolution of the identity.*

Proof. Let $E(\cdot)$ be a resolution of the identity of class Γ for T. By Theorem 6.5, T has a resolution of the identity $F(\cdot)$ of class X^*. Since $\Gamma \subseteq X^*$, $E(\cdot)$ and $F(\cdot)$ are both resolutions of the identity of class Γ for T. By Theorem 5.13, $E(\cdot) = F(\cdot)$ and the proof is complete.

The following result gives the *canonical reduction* of a spectral operator.

THEOREM 6.8. *Let $T \in L(X)$. Then T is spectral if and only if it is the sum $T = S + N$ of a scalar-type spectral operator S on X and a quasinilpotent operator N such that $SN = NS$. Furthermore this decomposition is unique. T and S have the same spectrum and the same resolution of the identity.*

Proof. The only part of the theorem which cannot immediately be deduced from Theorems 5.13 and 5.15 is the following statement: if S is a scalar-type spectral operator and N is a quasinilpotent operator such that $SN = NS$ then $S + N$ is spectral. However, if $E(\cdot)$ is the resolution of the identity for S then, by Theorem 6.6, $NE(\tau) = E(\tau)N$ $(\tau \in \Sigma_p)$, and so the desired conclusion follows from Theorem 5.15.

Note. The decomposition, given in Theorem 6.8, of a spectral operator $T = S + N$ into a sum of a scalar-type spectral operator S and a quasinilpotent N commuting with S is called the *canonical decomposition* of T. The operator S is called the *scalar part* of T, and N is called the *radical part* of T.

THEOREM 6.9. *Let T be a spectral operator on X with resolution of the identity $E(\cdot)$. Then T^* is a prespectral operator with resolution of the identity $E^*(\cdot)$ of class X.*

Proof. The map $\tau \to E^*(\tau)$ is certainly a spectral measure of class (Σ_p, X). Moreover,

$$T^*E^*(\tau) = E^*(\tau)\,T^* \qquad (\tau \in \Sigma_p).$$

It remains only to show that $\sigma(T^*|E^*(\tau)\,X^*) \subseteq \bar{\tau}(\tau \in \Sigma_p)$. If $\tau \in \Sigma_p$ and $\lambda \in \mathbf{C}\backslash\bar{\tau}$ then the restriction of $\lambda I - T$ to $E(\tau)X$ has an inverse R_τ. Define P_τ, in $L(X)$, by $P_\tau = R_\tau E(\tau)$. Then clearly $E(\tau)\,P_\tau = P_\tau = P_\tau E(\tau)$. Hence $P_\tau^* E^*(\tau) = E^*(\tau)P_\tau^*$, so that P_τ^* maps $E^*(\tau)X^*$ into itself. Also $(\lambda I - T)\,P_\tau = E(\tau)$ and

$$P_\tau(\lambda I - T) = P_\tau E(\tau)\,(\lambda I - T) = P_\tau(\lambda I - T)\,E(\tau) = E(\tau).$$

Hence

$$P_\tau^*(\lambda I^* - T^*) = (\lambda I^* - T^*)\,P_\tau^* = E^*(\tau).$$

Consequently, the restriction of P_τ^* to $E^*(\tau)X^*$ is the inverse of the restriction of $\lambda I^* - T^*$ to $E^*(\tau)\,X^*$. Hence $\lambda \in \rho(T^*|E^*(\tau)\,X^*)$. This shows that $\sigma(T^*|E^*(\tau)\,X^*) \subseteq \bar{\tau}(\tau \in \Sigma_p)$ and the proof is complete.

We now prove some theorems which give certain conditions on an operator on a weakly complete Banach space sufficient to ensure that the operator is spectral. The basic result is given by the following theorem.

THEOREM 6.10. *Let X be weakly complete and let $S \in L(X)$. Suppose that there is a compact subset K of $\mathbf{C}$ and a continuous algebra homomorphism ψ from $C(K)$ into $L(X)$ such that $\psi(f_0) = I$ and $\psi(f_1) = S$, where*

$$f_0(\lambda) = 1,\; f_1(\lambda) = \lambda \qquad (\lambda \in K).$$

Then S is a scalar-type spectral operator.

Proof. Let Σ_0 denote the σ-algebra of Borel subsets of K. By Theorem 5.21 there is a spectral measure $F(\cdot)$ of class (Σ_0, X) such that

$$\psi(f)^* = \int_K f(\lambda)\,F(d\lambda) \qquad (f \in C(K)).$$

Let τ be a compact subset of K and let $d(z, \tau)$ denote the distance of z from τ. If $z \in K\backslash\tau$ then $d(z, \tau) > 0$ since τ is compact. For each positive integer n define

f_n in $C(K)$ by

$$f_n(z) = 1 - \min(1 - n^{-1}, nd(z, \tau)) \qquad (z \in K).$$

Observe that for each n

$$|f_n(z)| \leqslant 1 \qquad (z \in K)$$
$$\lim_{n \to \infty} f_n(z) = \chi(\tau; z) \qquad (z \in K).$$

Let $x \in X$, $y \in X^*$ and $\mu(\delta) = \langle x, F(\delta) y\rangle (\delta \in \Sigma_0)$. It follows from Lebesgue's theorem of dominated convergence that

$$\lim_{n \to \infty} \int_K f_n(\lambda) \mu(d\lambda) = \int_K \chi(\tau; \lambda) \, \mu(d\lambda).$$

Hence, by Proposition 5.8,

$$\lim_{n \to \infty} \langle x, \psi(f_n)^* y\rangle = \langle x, F(\tau) y\rangle \qquad (x \in X, y \in X^*).$$

Therefore, for each x in X, $\{\psi(f_n)x\}$ is a weak Cauchy sequence. Since X is weakly complete there is a vector $E(\tau)x$ in X to which $\{\psi(f_n)x\}$ converges weakly. Hence

$$\langle E(\tau)x, y\rangle = \langle x, F(\tau) y\rangle \qquad (x \in X, y \in X^*).$$

By the uniform boundedness theorem, $E(\tau) \in L(X)$ and $E(\tau)^* = F(\tau)$. Consider the class Σ of Borel subsets τ of K such that there is $E(\tau)$ in $L(X)$ with $E(\tau)^* = F(\tau)$. Clearly Σ is a Boolean algebra of subsets of K. We show that Σ is a σ-algebra. Let $\{\tau_n\}$ be a sequence of sets in Σ with union τ. Observe that, for each n,

$$\langle E(\tau_n)x, y\rangle = \langle x, F(\tau_n) y\rangle \qquad (x \in X, y \in X^*)$$

and

$$\lim_{n \to \infty} \langle x, F(\tau_n) y\rangle = \langle x, F(\tau) y\rangle \qquad (x \in X, y \in X^*).$$

Hence, for each x in X, $\{(E(\tau_n)x\}$ is a weak Cauchy sequence, and so there is a vector $E(\tau)x$ in X to which $\{E(\tau_n)x\}$ converges weakly. Therefore

$$\langle E(\tau)x, y\rangle = \langle x, F(\tau)y\rangle \qquad (x \in X, y \in X^*).$$

As before, $E(\tau) \in L(X)$ and $E(\tau)^* = F(\tau)$. It follows that Σ is a σ-algebra of subsets of K containing all closed subsets and so $\Sigma = \Sigma_0$. Therefore

$$\langle E(\tau)x, y\rangle = \langle x, F(\tau)y\rangle \qquad (\tau \in \Sigma_0, x \in X, y \in X^*)$$

and so $E(\cdot)$ is a spectral measure of class (Σ_0, X). Also

$$S = \int_K \lambda E(d\lambda)$$

and so, by Proposition 5.8, S is a scalar-type operator of class X^*. Hence, by Theorem 6.5, S is a scalar-type spectral operator and the proof is complete.

THEOREM 6.11. *Let X be weakly complete and let T be a prespectral operator on X. Then T is spectral.*

Proof. Let $E(\cdot)$ be a resolution of the identity for T. Define

$$S = \int_{\sigma(T)} \lambda E(d\lambda), \quad N = T - S.$$

By Theorem 5.15, N is a quasinilpotent operator commuting with S, $\sigma(T) = \sigma(S)$, and S is a scalar-type operator. If we define

$$\psi(f) = \int_{\sigma(S)} f(\lambda)\, E(d\lambda) \qquad (f \in C(\sigma(S)))$$

then, by Proposition 5.9, ψ is a bicontinuous algebra isomorphism of $C(\sigma(S))$ into $L(X)$ with $\psi(f_0) = I$ and $\psi(f_1) = S$, in the notation of the statement of Theorem 6.10. Hence S is a scalar-type spectral operator. The desired conclusion now follows from Theorem 6.8.

THEOREM 6.12. *Let X be weakly complete and let T, in $L(X)$, have totally disconnected spectrum. In order that T be a spectral operator it is necessary and* *sufficient that the set $\{E(\delta): \delta$ open-and-closed in $\sigma(T)\}$ of spectral projections for T be uniformly bounded in norm.*

Proof. If T is spectral with resolution of the identity $F(\cdot)$ then, by Proposition 5.27, $E(\delta) = F(\delta)$ for each open-and-closed subset δ of $\sigma(T)$. Hence the condition is necessary.

Conversely, suppose that $\|E(\delta)\| \leqslant M < \infty$ for each open-and-closed subset δ of $\sigma(T)$. By Theorem 5.36, T^* is prespectral with resolution of the identity $F(\cdot)$ of class X such that

$$E(\delta)^* = F(\delta) \qquad (\delta \text{ open-and-closed in } \sigma(T)).$$

Define

$$\psi(f) = \int_{\sigma(T)} f(\lambda)\, F(d\lambda) \qquad (f \in C(\sigma(T))).$$

By Proposition 5.9, ψ is a bicontinuous algebra isomorphism of $C(\sigma(T))$ into $L(X^*)$. Since the topology of $\sigma(T)$ has a base of open-and-closed subsets, then by the Stone-Weierstrass theorem, the algebra $\mathscr{A}$ of finite linear combinations of characteristic functions of open-and-closed subsets of $\sigma(T)$ is norm dense in $C(\sigma(T))$. Given any f in $C(\sigma(T))$ we can find a sequence $\{f_n\}$ in $\mathscr{A}$ converging uniformly on $\sigma(T)$ to f. For each n, there is $\psi_0(f_n)$ in $L(X)$ such that $\psi_0(f_n)^* = \psi(f_n)$. Now, $\{\psi_0(f_n)\}$ is a Cauchy sequence in $L(X)$ and so it converges to an operator which we denote by $\psi_0(f)$. Observe that

$$\psi_0(f)^* = \psi(f) \qquad (f \in C(\sigma(T))).$$

For λ in $\sigma(T)$ define $f_0(\lambda) = 1$ and $f_1(\lambda) = \lambda$. Let $\psi_0(f_1) = S$ and $N = T - S$. Then $S^* + N^*$ is the Jordan decomposition of T^* and $\sigma(S) = \sigma(S^*) = \sigma(T)$. Observe that ψ_0 is a bicontinuous algebra isomorphism of $C(\sigma(S))$ into $L(X)$ with $\psi_0(f_0) = I$. It follows from Theorem 6.10 that S is a scalar-type spectral operator. Since N is a quasinilpotent operator commuting with S the desired conclusion follows from Theorem 6.8.

THEOREM 6.13. *Let X be weakly complete and let $S \in L(X)$. Let $\sigma(S)$ be an R-set. Then S is a scalar-type spectral operator if and only if there is a positive real* *constant M such that for each rational function g with poles outside $\sigma(S)$*

$$\|g(S)\| \leqslant M \sup\{|g(\lambda)| : \lambda \in \sigma(S)\}.$$

Proof. Necessity is obvious, since if $E(\cdot)$ is the resolution of the identity for S then

$$g(S) = \int_{\sigma(S)} g(\lambda)\, E(d\lambda).$$

Conversely, suppose that the condition is satisfied. Let g be a rational function with poles outside $\sigma(S)$. For brevity write

$$\|g\| = \sup\{|g(\lambda)| : \lambda \in \sigma(S)\}.$$

By the spectral mapping theorem, $g(\sigma(S)) = \sigma(g(S))$. Also the spectral radius of $g(S)$ is $\leqslant \|g(S)\|$ and so we have

$$\|g\| \leqslant \|g(S)\| \leqslant M\|g\|.$$

The map $r \to r(S)$, which sends a rational function in $C(\sigma(S))$ into an element of $L(X)$, is well-defined. Since $\sigma(S)$ is an R-set, this map can be extended to a bicontinuous algebra isomorphism from $C(\sigma(S))$ into a subalgebra of $L(X)$. It now follows from Theorem 6.10 that S is a scalar-type spectral operator.

We now prove sufficient conditions for a net of scalar-type spectral operators to converge to a scalar-type spectral operator.

THEOREM 6.14. *Let X be weakly complete and let $\{S_\alpha : \alpha \in A\}$ be a net of scalar-type spectral operators on X converging in the strong operator topology to S. Suppose that all the spectra $\{\sigma(S_\alpha) : \alpha \in A\}$ and $\sigma(S)$ lie in a compact set K which is an R-set. Suppose also that there is a real constant M such that*

$$\|E_\alpha(\delta)\| \leqslant M \qquad (\delta \in \Sigma_p,\ \alpha \in A)$$

where $E_\alpha(\cdot)$ is the resolution of the identity of S_α. Then S is a scalar-type spectral operator.

Proof. Observe that if $\lambda \in \mathbf{C}\backslash K$ and $\alpha \in A$ then

$$(\lambda I - S_\alpha)^{-1} = \int_K (\lambda - \zeta)^{-1} E_\alpha(d\zeta)$$

and so it follows that

$$\|(\lambda I - S_\alpha)^{-1}\| \leqslant 4M \sup\{|\zeta - \lambda|^{-1} : \zeta \in K\}.$$

Hence it follows from the resolvent equation

$$(\lambda I - S_\alpha)^{-1} - (\lambda I - S)^{-1} = (\lambda I - S_\alpha)^{-1}(S_\alpha - S)(\lambda I - S)^{-1}$$

that for λ in $\mathbf{C}\backslash K$, $(\lambda I - S_\alpha)^{-1} \to (\lambda I - S)^{-1}$ in the strong operator topology. Now let r be a rational function in $C(K)$. Then $r(S_\alpha)$ is a finite product of terms of the form $\lambda I - S_\alpha$ and $(\lambda I - S_\alpha)^{-1}$. Thus, since multiplication is a continuous function of both variables on bounded subsets of $L(X)$ we see that $r(S_\alpha) \to r(S)$ in the strong operator topology. Now

$$\begin{aligned} \|r(S_\alpha)\| &= \left\| \int_K r(\lambda)\, E_\alpha(d\lambda) \right\| \\ &\leqslant 4M \sup\{|r(\zeta)| : \zeta \in K\} \end{aligned}$$

and since $r(S_\alpha) \to r(S)$ strongly, we have

$$\|r(S)\| \leqslant 4M \sup\{|r(\zeta)| : \zeta \in K\}.$$

Since K is an R-set, the map $r \to r(S)$ can be extended uniquely to a continuous algebra homomorphism ψ of $C(K)$ into $L(X)$ such that $\psi(f_0) = I$ and $\psi(f_1) = S$, where

$$f_0(z) = 1, \; f_1(z) = z \qquad (z \in K).$$

The desired conclusion now follows from Theorem 6.10.

Note 6.15. The hypothesis $\sigma(S) \subseteq K$ in Theorem 6.14 can be replaced by the following hypothesis: $\mathbf{C}\backslash K$ is connected. In this case K is nowhere dense with connected complement and so by Theorem 5.37 the polynomials are norm dense in $C(K)$. For each polynomial p and x in X we have

$$\|p(S_\alpha)x\| \leqslant 4M \sup\{|p(\lambda)|: \lambda \in K\} \|x\| \qquad (\alpha \in A)$$

and so

$$\|p(S)x\| \leqslant 4M \sup\{|p(\lambda)|: \lambda \in K\}\|x\|.$$

Hence the map $p \to p(S)$ can be uniquely extended to a continuous algebra homomorphism ψ of $C(K)$ into $L(X)$ as before.

We shall discuss this problem further in the next chapter.

Observe that a reflexive Banach space is weakly complete. The following result therefore follows immediately from Theorems 6.9 and 6.11.

THEOREM 6.16. *Let X be reflexive and let $T \in L(X)$. Then T is spectral if and only if T^* is spectral.*

The next result follows from Theorems 5.40, 5.41 and 6.16.

THEOREM 6.17. *Let X be reflexive and let $S \in L(X)$. Then S is a scalar-type spectral operator if and only if the following condition is satisfied. There are operators R and J on X such that*

(i) $S = R + iJ$,

(ii) $RJ = JR$,

(iii) *there is an equivalent norm on X with respect to which the operators $\{R^m J^n: m, n = 0, 1, 2, \ldots\}$ are all hermitian.*

In order to proceed further in this direction, we require some deep results on weakly compact linear maps.

Definition 6.18. Let X and Y be complex Banach spaces and let T be a linear map from X into Y. Let $B = \{x \in X: \|x\| \leqslant 1\}$. Then T is said to be *weakly compact* if the weak closure of TB is compact in the weak topology of Y.

Thus, by the Eberlein–Šmulian theorem (see, for example, V.6.1 of [**5**; p. 430–3]) T is weakly compact if and only if it maps bounded sets into weakly sequentially compact sets. For a full discussion of weakly compact linear maps the reader is referred to [**5**; p. 482–5].

We require also the following results on linear maps of $C(K)$, where K is a compact Hausdorff space.

THEOREM 6.19. *Let X be weakly complete. Let K be a compact Hausdorff space and let T be a bounded linear map from $C(K)$ into X. Then T is weakly compact.*

For a proof of this result see Theorem VI.7.6 of [**5**; p. 494–6].

THEOREM 6.20. *Let K be a compact Hausdorff space and let Σ_0 be the σ-algebra of Borel subsets of K. Let T be a weakly compact linear map from $C(K)$ into X. Then there is a vector-valued measure $\boldsymbol{\mu}$ defined on Σ_0 with values in X such that*

(i) $\langle \boldsymbol{\mu}(\cdot), y \rangle$ *is a regular measure, for each y in X^*;*

(ii) $Tf = \displaystyle\int_K f(\lambda)\, \boldsymbol{\mu}(d\lambda) \qquad (f \in C(K)).$

Conversely, if $\boldsymbol{\mu}$ is a vector-valued measure on Σ_0 with values in X satisfying (i) *then the linear map T defined by* (ii) *is a weakly compact linear map from $C(K)$ into X.*

For a proof of this result see Theorem VI.7.3 of [**5**; p. 493–4].

In the case of a weakly complete Banach space X, the following stronger version of Proposition 5.43 holds.

THEOREM 6.21. *Let X be weakly complete. Let $\mathbf{B}$ be a bounded Boolean algebra of projections on X. Let Ω be the Stone representation space of $\mathbf{B}$ and let Σ denote the σ-algebra of Borel subsets of Ω. There is a spectral measure $E(\cdot)$ of class (Σ, X^*) such that $E(\tau)$ is the projection corresponding to the open-and-closed subset τ of Ω under the isomorphism of the Stone representation theorem.*

Proof. Let $\mathscr{A}$ be the closed subalgebra of $L(X)$ generated by $\mathbf{B}$. We observe that, as in Proposition 5.43, the isomorphism of the Stone representation theorem can be extended to a bicontinuous algebra isomorphism ψ from $C(\Omega)$ onto $\mathscr{A}$. By Theorem 5.21, there is a spectral measure $F(\cdot)$ of class (Σ, X) with values in $L(X^*)$ such that

$$\psi(f)^* = \int_\Omega f(\lambda)\, F(d\lambda) \qquad (f \in C(\Omega)).$$

Let $x \in X$. Since X is weakly complete the map $f \to \psi(f)x$ of $C(\Omega)$ into X is weakly compact by Theorem 6.19. Hence by Theorem 6.20 there is a vector-valued measure $\boldsymbol{\mu}_x(\cdot)$ such that

$$\psi(f)x = \int_\Omega f(\lambda)\, \boldsymbol{\mu}_x(d\lambda) \qquad (f \in C(\Omega)).$$

Define

$$E(\tau)x = \boldsymbol{\mu}_x(\tau) \qquad (\tau \in \Sigma,\, x \in X).$$

Let $y \in X^*$. Let $\mu_1(\tau) = \langle E(\tau)x, y\rangle$ and $\mu_2(\tau) = \langle x, F(\tau)y\rangle$ $(\tau \in \Sigma)$. Observe that

$$\int_\Omega f(\lambda)\,\mu_1(d\lambda) = \int_\Omega f(\lambda)\,\mu_2(d\lambda) \qquad (f \in C(\Omega)).$$

Since μ_1 and μ_2 are both regular measures they are identical, by the Riesz representation theorem; hence

$$\langle E(\tau)x, y\rangle = \langle x, F(\tau)y\rangle \qquad (x \in X,\, y \in X^*,\, \tau \in \Sigma).$$

By the uniform boundedness theorem for each τ in Σ, $E(\tau) \in L(X)$ and $E(\tau)^* = F(\tau)$. Hence $E(\cdot)$ is a spectral measure of class (Σ, X^*). The other statement of theorem follows easily.

THEOREM 6.22. *Let X be weakly complete. Let* **B** *be a Boolean algebra of projections on X. A necessary and sufficient condition for every operator in the closed subalgebra $\mathscr{A}$ of $L(X)$ generated by* **B** *to be a scalar-type spectral operator is that* **B** *be bounded.*

Proof. Let $S \in \mathscr{A}$. If S is a scalar-type spectral operator then, by Theorem 6.9, S^* is a scalar-type operator of class X. Hence, by Theorem 5.46, **B** is bounded. Conversely if **B** is bounded then in the notation of Theorem 6.21

$$\psi(f) = \int_\Omega f(\lambda)\,E(d\lambda) \qquad (f \in C(\Omega))$$

where $E(\cdot)$ is a spectral measure of class (Σ, X^*). The desired conclusion now follows from Proposition 5.8 and Theorem 6.5.

This result generalizes a theorem of Lumer. The last theorem in this chapter is due to Spain [**1**]. A preliminary definition is required.

Definition 6.23. Let $S \in L(X)$. We say that S possesses a *C-operational calculus* ψ if there is a bicontinuous algebra isomorphism ψ from $C(\sigma(S))$ into a subalgebra of $L(X)$ such that $\psi(f_0) = I$ and $\psi(f_1) = S$ where

$$f_0(\lambda) = 1, \quad f_1(\lambda) = \lambda \qquad (\lambda \in \sigma(S)).$$

Observe that, by Proposition 5.9, every scalar-type operator possesses a C-operational calculus. An example, will be given later to show that the converse of this result fails. Note that it follows from Theorem 6.10 and Proposition 5.9 that if X is weakly complete and $S \in L(X)$ then S is a scalar-type spectral operator if and only if S possesses a C-operational calculus.

We observe that if S possesses a C-operational calculus ψ then, by Theorem 5.21, S^* is a scalar-type operator with resolution of the identity $F(\cdot)$ of class X such that

$$\psi(f)^* = \int_{\sigma(S)} f(\lambda)\, F(d\lambda) \qquad (f \in C(\sigma(S))). \tag{1}$$

It follows from this and Theorem 5.13 that if S possesses a C-operational calculus ψ then ψ is uniquely determined.

THEOREM 6.24. *Let $S \in L(X)$. The following statements are equivalent.*

(i) *S is a scalar-type spectral operator.*

(ii) *S possesses a C-operational calculus ψ and for each τ in Σ_p there exists a projection $E(\tau)$ in $L(X)$ with $E(\tau)^* = F(\tau)$, where $F(\cdot)$ and ψ are related by* (1).

(iii) *S possesses a C-operational calculus ψ and for each x in X the map $f \to \psi(f)x$ from $C(\sigma(S))$ into X is weakly compact.*

Proof. (i) $\Rightarrow$ (ii). Let S be a scalar-type spectral operator with resolution of the identity $E(\cdot)$. Define

$$\psi(f) = \int_{\sigma(S)} f(\lambda)\, E(d\lambda) \qquad (f \in C(\sigma(S))).$$

By Proposition 5.9, ψ is a C-operational calculus for S. Hence, by Theorem 5.21, S^* is a scalar-type operator with resolution of the identity $F(\cdot)$ of class X satisfying (1). However, by Theorem 6.9, S^* is a scalar-type operator with a resolution of the identity $E^*(\cdot)$ of class X. By Theorem 5.13, $F(\cdot) = E^*(\cdot)$.

(ii) $\Rightarrow$ (iii). $E(\cdot)$ is clearly a spectral measure of class (Σ_p, X^*). As in Theorem

6.5, $E(\cdot)$ is countably additive in the strong operator topology. By equation (1)

$$\psi(f) = \int_{\sigma(S)} f(\lambda)\, E(d\lambda) \qquad (f \in C(\sigma(S))).$$

For each x in X, $E(\cdot)x$ is a vector-valued measure on the σ-algebra of Borel subsets of $\sigma(S)$ and so by Theorem 6.20 the map $f \to \psi(f)x$ from $C(\sigma(S))$ into X is weakly compact.

(iii) $\Rightarrow$ (i). By Theorem 6.20, for each x in X there is a unique vector-valued measure $\boldsymbol{\mu}_x(\cdot)$ defined on the σ-algebra Σ of Borel subsets of $\sigma(S)$ such that

$$\psi(f)x = \int_{\sigma(S)} f(\lambda)\, \boldsymbol{\mu}_x(d\lambda) \qquad (f \in C(\sigma(S))).$$

For each τ in Σ and x in X define

$$E(\tau)x = \boldsymbol{\mu}_x(\tau).$$

Let $y \in X^*$. Let $\mu_1(\tau) = \langle E(\tau)x, y\rangle$ and $\mu_2(\tau) = \langle x, F(\tau)y\rangle \quad (\tau \in \Sigma)$. Observe that

$$\int_{\sigma(S)} f(\lambda)\, \mu_1(d\lambda) = \int_{\sigma(S)} f(\lambda)\, \mu_2(d\lambda) \qquad (f \in C(\sigma(S))).$$

Since μ_1 and μ_2 are both regular measures they are identical by the Riesz representation theorem. Hence

$$\langle E(\tau)x, y\rangle = \langle x, F(\tau)y\rangle \qquad (x \in X,\ y \in X^*,\ \tau \in \Sigma).$$

By the uniform boundedness theorem, $E(\tau) \in L(X)$ and $E(\tau)^* = F(\tau)(\tau \in \Sigma)$. Hence $E(\cdot)$ is a spectral measure of class (Σ, X^*). By Proposition 5.8, S is a scalar-type operator of class X^*. It now follows from Theorem 6.5 that S is a scalar-type spectral operator and the proof is complete.

Example 6.26. Let $X = C[0, 1]$. Define S, in $L(X)$, by

$$(Sf)(t) = tf(t) \qquad (f \in X,\ 0 \leqslant t \leqslant 1).$$

Then S possesses a C-operational calculus ψ given by

$$\psi(g)f = gf \qquad (f, g \in X).$$

Clearly $\sigma(S) = [0, 1]$. Suppose that $P^2 = P \in L(X)$ and $SP = PS$. Let $f_0(t) = 1$ $(0 \leqslant t \leqslant 1)$. By the Stone-Weierstrass theorem, $Pf = (Pf_0)f$, for all f in X, so that $(Pf_0)^2 = Pf_0$. Thus $P = 0$ or $P = I$. This shows that S is not a prespectral operator.

7. Normal Operators

The main purpose of this chapter is to prove the spectral theorem for a normal operator. This states that a normal operator is a scalar-type spectral operator with a resolution of the identity all of whose values are self-adjoint projections. The proof given here is due to Whitley [**1**]. Rosenblum's proof of Fuglede's theorem is given. It is shown that an operator on a Hilbert space is hermitian in the sense of Chapter 4 if and only if it is self-adjoint.

Definition 7.1. Let $T \in L(H)$. Then T is said to be *normal* if $TT^* = T^*T$.

Definition 7.2. Let $T \in L(H)$. Then T is *self-adjoint* if $T = T^*$.

Definition 7.3. Let $T \in L(H)$. Then T is said to be *positive* if $\langle Tx, x \rangle \geqslant 0$ for every x in H.

A positive operator is self-adjoint.

Definition 7.4. Let $T \in L(H)$. Then T is *unitary* if $TT^* = T^*T = I$.

Observe that every T in $L(H)$ can be expressed in the form $T = A + iB$ where A and B are self-adjoint operators. It suffices to take

$$A = \tfrac{1}{2}(T + T^*), \; B = \frac{1}{2i}(T - T^*).$$

Clearly T is normal if and only if $AB = BA$. The following result gives another characterization of normal operators.

PROPOSITION 7.5. *Let $T \in L(H)$. Then T is normal if and only if*

$$\|Tx\| = \|T^*x\| \qquad (x \in H).$$

Proof. Since $\|Tx\|^2 = \langle Tx, Tx\rangle = \langle T^*Tx, x\rangle$ and similarly $\|T^*x\|^2 = \langle T^*x, T^*x\rangle = \langle TT^*x, x\rangle$ for all x in H, the identity of $\|Tx\|$ and $\|T^*x\|$ is equivalent to the identity of $\langle T^*Tx, x\rangle$ and $\langle TT^*x, x\rangle$ for all x in H. This last condition is equivalent to the normality of T.

Next, we give a characterization of unitary operators.

PROPOSITION 7.6. *Let $T \in L(H)$. Then T is unitary if and only if T is isometric and invertible in $L(H)$.*

Proof. If T is unitary, then T is invertible in $L(H)$. Also

$$\|Tx\|^2 = \langle Tx, Tx\rangle = \langle T^*Tx, x\rangle = \langle x, x\rangle = \|x\|^2 \qquad (x \in H)$$

and so T is isometric. Conversely if T is isometric and invertible in $L(H)$ we have

$$\langle Tx, Ty\rangle = \langle x, y\rangle \qquad (x, y \in H)$$

and so $T^*T = I$. Hence $TT^*T = T$ and so $TT^* = I$ since $TH = H$.

We shall use the term 'reducing subspace' in a different sense from Chapter 1 for operators on a Hilbert space.

Definition 7.7. Let $A \in L(H)$. A closed subspace Y of H is said to *reduce* A if both Y and $Y^\perp$ are invariant under A.

PROPOSITION 7.8. *Let $A \in L(H)$. A necessary and sufficient condition that a closed subspace Y of H be invariant under A is that $Y^\perp$ be invariant under A^*.*

Proof. By symmetry it is enough to prove that the condition is necessary. If $AY \subseteq Y$, if $x \in Y$ and $y \in Y^\perp$, then $\langle x, A^*y\rangle = \langle Ax, y\rangle = 0$, so that $A^*y \in Y^\perp$. Consequently, $Y^\perp$ is invariant under A^*.

PROPOSITION 7.9. *Let $A \in L(H)$ and let Y be a closed subspace of H. Then Y reduces A if and only if Y is invariant under both A and A^*.*

Proof. This follows immediately from Proposition 7.8.

Next, we prove some spectral properties of normal operators.

THEOREM 7.10. *Let A be a normal operator on H. Then $\sigma_a(A) = \sigma(A)$.*

Proof. In view of Theorem 1.16(i) it is sufficient to prove that $\sigma(A) \subseteq \sigma_a(A)$. If $\lambda \in \mathbf{C} \backslash \sigma_a(A)$, then there exists a positive real number ε such that

$$\|Ay - \lambda y\| \geqslant \varepsilon \|y\| \qquad (y \in H).$$

Since $(A - \lambda I)^* = A^* - \bar{\lambda} I$, it follows that $A - \lambda I$ is normal. By Proposition 7.5,

$$\|A^* y - \bar{\lambda} y\| \geqslant \varepsilon \|y\| \qquad (y \in H).$$

In order to prove that $\lambda \in \rho(A)$ it is enough to prove that the range of $A - \lambda I$ is dense in H, or, equivalently, that the orthogonal complement in H of the range is $\{0\}$. Clearly, however, if a vector y is orthogonal to the range of $A - \lambda I$, then

$$0 = \langle (A - \lambda I)x, y \rangle = \langle x, (A^* - \bar{\lambda} I)y \rangle \qquad (x \in H)$$

and hence $A^* y - \bar{\lambda} y = 0$. Since $\|A^* y - \bar{\lambda} y\| \geqslant \varepsilon \|y\|$ it follows that $y = 0$ and the proof is complete.

PROPOSITION 7.11. *If A is a self-adjoint operator on H then $\sigma(A)$ is real.*

Proof. If λ is not real and $x \neq 0$ then

$$\begin{aligned} 0 < |\lambda - \bar{\lambda}| \, \|x\|^2 &= |\langle (A - \lambda I)x, x \rangle - \langle (A - \bar{\lambda} I)x, x \rangle| \\ &= |\langle (A - \lambda I)x, x \rangle - \langle x, (A - \lambda I)x \rangle| \\ &\leqslant 2\|Ax - \lambda x\| \, \|x\|. \end{aligned}$$

The desired conclusion follows from the fact that for self-adjoint operators the spectrum and the approximate point spectrum are the same.

THEOREM 7.12. *If A is a normal operator on H then $\|A\| = v(A)$.*

Proof. We show first that $\|A^*A\| = \|A\|^2$. (This equation does not depend on normality.) We have $\|A^*A\| \leqslant \|A^*\| \, \|A\| = \|A\|^2$. Conversely, if $\|x\| = 1$

$$\|A^*Ax\| = \|A^*Ax\| \, \|x\| \geqslant \langle A^*Ax, x \rangle = \langle Ax, Ax \rangle = \|Ax\|^2.$$

Hence $\|A^*A\| \geqslant \|A\|^2$ and so $\|A^*A\| = \|A\|^2$.

Since A is normal, Proposition 7.5 shows that

$$\|A^2x\| = \|A(Ax)\| = \|A^*(Ax)\| \qquad (x \in H).$$

Hence $\|A^2\| = \|A^*A\| = \|A\|^2$. This argument shows that if m is a power of 2

$$\|A^m\| = \|A\|^m.$$

It follows from Proposition 1.8 that $\nu(A) = \|A\|$.

PROPOSITION 7.13. *Let A be a self-adjoint operator on H. If p is a real polynomial, then*

$$\|p(A)\| = \sup\{|p(\lambda)| : \lambda \in \sigma(A)\}.$$

The map $p \to p(A)$ can be extended to an isometric algebra isomorphism $f \to f(A)$ of $C_R(\sigma(A))$ into a subalgebra of self-adjoint operators on H.

Proof. By Theorem 7.12 and the spectral mapping theorem

$$\begin{aligned}\|p(A)\| &= \sup\{|\lambda| : \lambda \in \sigma(p(A))\}\\ &= \sup\{|\lambda| : \lambda \in p(\sigma(A))\}\\ &= \sup\{|p(\lambda)| : \lambda \in \sigma(A)\}.\end{aligned}$$

The other statement follows from the Weierstrass polynomial theorem.

PROPOSITION 7.14. *Let $E \in L(H)$, $E^2 = E$ and $E \neq 0$. The following statements are equivalent.*

(i) *E is self-adjoint.*

(ii) $\|E\| = 1$.

(iii) *EH and $(I - E)H$ are orthogonal subspaces.*

Proof. Suppose that (i) holds. If x is a non-zero vector in H, then

$$\langle Ex, x\rangle = \langle E^2x, x\rangle = \langle Ex, E^*x\rangle = \langle Ex, Ex\rangle = \|Ex\|^2.$$

Therefore $\|Ex\|^2 \leqslant \|Ex\| \, \|x\|$ and so $\|Ex\| \leqslant \|x\|$. Since E is idempotent and $E \neq 0$ it follows that $\|E\| = 1$. Thus (i) implies (ii).

Suppose that (ii) holds. Let $M = EH$ and $N = (I - E)H$, For every x in H, $y = Ex - x \in N$. If $x \in N^{\perp}$, then $Ex = x + y$ with $\langle x, y \rangle = 0$. It follows then that

$$\|x\|^2 \geqslant \|Ex\|^2 = \|x\|^2 + \|y\|^2$$

so that $y = 0$. Thus we have proved that $x \in N^{\perp}$ implies $x = Ex$; that is $N^{\perp} \subseteq M$. Conversely, let $z \in M$ so that $z = Ez$. Then we have the orthogonal decomposition $z = y + x$, where $y \in N$ and $x \in N^{\perp}$. Hence $z = Ez = Ey + Ex = Ex = x$. This shows that $M = EH \subseteq N^{\perp}$. We have proved that $M = N^{\perp}$ and so (ii) implies (iii).

Suppose that (iii) holds. Let $z_r = x_r + y_r$, where $x_r \in EH$ and $y_r \in (I - E)H$, for $r = 1, 2$. Then

$$\langle Ez_1, z_2 \rangle = \langle x_1, z_2 \rangle = \langle x_1, x_2 \rangle = \langle z_1, x_2 \rangle = \langle z_1, Ez_2 \rangle$$

and so E is self-adjoint. Thus (iii) implies (i) and the proof is complete.

We have now assembled all the elementary results necessary to prove the spectral theorem for a normal operator. However we require two other preliminary results.

LEMMA 7.15. *Let T be a normal operator on H and let $0 \in \sigma(T)$. Let ε in $(0, 1]$ be given. There is a closed non-zero subspace M of H with the property that any operator which commutes with T^*T is reduced by M and $\|T|M\| \leqslant \varepsilon$.*

Proof. Let $A = T^*T$. Since $0 \in \sigma(T)$ then by Theorem 7.10 there is a sequence $\{x_n\}$ of unit vectors with $Tx_n \to 0$. Thus $Ax_n \to 0$ and the self-adjoint operator A has 0 in its spectrum. Define

$$f(t) = \begin{cases} 1 & (|t| < \varepsilon/2) \\ 2(1 - |\varepsilon^{-1}t|) & (\varepsilon/2 \leqslant |t| \leqslant \varepsilon) \\ 0 & (|t| \geqslant \varepsilon). \end{cases}$$

Since f is continuous on $\mathbf{R}$ we can define $f(A)$ as in Proposition 7.13.

Let M be the closed subspace $\{x : f(A)x = x\}$ of H. By Proposition 7.13, if B is an operator which commutes with A then B commutes with $f(A)$.

Consequently if $x \in M$ then $Bx = Bf(A)x = f(A)Bx$, which shows that M is invariant under B. Since B^* also commutes with A, M is invariant under B^* and so reduces B.

If x is a unit vector in M then

$$\|Ax\| = \|Af(A)x\| \leqslant \|Af(A)\| = \sup\{|\lambda f(\lambda)| : \lambda \in \sigma(A)\} \leqslant \varepsilon.$$

Thus $\|Tx\|^2 = \langle Ax, x\rangle \leqslant \varepsilon$ and so $\|T|M\| \leqslant \varepsilon^{1/2} \leqslant \varepsilon$.

It remains to show that $M \neq \{0\}$. Observe that

$$\|(I - f(A))f(2A)\| = \sup\{|(1 - f(\lambda))(f(2\lambda))| : \lambda \in \sigma(A)\} = 0$$

since $f(\lambda) = 1$ whenever $f(2\lambda) \neq 0$. Hence every element in the range of the operator $f(2A)$ lies in M and this range is not $\{0\}$ because

$$\|f(2A)\| = \sup\{|f(2\lambda)| : \lambda \in \sigma(A)\} \geqslant f(0) = 1.$$

LEMMA 7.16. *Let T be a normal operator on H and let p be a polynomial in two variables. Then*

$$\sigma(p(T, T^*)) = \{p(\lambda, \bar{\lambda}) : \lambda \in \sigma(T)\}.$$

Proof. We write $p(w, z) = \sum_{n,m} a_{nm} w^n z^m$. Let $\lambda \in \sigma(T)$. By Theorem 7.10, there is a sequence $\{x_r\}$ of unit vectors such that $(\lambda I - T)x_r \to 0$. Since T is normal it follows from Proposition 7.5 that $(\bar{\lambda} I - T^*)x_r \to 0$. Hence

$$\begin{aligned}
[p(T, T^*) - p(\lambda, \bar{\lambda})I]x_r &= \sum_{n,m} a_{nm}(T^n T^{*m} - \lambda^n \bar{\lambda}^m I)x_r \\
&= \sum_{n,m} a_{nm}[T^n(T^{*m} - \bar{\lambda}^m I)x_r + \bar{\lambda}^m(T^n - \lambda^n I)x_r] \\
&= \sum_{n,m} a_{nm}[T^n(T^{*m-1} + \ldots + \bar{\lambda}^{m-1}I)(T^* - \bar{\lambda}I)x_r] \\
&\quad + \sum_{n,m} a_{nm}\bar{\lambda}^m(T^{n-1} + \ldots + \lambda^{n-1}I)(T - \lambda I)x_r
\end{aligned}$$

which converges to 0 as $r \to \infty$. We conclude that $p(\lambda, \bar{\lambda}) \in \sigma(p(T, T^*))$.

Now choose μ in $\sigma(p(T, T^*))$. The operator $B = p(T, T^*) - \mu I$ is normal and $0 \in \sigma(B)$. By Lemma 7.15, for each positive integer n there is a closed non-zero subspace M_n which reduces B such that $\|B|M_n\| \leqslant n^{-1}$ and, since T commutes with B^*B, M_n reduces T. It follows that $T|M_n$ is normal. By Theorem 7.12, $\sigma(T|M_n)$ is non-empty and so we can choose λ_n in $\sigma(T|M_n)$. Hence by Theorem 7.10 there is a unit vector y_n in M_n with $\|(\lambda_n I - T)y_n\| \leqslant n^{-1}$. The sequence $\{\lambda_n\}$ is bounded by $\|T\|$ and so contains a subsequence converging to λ. By re-indexing we may suppose that $\{\lambda_n\}$ converges to λ.

The point λ is in $\sigma(T)$, since $\|(\lambda I - T)y_n\| \leqslant |\lambda_n - \lambda| + \|(\lambda_n I - T)y_n\|$ and so $(\lambda I - T)y_n \to 0$. From above, if $(\lambda I - T)y_n \to 0$ then

$$[p(T, T^*) - p(\lambda, \bar{\lambda})I]y_n \to 0.$$

Since $y_n \in M_n$ it follows that $[p(T, T^*) - \mu I]y_n \to 0$. Consequently $\mu = p(\lambda, \bar{\lambda})$ and the proof of the lemma is complete.

LEMMA 7.17. *Let T be a normal operator on H. There is an isometric algebra isomorphism ψ of $C(\sigma(T))$ into a subalgebra of $L(H)$ consisting of normal operators such that*

(i) *ψ maps the polynomial $p(\lambda, \bar{\lambda})$ into $p(T, T^*)$,*
(ii) $\psi(\bar{f}) = \psi(f)^* \qquad (f \in C(\sigma(T)))$.

Proof. This follows easily from the Stone-Weierstrass theorem, Theorem 7.12 and Lemma 7.16.

We now prove the spectral theorem for a normal operator.

THEOREM 7.18. *Let T be a normal operator on H. Then T is a scalar-type spectral operator. The values of the resolution of the identity of T are self-adjoint projections.*

Proof. Let ψ be the isometric algebra isomorphism of $C(\sigma(T))$ into a subalgebra of $L(H)$ given by Lemma 7.17. Let $x, y \in H$. The map $f \to \langle\psi(f)x, y\rangle$ is a bounded linear functional on $C(\sigma(T))$ and so, by the Riesz representation theorem, there is a uniquely determined regular Borel measure $\mu(\cdot, x, y)$ such that, for all f in $C(\sigma(T))$,

$$\langle\psi(f)x, y\rangle = \int_{\sigma(T)} f(\lambda)\mu(d\lambda, x, y). \tag{1}$$

Let Σ_0 denote the σ-algebra of Borel subsets of $\sigma(T)$. Since $\mu(\tau, x, y)$ is uniquely determined by τ, x and y it is, for each τ in Σ_0, bilinear in x and y. Also, since

$$\langle \psi(f)x, y \rangle = \langle x, \psi(f)^* y \rangle = \overline{\langle \psi(f)^* y, x \rangle} \qquad (f \in C(\sigma(T))),$$

$\mu(\tau, x, y)$ is symmetric. Since $|\mu(\tau, x, y)| \leqslant \|x\| \, \|y\|$, these bilinear functionals are bounded. It follows that for each τ in Σ_0 there is a unique self-adjoint operator $E(\tau)$ on H such that

$$\mu(\tau, x, y) = \langle E(\tau)x, y \rangle \qquad (x, y \in H). \tag{2}$$

It will now be shown that $E(\cdot)$ is a spectral measure. By putting $f = 1$ in equation (1) we see that $I = E(\sigma(T))$ and, since $E(\cdot)$ is additive, that

$$I - E(\delta) = E(\sigma(T)) - E(\delta) = E(\sigma(T) \backslash \delta) \qquad (\delta \in \Sigma_0).$$

To prove that $E(\cdot)$ is a spectral measure it will therefore suffice to show that $E(\delta \cap \tau) = E(\delta)E(\tau)$ for every pair δ, τ of Borel subsets of $\sigma(T)$. For fixed x, y in H and g in $C(\sigma(T))$ define

$$\nu(\tau) = \int_\tau g(\lambda)\mu(d\lambda, x, y) \qquad (\tau \in \Sigma_0).$$

If $f \in C(\sigma(T))$, then

$$\int_{\sigma(T)} f(\lambda)\nu(d\lambda) = \int_{\sigma(T)} f(\lambda)g(\lambda)\mu(d\lambda, x, y) = \langle \psi(f)\psi(g)x, y \rangle$$

$$= \int_{\sigma(T)} f(\lambda)\mu(d\lambda, \psi(g)x, y).$$

Since $\nu(\cdot)$ and $\mu(\cdot, \psi(g)x, y)$ are regular measures

$$\nu(\tau) = \int_{\sigma(T)} g(\lambda)\chi(\tau; \lambda)\mu(d\lambda, x, y) = \langle E(\tau)\psi(g)x, y \rangle$$

$$= \int_{\sigma(T)} g(\lambda)\mu(d\lambda, E(\tau)x, y) \qquad (\tau \in \Sigma_0).$$

Since g is an arbitrary element of $C(\sigma(T))$,

$$\langle E(\tau \cap \delta)x, y\rangle = \int_{\tau \cap \delta} \mu(d\lambda, x, y) = \int_{\delta} \chi(\tau; \lambda)\mu(d\lambda, x, y)$$

$$= \langle E(\delta)E(\tau)x, y\rangle \qquad (\delta, \tau \in \Sigma_0).$$

It follows that for all pairs of Borel subsets δ, τ of $\sigma(T)$

$$E(\delta)E(\tau) = E(\delta \cap \tau) = E(\tau)E(\delta)$$

and this completes the proof that $E(\cdot)$ is a spectral measure.

Let $\{\tau_n\}$ be a sequence of pairwise disjoint sets in Σ_0 with union τ. Then

$$\langle E(\tau_r)x, E(\tau_k)x\rangle = \langle E(\tau_k)E(\tau_r)x, x\rangle = 0 \qquad (r \neq k)$$

and so, for every x in H, $\{E(\tau_n)x\}$ is an orthogonal sequence of vectors. Since

$$\sum_{n=1}^{\infty} \|E(\tau_n)x\|^2 = \sum_{n=1}^{\infty} \langle E(\tau_n)x, x\rangle = \langle E(\tau)x, x\rangle = \|E(\tau)x\|^2,$$

it follows that $E(\cdot)$ is countably additive on Σ_0 in the strong operator topology. Define

$$E(\tau) = E(\tau \cap \sigma(T)) \qquad (\tau \in \Sigma_p).$$

Then, as in the proof of Proposition 5.8, $\sigma(T|E(\tau)H) \subseteq \bar{\tau}$ for each τ in Σ_p, since it follows from equation (1) and the definition of the integral with respect to a spectral measure that

$$T = \int_{\sigma(T)} \lambda E(d\lambda)$$

and this completes the proof.

Note 7.19. Let $E(\cdot)$ be a spectral measure defined on the σ-algebra of Borel subsets of K, a compact Hausdorff space. Suppose that the values of $E(\cdot)$

are self-adjoint projections on H. If $f \in C(K)$ and

$$T = \int_K f(\lambda) E(d\lambda)$$

then

$$T^* = \int_K \overline{f(\lambda)} E(d\lambda)$$

and $TT^* = T^*T$. These facts follow readily from the definition of the integral with respect to a spectral measure.

THEOREM 7.20. *Let T be a normal operator on H. Then T is unitary, self-adjoint, or positive if and only if $\sigma(T)$ lies in the unit circle, the real line or the non-negative real line respectively.*

Proof. By Lemma 7.17, $TT^* = T^*T = I$ if and only if $\lambda\bar{\lambda} = 1$ for every λ in $\sigma(T)$ and $T = T^*$ if and only if $\lambda = \bar{\lambda}$ for every λ in $\sigma(T)$. Now let $E(\cdot)$ be the resolution of the identity for the self-adjoint operator T. If $\sigma(T)$ is non-negative then

$$\langle Tx, x \rangle = \int_{\sigma(T)} \lambda \langle E(d\lambda)x, x \rangle \geqslant 0$$

which shows that T is positive. Conversely, if an open interval δ of negative numbers intersects the spectrum of T then, by Note 5.7, $E(\delta) \neq 0$. If x is a non-zero vector in $E(\delta)H$, then

$$\langle Tx, x \rangle = \int_{\sigma(T)} \lambda \langle E(d\lambda)E(\delta)x, x \rangle < 0$$

which shows that T is not positive. The proof is complete.

The following result was first proved by Fuglede in 1950. The simple proof given here is due to M. Rosenblum.

THEOREM 7.21. *Let T be a normal operator on H and let $A \in L(H)$. If $AT = TA$ then $AT^* = T^*A$.*

Proof. Define

$$U = \tfrac{1}{2}(T + T^*),\ V = \frac{1}{2i}(T - T^*).$$

Then U, V are self-adjoint operators and $T = U + iV$, $T^* = U - iV$. Since A commutes with all powers of T, we have for any λ in **C**.

$$\exp(\lambda T^*)A\exp(-\lambda T^*) = \exp(\lambda T^*)\exp(-\bar{\lambda}T)A\exp(\bar{\lambda}T)\exp(-\lambda T^*).$$

Now $TT^* = T^*T$, whence

$$\exp(\lambda T^*)A\exp(-\lambda T^*) = \exp(\lambda T^* - \bar{\lambda}T)\,A\exp(\bar{\lambda}T - \lambda T^*).$$

Put $\lambda = \alpha + i\beta$, where α, β are real. Then

$$\lambda T^* - \bar{\lambda}T = 2i(\beta U - \alpha V) = iW_\lambda,$$

where W_λ is self-adjoint. Hence

$$\|\exp(iW_\lambda)\| = \|\exp(-iW_\lambda)\| = 1$$

and so

$$\|\exp(\lambda T^*)A\exp(-\lambda T^*)\| = \|\exp(iW_\lambda)A\exp(-iW_\lambda)\| \leqslant \|A\|.$$

The map $\lambda \to \exp(\lambda T^*)\,A\exp(-\lambda T^*)$ is therefore a bounded entire function and so by Liouville's theorem is constant. Hence

$$A = \exp(\lambda T^*)\,A\exp(-\lambda T^*)$$
$$\exp(\lambda T^*)\,A = A\exp(\lambda T^*) \qquad (\lambda \in \mathbf{C}).$$

On equating the coefficients of λ on both sides we obtain $AT^* = T^*A$.

We conclude this chapter with some results on limits of sequences of normal operators. As an application of these results we prove that an operator on a Hilbert space is hermitian in the sense of Chapter 4 if and only if it is self-adjoint.

PROPOSITION 7.22 *Let $\{T_n\}$ be a sequence of normal operators on H converging in the norm of $L(H)$ to T. Then T is normal.*

Proof. Observe that $\{T_n^*\}$ converges to T^*. Since $L(H)$ is a Banach algebra the equations $T_n T_n^* = T_n^* T_n$ for each n imply that $TT^* = T^*T$.

THEOREM 7.23. *Let $A \in L(H)$. Then A is hermitian if and only if A is self-adjoint.*

Proof. If A is self-adjoint then $\sigma(A)$ is real and so

$$\| \exp(i\alpha A) \| = 1 \qquad (\alpha \in \mathbf{R}). \tag{3}$$

Hence A is hermitian, by Theorem 4.7. Conversely, if A is hermitian then (3) shows that, for each real α, $\exp i\alpha A$ is an invertible isometry and so is unitary, by Proposition 7.6. The function ϕ defined by

$$\phi(\lambda) = \frac{\exp i\lambda A - I}{i\lambda} \qquad (\lambda \neq 0)$$

$$\phi(0) = A$$

is entire. For each positive integer n, $\phi(n^{-1})$ is normal and $\lim_{n\to\infty} \phi(n^{-1}) = A$ in the norm of $L(H)$. Hence, by Proposition 7.22, A is a normal operator. Also, $\sigma(A)$ is real and so, by Theorem 7.20, A is self-adjoint.

The final theorem in this chapter is the analogue for normal operators of Theorem 6.14.

THEOREM 7.24. *Let $\{S_\alpha : \alpha \in A\}$ be a net of normal operators on H converging in the strong operator topology to S. Suppose that all the spectra $\{\sigma(S_\alpha): \alpha \in A\}$ and $\sigma(S)$ lie in a compact set K which is an R-set. Then S is a normal operator.*

Proof. For each α in A let $E_\alpha(\cdot)$ be the resolution of the identity for S_α. By Proposition 7.14, $\| E_\alpha(\tau) \| \leqslant 1$ for every τ in Σ_p and so, by Theorem 6.14, S is a scalar-type spectral operator. Since each S_α is normal

$$\begin{aligned} \| r(S_\alpha) \| &\leqslant \sup\{|r(\lambda)| : \lambda \in \sigma(S_\alpha)\} \\ &\leqslant \sup\{|r(\lambda)| : \lambda \in K\} \end{aligned}$$

for each rational function r with poles outside K, by Lemma 7.17. As in the proof of Theorem 6.14 it follows that

$$\|r(S)\| \leqslant \sup\{|r(\lambda)| : \lambda \in K\}.$$

Let $E(\cdot)$ be the resolution of the identity for S. Since K is an R-set, we can extend the map $r \to r(S)$ to a continuous algebra homomorphism ψ of $C(K)$ into $L(X)$ of norm 1. The argument of the proof of Theorem 6.10 shows that if τ is a closed subset of K, there is a sequence $\{f_n\}$ of functions in $C(K)$ with $\|f_n\| \leqslant 1$ such that $\{\psi(f_n)\}$ converges to $E(\tau)$ in the weak operator topology. Hence if τ is closed and $x, y \in H$, then $|\langle E(\tau)x, y\rangle| \leqslant \|x\| \, \|y\|$. Since the measure $\langle E(\cdot)x, y\rangle$ is regular with support contained in K,

$$|\langle E(\tau)x, y\rangle| \leqslant \|x\| \, \|y\| \qquad (\tau \in \Sigma_p;\ x, y \in H).$$

Therefore $\|E(\tau)\| \leqslant 1 (\tau \in \Sigma_p)$. Since $E(\tau)$ is idempotent, each projection $E(\tau)$ has norm 1 or 0. By Proposition 7.14, the values of the resolution of the identity $E(\cdot)$ are self-adjoint projections. It follows from Note 7.19 that S is a normal operator.

Example 7.25. On the complex Hilbert space l^2, define operators T and T_n $(n = 1, 2, \ldots)$ by

$$T_n x = \{x_n, x_1, \ldots, x_{n-1}, x_{n+1}, x_{n+2}, \ldots\},$$
$$Tx = \{0, x_1, \ldots, x_{n-1}, x_n, x_{n+1}, \ldots\},$$

where $x = \{x_1, x_2, \ldots\}$. Then $\{T_n\}$ is a sequence of unitary operators converging to T in the strong operator topology. However, T is not a normal operator. Indeed it is not even a spectral operator. Observe that it was shown in Theorem 5.47 that the spectrum and approximate point spectrum of a spectral operator coincide. Also, it was shown in Example 1.32 that $\sigma_a(T) \neq \sigma(T)$. Therefore, the hypothesis $\sigma(S) \subseteq K$ cannot be omitted in Theorems 6.14 and 7.24.

8. Spectral Operators on Hilbert Space

The main result of this chapter is due to Sz.-Nagy. It states that a bounded multiplicative abelian group of operators on a Hilbert space is equivalent to a group of unitary operators. From this theorem we deduce the following results of Wermer. Every scalar-type spectral operator on a Hilbert space is similar to a normal operator. The sum and product of two commuting spectral operators on a Hilbert space are also spectral operators.

THEOREM 8.1. *Let* $\mathbf{G}$ *be a bounded multiplicative abelian group of operators on* H *with identity element* I. *There is a self-adjoint operator* B, *invertible in* $L(H)$, *such that for every* T *in* $\mathbf{G}$ *the operator* BTB^{-1} *is unitary.*

Proof. Let $\mathscr{L}$ be the vector space of complex functions on $H \times H$ and let $\mathscr{B}$ consist of those f in $\mathscr{L}$ which are bilinear, symmetric and have $f(x, x) \geqslant 0$, for all x in H. Let $\mathscr{L}$ be given its weak product topology so that, by definition, the sets

$$\{f \in \mathscr{L} : |f(x, y) - g(x, y)| < \varepsilon\},$$

where $x, y \in H$ and $\varepsilon > 0$, form a subbase for the neighbourhoods of a point g in $\mathscr{L}$. It is easily seen that $\mathscr{B}$ is a closed set in $\mathscr{L}$. Let $\mathscr{R}$ be the smallest closed convex subset of $\mathscr{B}$ containing all functions f having the form $f(x, y) = \langle Tx, Ty \rangle$, where $T \in \mathbf{G}$. If M is an upper bound for the norms of the operators in $\mathbf{G}$, then $|\langle Tx, Ty \rangle| \leqslant M^2 \|x\| \, \|y\|$ and so

$$|f(x, y)| \leqslant M^2 \|x\| \, \|y\| \qquad (f \in \mathscr{R}). \tag{1}$$

Also, since $\|x\|^2 = \langle T^{-1} Tx, T^{-1}Tx \rangle \leqslant M^2 \|Tx\|^2$, we have

$$\frac{\|x\|^2}{M^2} \leqslant f(x, x) \qquad (f \in \mathscr{R}). \tag{2}$$

Since $\mathscr{R}$ is closed in $\mathscr{B}$ and $\mathscr{B}$ is closed in $\mathscr{L}$, it follows from (1) and Tychonoff's theorem that $\mathscr{R}$ is a compact set. For each T in **G** we define the continuous linear map J_T of $\mathscr{L}$ into itself by

$$(J_T f)(x, y) = f(Tx, Ty) \qquad (x, y \in H).$$

Since $J_T J_U = J_{TU}$ and since **G** is an abelian group, the collection $\{J_T\}$ is itself an abelian group of continuous linear mappings of $\mathscr{L}$ into itself. Moreover each J_T maps $\mathscr{R}$ into $\mathscr{R}$. It follows from the Markov-Kakutani fixed point theorem (see for example Theorem V.10.6 of [**5**; p. 456]) that there is an f_0 in $\mathscr{R}$ with $J_T f_0 = f_0$, for every T in **G**.

There is a bounded self-adjoint operator A on H such that $f_0(x, y) = \langle Ax, y\rangle$, for all x, y in H. Therefore

$$\langle Ax, y\rangle = \langle ATx, Ty\rangle = \langle T^*ATx, y\rangle \qquad (T \in \mathbf{G}),$$

so that $A = T^*AT$, for every T in **G**. Since $f_0 \in \mathscr{B}$ we have

$$\langle Ax, x\rangle = f_0(x, x) \geqslant 0 \qquad (x \in H)$$

and so, by Theorem 7.20, the spectrum of A is non-negative. If $E(\cdot)$ is the resolution of the identity for A, then the operator

$$B = \int_{\sigma(A)} \lambda^{1/2}\, E(d\lambda)$$

is self-adjoint and $B^2 = A$. In view of (2) we have

$$\frac{\|x\|^2}{M^2} \leqslant \langle Ax, x\rangle = \langle B^2 x, x\rangle = \|Bx\|^2 \qquad (x \in H).$$

Let $\{y_n\}$ be a Cauchy sequence of elements of BH. Thus $y_n = Bx_n$, for some x_n in H. Now $\{x_n\}$ is a Cauchy sequence in H and so converges to x, say. By the continuity of B, $\{y_n\}$ converges to Bx, and so the range of B is closed. Also B is one-to-one. To show that B is invertible in $L(H)$ it will therefore suffice to show that 0 is the only vector orthogonal to BH. If y is orthogonal to BH then

$$0 = \langle B^2 y, y\rangle = \langle By, By\rangle$$

and so $By = 0$. Since B is one-to-one, $y = 0$. This proves that B is self-adjoint and invertible in $L(H)$. Now, since $T^*AT = A$ we have $B^2T = (T^*)^{-1}B^2$, and so

$$BTB^{-1} = B^{-1}(T^*)^{-1}B = ((BTB^{-1})^*)^{-1},$$

which proves that BTB^{-1} is a unitary operator.

PROPOSITION 8.2. *Let $\mathbf{B}_1, \ldots, \mathbf{B}_k$ be a finite collection of commuting bounded Boolean algebras of projections on H. There is a self-adjoint operator B, invertible in $L(H)$, such that BEB^{-1} is a self-adjoint projection, for every E in the Boolean algebra of projections generated by $\mathbf{B}_1, \ldots, \mathbf{B}_k$.*

Proof. For E in $\mathbf{B}_r$, put $F(E) = I - 2E$. Then

$$F(E)^2 = I - 4E + 4E^2 = I,$$

$$F(E_1)F(E_2) = I - 2E_1 - 2E_2 + 4E_1E_2$$

$$= F[(E_1 \wedge (I - E_2)) \vee (E_2 \wedge (I - E_1))].$$

Thus the collection $\mathbf{G}_r$ of all $F(E)$ with E in $\mathbf{B}_r$ forms a bounded multiplicative abelian group of operators on H. Clearly all the elements of $\mathbf{G}_r$ commute with all the elements of $\mathbf{G}_s$. Thus the set $\mathbf{G}$ of products $g_1 \ldots g_k$ with $g_r \in \mathbf{G}_r$ is a bounded multiplicative abelian group of operators on H. It follows immediately from the preceding theorem that there is a self-adjoint operator B, invertible in $L(H)$, such that $BF(E)B^{-1} = U$ is unitary, for every E in $\mathbf{B}_r$ $(r = 1, \ldots, k)$. Since $F(E)^2 = I$, we have $U^2 = I$, so that $U^* = U^{-1} = U$. Thus U is self-adjoint and therefore the operator

$$BEB^{-1} = \tfrac{1}{2}B(I - F(E))B^{-1} = \tfrac{1}{2}(I - U)$$

is also self-adjoint, for every E in $\mathbf{B}_r$ $(r = 1, \ldots, k)$. From this the desired conclusion follows easily.

THEOREM 8.3. *Let $S_1, \ldots, S_k$ be commuting scalar-type spectral operators on H. There is a self-adjoint operator B, invertible in $L(H)$, such that the operators BS_rB^{-1} $(r = 1, \ldots, k)$ are all normal.*

Proof. Let $E_r(\cdot)$ be the resolution of the identity for S_r. By Proposition 8.2

there is a self-adjoint operator B, invertible in $L(H)$, such that the projections $BE_r(\delta)B^{-1}$ $(r = 1, \ldots, k; \delta \in \Sigma_p)$ are all self-adjoint. Thus the operators

$$BS_rB^{-1} = \int_{\mathbf{C}} \lambda BE_r(d\lambda)\, B^{-1} \qquad (r = 1, \ldots, k)$$

are normal.

THEOREM 8.4. *Let T_1, T_2 be commuting spectral operators on H. Then $T_1 + T_2$ and T_1T_2 are spectral operators.*

Proof. Let S_r, N_r be respectively the scalar and radical parts of T_r $(r = 1, 2)$. It follows from Theorem 6.6. that the operators S_1, S_2, N_1 and N_2 all commute with each other. Thus the sum and product of T_1 and T_2 have the forms

$$T_1 + T_2 = S_1 + S_2 + N,\ T_1T_2 = S_1S_2 + Q,$$

where N, Q are quasinilpotents commuting with S_1 and S_2. To see that $T_1 + T_2$ and T_1T_2 are spectral operators it therefore suffices, in view of Theorem 6.8, to show that $S_1 + S_2$ and S_1S_2 are scalar-type spectral operators. By Theorem 8.3 there is a self-adjoint operator B, invertible in $L(H)$, such that BS_1B^{-1} and BS_2B^{-1} are normal operators. These operators commute and so, by Fuglede's theorem (7.21), their sum and product are also normal operators, from which it is evident that $S_1 + S_2$ and S_1S_2 are scalar-type spectral operators. The proof is complete.

COROLLARY 8.5. *Let S_1, S_2 be commuting scalar-type spectral operators on H. Then $S_1 + S_2$ and S_1S_2 are scalar-type spectral operators.*

9. McCarthy's Example

It was shown in the last chapter that the sum and product of two commuting spectral operators on a Hilbert space are also spectral operators. The following example, due to McCarthy [**1**; p. 303–4] shows that the analogous result for a separable reflexive Banach space may fail. The construction relies heavily on an earlier example of two commuting scalar-type spectral operators whose sum is not spectral, due to Kakutani [**1**].

Example 9.1. Suppose that $N = 2^n$, for some positive integer n. Let S be the finite set of integers $\{1, \ldots, N\}$. Let X_N denote the space of functions h of the form

$$h(s, t) = \sum_{r=1}^{m} f_r(s)\, g_r(t) \qquad (s, t \in S), \tag{1}$$

where $f_r \in C(S)$, $g_r \in C(S)$, for $r = 1, \ldots, m$, under the norm

$$|||h||| = \inf \sum_{r=1}^{m} \|f_r\|\, \|g_r\|,$$

where the infimum is taken over all representations of h in the form (1) and $\|\ \|$ denotes the supremum norm. X_N is a finite-dimensional complex Banach space. The dimension of X_N is N^2. The elements of X_N may be thought of in a natural way as $N \times N$ matrices with entries $x(s, t)$ for s, t in S.

For $r = 1, \ldots, N$ define projections E_r and F_r by

$$(E_r x)(s, t) = \begin{cases} x(s, t) & \text{if } s = r \\ 0 & \text{if } s \neq r \end{cases},$$

$$(F_r x)(s, t) = \begin{cases} x(s, t) & \text{if } t = r \\ 0 & \text{if } t \neq r \end{cases}.$$

Observe that E_r projects onto the rth row and F_r projects onto the rth column of an element of X_N. Let $\mathscr{E}_N$ and $\mathscr{F}_N$ be the commuting Boolean algebras of projections, each of bound 1, generated by the projections $\{E_r: 1 \leqslant r \leqslant N\}$ and $\{F_r: 1 \leqslant r \leqslant N\}$ respectively. There is a projection G in the Boolean algebra of projections generated by $\mathscr{E}_N$ and $\mathscr{F}_N$ such that $2G - I$ takes the element e of X_N, all of whose entries are 1, into the element ρ whose entries are given by the $2^n \times 2^n$ matrix M_n defined inductively in the following way.

$$M_1 = \begin{bmatrix} 1 & 1 \\ 1 & -1 \end{bmatrix}.$$

$$M_2 = \begin{bmatrix} M_1 & M_1 \\ M_1 & -M_1 \end{bmatrix} = \begin{bmatrix} 1 & 1 & 1 & 1 \\ 1 & -1 & 1 & -1 \\ 1 & 1 & -1 & -1 \\ 1 & -1 & -1 & 1 \end{bmatrix}.$$

Assume that we have defined M_{n-1} for some positive integer $n \geqslant 2$. Define

$$M_n = \begin{bmatrix} M_{n-1} & M_{n-1} \\ M_{n-1} & -M_{n-1} \end{bmatrix}.$$

We put a measure μ on S which assigns to each point the measure N^{-1}. Observe that the functions $\{\rho(s, t): 1 \leqslant t \leqslant N\}$ form an orthonormal basis for $L^2(S, \mu)$. Consequently, by Bessel's inequality, for any y in $L^2(S, \mu)$,

$$\begin{aligned} \int_S \left| \int_S \rho(s, t)\, y(s)\, \mu(ds) \right|^2 \mu(dt) &= \frac{1}{N} \sum_{m=1}^{N} \left| \int_S \rho(s, m)\, y(s)\, \mu(ds) \right|^2 \\ &\leqslant N^{-1} \|y\|_2^2, \end{aligned}$$

where $\| \; \|_2$ denotes the norm of $L^2(S, \mu)$. From this and Hölder's inequality we obtain

$$\left| \int_S \int_S \rho(s, t)\, y(s)\, z(t)\, \mu(ds)\, \mu(dt) \right|^2$$

$$\leqslant \left\{ \int_S \left| \int_S \rho(s, t)\, y(s)\, \mu(ds) \right| |z(t)|\, \mu(dt) \right\}^2$$

$$\leqslant \int_S \left| \int_S \rho(s, t)\, y(s)\, \mu(ds) \right|^2 \mu(dt) . \int_S |z(t)|^2\, \mu(dt)$$

$$\leqslant N^{-1} \|y\|_2^2 \, \|z\|_2^2 \leqslant N^{-1} \|y\|^2 \, \|x\|^2 \qquad (y, z \in C(S)).$$

It follows that

$$\left| \int_S \int_S \rho(s, t)\, h(s, t)\, \mu(ds)\, \mu(dt) \right| \leqslant N^{-1/2}\, |||h||| \qquad (h \in X_N). \tag{2}$$

Hence on putting $h = \rho$ in (2) we obtain $|||\rho||| \geqslant N^{1/2}$. Also, since $|||e||| = 1$, it follows that $|||I - 2G||| \geqslant N^{1/2}$. Therefore $|||G||| \geqslant (N^{1/2} - 1)/2$.

For each positive integer n take one copy X_N of the example above, where $N = 2^n$, and form the l^2 direct sum of the X_N, which we call X. Elements of X are sequences $\{x_n\}$ where $x_n \in X_N$ and

$$\|\{x_n\}\| = \left[\sum_{n=1}^{\infty} |||x_n|||_N^2 \right]^{1/2} < \infty,$$

where $||| \; |||_N$ denotes the norm of X_N.

The algebras $\mathscr{E}_N$ and $\mathscr{F}_N$ on X_N have a natural extension to all of X defined by $\mathscr{E}_N X_M = \mathscr{F}_N X_M = 0\,(M \neq N)$. Let $\mathscr{E}$ (respectively $\mathscr{F}$) be the commuting Boolean algebra of projections on X, of bound 1, generated by all the $\mathscr{E}_N$ (respectively $\mathscr{F}_N$). Observe that the Boolean algebra of projections generated by $\mathscr{E}$ and $\mathscr{F}$ contains a projection of norm at least $(N^{1/2} - 1)/2$ on the subspace X_N. It follows that the Boolean algebra generated by $\mathscr{E}$ and $\mathscr{F}$ is not bounded. Since X is an l^2 direct sum of finite-dimensional (hence reflexive) spaces, X must itself be reflexive and also separable.

Now let S_1 and S_2 be operators on X defined by

$$S_1 \sum_{n=1}^{\infty} \oplus\, x_n(s, t) = \sum_{n=1}^{\infty} \oplus\, 2^{-n} 3^{-s} x_n(s, t),$$

$$S_2 \sum_{n=1}^{\infty} \oplus\, x_n(s, t) = \sum_{n=1}^{\infty} \oplus\, 5^{-t} x_n(s, t).$$

Then S_1 and S_2 are commuting scalar-type spectral operators on X. The operator $S_1 S_2$ has eigenvalues of multiplicity one at the distinct points

$$\{2^{-n}3^{-s}5^{-t} : 1 \leqslant s, t \leqslant 2^n < \infty\}.$$

There are no other non-zero points in $\sigma(S_1 S_2)$. The projection $E(2^{-m}3^{-r}5^{-k})$ corresponding to the eigenvalue $2^{-m}3^{-r}5^{-k}$ is given by

$$E(2^{-m}3^{-r}5^{-k})\left(\sum_{n=1}^{\infty} \oplus x_n(s, t)\right) = \sum_{n=1}^{\infty} \oplus \delta_{mn}\delta_{rs}\delta_{kt}x_n(s, t).$$

Thus the Boolean algebra of projections generated by these spectral projections contains both $\mathscr{E}$ and $\mathscr{F}$. Hence it is unbounded. Therefore $S_1 S_2$ cannot be a spectral operator.

The operator $S_1 + S_2$ has eigenvalues of multiplicity one at the distinct points

$$\{2^{-n}3^{-s} + 5^{-t} : 1 \leqslant s, t \leqslant 2^n < \infty\}.$$

There are no other non-zero points in $\sigma(S_1 + S_2)$. The spectral projection $E(m, r, k)$ corresponding to the eigenvalue $2^{-m}3^{-r} + 5^{-k}$ is given by

$$E(m, r, k) \sum_{n=1}^{\infty} \oplus x_n(s, t) = \sum_{n=1}^{\infty} \oplus \delta_{mn}\delta_{rs}\delta_{kt}x_n(s, t).$$

Thus the Boolean algebra of projections generated by these spectral projections contains both $\mathscr{E}$ and $\mathscr{F}$. Hence it is unbounded. Therefore $S_1 + S_2$ is not a spectral operator.

Observe that it follows immediately from the definitions of S_1 and S_2 that

$$\|\exp i\alpha S_1\| = \|\exp i\alpha S_2\| = 1 \qquad (\alpha \in \mathbf{R}).$$

Therefore, by Theorem 4.7, S_1 and S_2 are hermitian operators. Hence, by Theorem 4.4, $S_1 + S_2$ is hermitian. Thus $S_1 + S_2$ is an example of a hermitian operator on a reflexive Banach space which is not a spectral operator.

10. Logarithms of Prespectral Operators

If T is an operator on X, an operator A on X such that $\exp A = T$ is called a *logarithm* of T. Also, if m is a positive integer and B is an operator on X which satisfies $B^m = T$, then B is called an mth *root* of T. In this chapter we study logarithms and mth roots of prespectral operators. As an application of this theory we show that a power-bounded invertible prespectral operator is of scalar-type.

Recall that in the last chapter it was shown that the sum and product of two commuting scalar-type spectral operators need not be spectral operators. However, if one of the operators has finite spectrum then the sum and product of the operators are indeed spectral. In order to prove this some preliminary results are required.

LEMMA 10.1. *Let $T \in L(X)$. Suppose that*

$$X = X_1 \oplus \ldots \oplus X_n$$

where each X_r is a closed subspace of X invariant under T and $T|X_r$ is prespectral of class $\Gamma_r (r = 1, \ldots, n)$. Then T is prespectral of class $\Gamma = \Gamma_1 \oplus \ldots \oplus \Gamma_n$. Assume that Γ_r is a subspace of X_r^* $(r = 1, \ldots, n)$.

Proof. Let $E_r(\cdot)$ be the resolution of the identity of class Γ_r for $T|X_r$. If $x = \sum_{r=1}^{n} x_r$ with $x_r \in X_r$, we define for every τ in Σ_p

$$E(\tau)x = \sum_{r=1}^{n} E_r(\tau)x_r.$$

Clearly $E(\cdot)$ is a spectral measure of class (Σ_p, Γ). Also $TE(\delta) = E(\delta)T$ for all

δ in Σ_p. Let $\tau \in \Sigma_p$ and $\lambda \in \mathbf{C} \backslash \bar{\tau}$. Then $\lambda I - T$ maps each of the spaces $E_r(\tau) X_r$ in a one-to-one manner onto all of itself. Hence $\lambda I - T$ maps $E(\tau) X$ in a one-to-one manner onto all of itself. This shows that $\sigma(T | E(\tau) X) \subseteq \bar{\tau}$ and completes the proof that T is a prespectral operator with resolution of the identity $E(\cdot)$ of class Γ.

LEMMA 10.2. *Let S be a scalar-type operator on X with resolution of the identity $E(\cdot)$ of class Γ. Let $P \in L(X)$. Suppose that $P^2 = P$ and $PE(\tau) = E(\tau)P$ $(\tau \in \Sigma_p)$. Then $S | PX$ is a scalar-type operator of class Γ.*

Proof. Define $G(\tau) = E(\tau) | PX$ $(\tau \in \Sigma_p)$. Observe that Γ is a total set for the Banach space PX. Hence $G(\cdot)$ is a spectral measure of class (Σ_p, Γ) with values in $L(PX)$. Also, by Lemma 5.34,

$$\sigma(S | G(\tau) PX) \subseteq \sigma(S | E(\tau) X) \subseteq \bar{\tau} \qquad (\tau \in \Sigma_p).$$

Clearly

$$S | PX = \int_{\mathbf{C}} \lambda G(d\lambda)$$

and so $S | PX$ is a scalar-type operator of class Γ.

THEOREM 10.3. *Let S_0 be a scalar-type spectral operator on X with finite spectrum. Assume that $S_0 \neq 0$.*

(i) *Let T be a prespectral operator on X with resolution of the identity $G(\cdot)$ of class Γ. Suppose that $G(\tau) S_0 = S_0 G(\tau)$ $(\tau \in \Sigma_p)$. Then $T + S_0$ is prespectral of class Γ.*

(ii) *Let S be a scalar-type operator on X with resolution of the identity $G(\cdot)$ of class Γ. Suppose that $G(\tau) S_0 = S_0 G(\tau)$ $(\tau \in \Sigma_p)$. Then $S + S_0$ is a scalar-type operator of class Γ.*

Proof. Let $E(\cdot)$ be the resolution of the identity of S_0 and let $\{\lambda_r : r = 1, \ldots, n\}$ be the non-zero points of $\sigma(S_0)$. Then

$$S_0 = \sum_{r=1}^{n} \lambda_r E(\lambda_r),$$

where

$$I = E(0) + \sum_{r=1}^{n} E(\lambda_r)$$

and

$$X = E(0)X \oplus E(\lambda_1)X \oplus \ldots \oplus E(\lambda_n)X. \tag{1}$$

Let Y be one of the $(n+1)$ subspaces on the right-hand side of (1) and let P be the projection of X onto Y. Then, by the commutativity theorem for spectral operators (6.6)

$$PG(\tau) = G(\tau)P \qquad (\tau \in \Sigma_p).$$

Let $T = S + N$ be the Jordan decomposition of T. By Lemma 10.2, $S|Y$ is a scalar-type operator with resolution of the identity $G(\cdot)|Y$ of class Γ. Therefore, by Proposition 5.8, the operator $(S + S_0)|Y$, which is $S|Y$ plus a scalar multiple of the identity on Y, is a scalar-type operator with resolution of the identity $H(\cdot)$ of class Γ and

$$\{H(\tau): \tau \in \Sigma_p\} \subseteq \{G(\tau)|Y: \tau \in \Sigma_p\}.$$

By Lemma 10.1, $S + S_0$ is a scalar-type operator of class Γ. Moreover, since $NS_0 = S_0N$ we have $NP = PN$ and $N|Y$ is quasinilpotent. Therefore

$$(N|Y)(G(\tau)|Y) = (G(\tau)|Y)(N|Y) \qquad (\tau \in \Sigma_p),$$
$$(N|Y)\,H(\tau) = H(\tau)(N|Y) \qquad (\tau \in \Sigma_p),$$

and so, by Theorem 5.15, $(S + S_0 + N)|Y$ is prespectral of class Γ. Another application of Lemma 10.1 suffices to complete the proof.

A similar argument establishes the following result.

THEOREM 10.4. *Let S_0 be a scalar-type spectral operator on X with finite spectrum. Assume that $S_0 \neq 0$.*

(i) *Let T be a prespectral operator on X with resolution of the identity $G(\cdot)$ of class Γ. Suppose that $G(\tau)S_0 = S_0G(\tau)(\tau \in \Sigma_p)$. Then S_0T is prespectral of class Γ.*

(ii) *Let S be a scalar-type operator on X with resolution of the identity $G(\cdot)$ of class Γ. Suppose that $G(\tau)S_0 = S_0G(\tau)(\tau \in \Sigma_p)$. Then S_0S is a scalar-type operator of class Γ.*

By virtue of the commutativity theorem for spectral operators (6.6), in the case $\Gamma = X^*$ Theorems 10.3 and 10.4 reduce to the following result of Foguel.

THEOREM 10.5. *Let S, S_0, in $L(X)$, be scalar-type spectral operators such that $SS_0 = S_0S$ and S_0 has finite spectrum. Then $S + S_0$ and SS_0 are scalar-type spectral operators.*

Prior to beginning the study of logarithms and mth roots of prespectral operators we require the following corollary to the minimal equation theorem.

PROPOSITION 10.6. *Let $T \in L(X)$. Suppose there is f in $\mathscr{F}(T)$ such that $f(T) = 0$ and $f'(\lambda) \neq 0$ for all λ in $\sigma(T)$. Then $\sigma(T)$ is a finite set and T is a scalar-type spectral operator of the form $\sum_{r=1}^{n} \lambda_r F(\lambda_r)$, where $F(\lambda_r)$ is the spectral projection corresponding to the open-and-closed subset $\{\lambda_r\}$ of $\sigma(T)$.*

Proof. First, in the minimal equation theorem (1.41) take f as given and $g = 0$. This shows that $\sigma(T)$ consists of a finite number of points, each of which is a simple pole of the resolvent of T. Let the non-zero points of $\sigma(T)$ be $\lambda_1, \ldots, \lambda_n$. Next, in the minimal equation theorem take

$$f(\lambda) = \lambda \text{ and } g(\lambda) = \sum_{r=1}^{n} \lambda_r \chi(\Omega_r, \lambda) \; (\lambda \in \mathbf{C})$$

where Ω_r is an open disc such that $\lambda_r \in \Omega_r$ but $\overline{\Omega}_r$ contains no other point of $\sigma(T)$. This completes the proof.

Our next main theorem shows that if a prespectral operator of class Γ is invertible in the Banach algebra $L(X)$ then it does possess a logarithm which is also prespectral of class Γ.

THEOREM 10.7. (i) *Let A, in $L(X)$, be an invertible prespectral operator of class Γ. Then there is T_0, in $L(X)$, such that $\exp T_0 = A$ and T_0 is prespectral of class Γ.*

(ii) *Let A, in $L(X)$, be an invertible scalar-type operator of class Γ. Then there is S_0, in $L(X)$, such that $\exp S_0 = A$ and S_0 is a scalar-type operator of class Γ.*

Proof. In (i), let $E(\cdot)$ be the resolution of the identity of class Γ for A. Define

$$S = \int_{\sigma(A)} \lambda E(d\lambda), \quad N = A - S, \; S_0 = \int_{\sigma(A)} \log \lambda E(d\lambda),$$

where log denotes the principal value of the logarithm defined on $\mathbf{C}\backslash\{0\}$. Observe that, since $0 \notin \sigma(A)$, log is a bounded Borel measurable function on $\sigma(A)$ and so the integral converges. Note that $\exp S_0 = S$. Define

$$Q = N \exp(-S_0).$$

Since N commutes with $E(\cdot)$ then $NS_0 = S_0 N$ and so Q is quasinilpotent. Let C denote the circle, centre the origin, radius $\frac{1}{2}$, described once counter-clockwise. The operator

$$N_0 = \frac{1}{2\pi i} \int_C (\lambda I - Q)^{-1} \log(1 + \lambda)\, d\lambda$$

is well-defined. Also $\sigma(N_0) = \{0\}$, and so N_0 is quasinilpotent. Further,

$$\exp N_0 = I + Q.$$

Since $N_0 S_0 = S_0 N_0$ and $NS_0 = S_0 N$, we obtain

$$\begin{aligned}
\exp(S_0 + N_0) &= \exp S_0 \exp N_0 = (\exp S_0)(I + Q) \\
&= (\exp S_0) + (\exp S_0)\, N \exp(-S_0) \\
&= S + N = A.
\end{aligned}$$

Define $T_0 = S_0 + N_0$. To complete the proof of (i) we show that T_0 is prespectral of class Γ. Now, by Proposition 5.8, S_0 is a scalar-type operator with resolution of the identity $F(\cdot)$ of class Γ given by

$$F(\tau) = E(g^{-1}(\tau)) \qquad (\tau \in \Sigma_p)$$

where g denotes the function log. Also we have successively

$$
\begin{aligned}
NE(\tau) &= E(\tau)N && (\tau \in \Sigma_p)\\
S_0E(\tau) &= E(\tau)S_0 && (\tau \in \Sigma_p)\\
QE(\tau) &= E(\tau)Q && (\tau \in \Sigma_p)\\
(\lambda I - Q)^{-1}E(\tau) &= E(\tau)(\lambda I - Q)^{-1} && (\tau \in \Sigma_p, \lambda \in \rho(Q))\\
N_0E(\tau) &= E(\tau)N_0 && (\tau \in \Sigma_p)\\
N_0F(\tau) &= F(\tau)N_0 && (\tau \in \Sigma_p)
\end{aligned}
$$

and so Theorem 5.15 shows that T_0 is prespectral of class Γ. If A is a scalar-type operator of class Γ then, in the notation above, $A = S$, $\exp S_0 = S$ and S_0 is a scalar-type operator of class Γ. This proves (ii).

THEOREM 10.8. *Let $m \geqslant 2$ be a positive integer.*

(i) *Let A, in $L(X)$, be an invertible prespectral operator of class Γ. Then there is A_0, in $L(X)$, such that $A_0^m = A$ and A_0 is prespectral of class Γ.*

(ii) *Let A, in $L(X)$, be an invertible scalar-type operator of class Γ. Then there is A_0, in $L(X)$, such that $A_0^m = A$ and A_0 is a scalar-type operator of class Γ.*

Proof. In order to prove (i) let T_0 be the operator constructed in Theorem 10.7(i). Define $A_0 = \exp(m^{-1}T_0)$. By Theorem 5.16, A_0 is prespectral of class Γ. Clearly $A_0^m = A$. If A is of scalar type, let S_0 be the operator constructed in Theorem 10.7(ii). Define $A_0 = \exp(m^{-1}S_0)$. By Proposition 5.8, A_0 is a scalar-type operator of class Γ. Clearly $A_0^m = A$ and so the proof is complete.

We now show that every logarithm and mth root of an invertible spectral operator is also a spectral operator.

THEOREM 10.9. (i) *Let A, in $L(X)$, be an invertible spectral operator. Then there is T_0, in $L(X)$, such that* $\exp T_0 = A$ *and T_0 is a spectral operator. Moreover, if T, in $L(X)$, is such that* $\exp T = A$ *then T is a spectral operator.*

(ii) *Let A, in $L(X)$, be an invertible scalar-type spectral operator. Then there is S_0, in $L(X)$, such that* $\exp S_0 = A$ *and S_0 is a scalar-type spectral operator. Moreover, if S, in $L(X)$, is such that* $\exp S = A$ *then S is a scalar-type spectral operator.*

Proof. In Theorem 10.7 take $\Gamma = X^*$. Observe that by the commutativity theorem for spectral operators (6.6), since $TA = AT$ then $TE(\tau) = E(\tau)T$ $(\tau \in \Sigma_p)$, where $E(\cdot)$ is the resolution of the identity for A. We have successively $TS_0 = S_0T$, $TN = NT$, $TQ = QT$,

$$T(\lambda I - Q)^{-1} = (\lambda I - Q)^{-1}T \qquad (\lambda \in \rho(Q)),$$

$TN_0 = N_0T$ and so $TT_0 = T_0T$. Hence, since $\exp T = A = \exp T_0$, we obtain

$$\exp(T - T_0) - I = 0.$$

By Proposition 10.6, $T - T_0$ is a scalar-type spectral operator with finite spectrum. Since $T - T_0$ commutes with T_0 it commutes with S_0 and N_0. By Theorem 10.5, $S_0 + (T - T_0)$ is scalar-type spectral and, since N_0 commutes with this operator, T is a spectral operator by Theorem 6.8. In case (ii) we have successively $N = 0$, $Q = 0$, $N_0 = 0$ and so S is a scalar-type spectral operator. This completes the proof.

THEOREM 10.10. *Let $m \geqslant 2$ be a positive integer.*

(i) *Let A, in $L(X)$, be an invertible spectral operator. Then there is A_0, in $L(X)$, such that $A_0^m = A$ and A_0 is a spectral operator. Moreover if B, in $L(X)$, is such that $B^m = A$ then B is a spectral operator.*

(ii) *Let A, in $L(X)$, be an invertible scalar-type spectral operator. Then there is A_0, in $L(X)$, such that $A_0^m = A$ and A_0 is a scalar-type spectral operator. Moreover if B, in $L(X)$, is such that $B^m = A$ then B is a scalar-type spectral operator.*

Proof. In Theorem 10.8 take $\Gamma = X^*$. Observe that by the commutativity theorem for spectral operators (6.6), since $BA = AB$ then $BE(\tau) = E(\tau)B$ $(\tau \in \Sigma_p)$ where $E(\cdot)$ is the resolution of the identity for A. In the notation of Theorem 10.7 we have successively

$$BS_0 = S_0B, BN = NB, BQ = QB,$$

$$B(\lambda I - Q)^{-1} = (\lambda I - Q)^{-1}B \qquad (\lambda \in \rho(Q))$$

$$BN_0 = N_0B, BT_0 = T_0B$$

and so $B\exp(m^{-1}T_0) = \exp(m^{-1}T_0)B$. Thus $BA_0 = A_0B$. Now $0 \notin \sigma(A_0)$, for otherwise by the spectral mapping theorem $0 \in \sigma(A)$, contradicting the hypothesis that A is invertible. Hence A_0 is invertible and $BA_0^{-1} = A_0^{-1}B$. Also $(BA_0^{-1})^m = I$. Again by the spectral mapping theorem, since I is invertible $0 \notin \sigma(BA_0^{-1})$. It follows from Proposition 10.6 that there is a finite family of projections $F(w_1), \ldots, F(w_n)$ such that

$$F(w_r)F(w_s) = 0 \qquad (r \neq s)$$

and

$$BA_0^{-1} = A_0^{-1}B = \sum_{i=1}^{n} w_i F(w_i);$$

i.e.

$$B = A_0 \sum_{i=1}^{n} w_i F(w_i).$$

The operators on the right-hand side commute. Hence $\sum_{i=1}^{n} w_i F(w_i)$ commutes with the scalar and radical parts of the spectral operator A_0. It follows from Theorem 10.5 that B is the sum of a scalar-type spectral operator and a commuting quasinilpotent. Hence, by Theorem 6.8, B is a spectral operator. In case (ii) we have successively $N = 0$, $Q = 0$, $N_0 = 0$ and so $B = A_0\left(\sum_{i=1}^{n} w_i F(w_i)\right)$ is a scalar-way spectral operator. This completes the proof.

Not every logarithm of a scalar-type operator of class Γ is a scalar-type operator of class Γ, as the following example shows.

Example 10.11. On the subspace of l^∞ consisting of convergent sequences, the map which assigns to each such sequence its limit is a linear functional of norm 1. Throughout, L denotes a fixed linear functional on l^∞ with $\|L\| = 1$, such that for each convergent sequence $\{\xi_n\}$

$$L(\{\xi_n\}) = \lim_n \{\xi_n\}.$$

Define operators S and A on l^∞ by

$$S\{\xi_n\} = \{\eta_n\},$$

where $\eta_1 = \xi_1$ and $\eta_n = ((n-1)/n)\xi_n$ $(n = 2, 3, 4, \ldots)$,

$$A\{\xi_n\} = \{L(\{\xi_n\}), 0, 0, \ldots\}.$$

Clearly $\|A\| = 1$ and $A^2 = 0$. Also

$$S\{\xi_n\} = \{\xi_n\} - \{\gamma_n\}$$

where $\gamma_1 = 0$ and $\gamma_n = \xi_n/n$ $(n = 2, 3, 4, \ldots)$. Since $L(\{\gamma_n\}) = 0$ then $AS\{\xi_n\} = A\{\xi_n\}$. It is easy to see that $SA\{\xi_n\} = A\{\xi_n\}$ and hence

$$AS = SA.$$

$\sigma(S)$ is the totally disconnected set consisting of 1 and the numbers $(n-1)/n$ for $n = 2, 3, 4, \ldots$. S is a scalar-type operator with resolution of the identity $E(\cdot)$ of class l^1 satisfying

$$E(\{1\})\{\xi_n\} = \{\xi_1, 0, 0, \ldots\}$$

$$E(\{(n-1)/n\})\{\xi_n\} = \{\delta_{nk}\xi_k\} \quad \text{for} \quad n = 2, 3, 4, \ldots.$$

Observe that

$$AE(\{1\})\{1, 1, 1, \ldots\} = \{0, 0, 0, \ldots\}$$

$$E(\{1\})A\{1, 1, 1, \ldots\} = \{1, 0, 0, \ldots\}$$

$$AE(\{1\}) \neq E(\{1\})A. \tag{2}$$

Define

$$F(\tau) = E(\tau) + AE(\tau) - E(\tau)A \qquad (\tau \in \Sigma_p).$$

Using the relations $A^2 = 0$ and $AE(\tau)A = 0 (\tau \in \Sigma_p)$ it is easily verified that $F(\cdot)$ is a homomorphism from Σ_p into a Boolean algebra of projections on l^∞ with $F(\sigma(S)) = I$. Clearly $\|F(\tau)\| \leqslant 3\,(\tau \in \Sigma_p)$. For each positive integer n let e_n in l^1 be given by $e_n = \{\delta_{nk}\}_{k=1}^{\infty}$, and let e_n^* be the corresponding linear functional on l^∞. Let Γ be the total linear manifold in $(l^\infty)^*$ generated by

$e_1^* - L$ and $\{e_n^* : n = 2, 3, 4, \ldots\}$. Since for each τ in Σ_p and x in l^∞

$$\langle F(\tau)x, e_n^* \rangle = \langle E(\tau)x, e_n^* \rangle \qquad (n = 2, 3, 4, \ldots)$$

$$\langle F(\tau)x, e_1^* - L \rangle = \chi(\tau; 1)\langle x, e_1^* - L \rangle,$$

it follows that $F(\cdot)$ is Γ-countably additive. Since $E(\cdot)$ and A commute with S, elementary algebra shows that

$$F(\tau)S = SF(\tau) \qquad (\tau \in \Sigma_p).$$

In order to prove that $F(\cdot)$ is a resolution of the identity for S it remains only to show that $\sigma(S|F(\tau)l^\infty) \subseteq \bar{\tau}(\tau \in \Sigma_p)$. By virtue of Lemma 5.34 it suffices to prove this inclusion when τ is a closed subset of $\sigma(S)$. Since $\sigma(S)$ is totally disconnected, it is enough to prove the inclusion for an open-and-closed subset τ of $\sigma(S)$. It is easy to see from the definition of $F(\cdot)$ that $E(\cdot)$ and $F(\cdot)$ agree on finite subsets of $\sigma(S)\backslash\{1\}$. Since every open-and-closed subset of $\sigma(S)$ is such a set or the complement in $\sigma(S)$ of such a set, $F(\cdot)$ and $E(\cdot)$ agree on open-and-closed subsets of $\sigma(S)$. Therefore

$$\sigma(S|F(\tau)l^\infty) = \sigma(S|E(\tau)l^\infty) \subseteq \tau$$

for τ open-and-closed in $\sigma(S)$. Hence S is a scalar-type operator with resolution of the identity $F(\cdot)$ of class Γ. By (2), $AE(\{1\}) \neq E(\{1\})A$. Hence $F(\{1\}) \neq E(\{1\})$. Since, by Theorem 5.33, these projections have the same range,

$$F(\{1\})E(\{1\}) \neq E(\{1\})F(\{1\}). \tag{3}$$

Now consider the operator $S + F(\{1\})$. Observe that

$$S + F(\{1\}) = \int_{\sigma(S)} g(\lambda)F(d\lambda)$$

where $g(\lambda) = \lambda + \chi(\{1\}; \lambda)\,(\lambda \in \mathbf{C})$ and so, by Proposition 5.8, $S + F(\{1\})$ is a scalar-type operator with resolution of the identity $G(\cdot)$ of class Γ given by

$$G(\tau) = F(g^{-1}(\tau)) \qquad (\tau \in \Sigma_p).$$

Let δ be the set consisting of 1, 2 and the numbers $(n-1)/n$ for $n = 2, 3, 4, \ldots$. Note that $G(\mathbf{C}\backslash\delta) = 0$. Also $G(\{2\}) = F(\{1\})$ and $G(\{(n-1)/n\}) = F(\{(n-1)/n\})$ $(n = 2, 3, 4, \ldots)$. Each of these projections is non-zero. By Note 5.7, $\sigma(S + F(\{1\}))$ is the complement of the largest open set ρ for which $G(\rho) = 0$. Hence $\sigma(S + F(\{1\})) = \delta$. Suppose now that $S + F(\{1\})$ is a prespectral operator with resolution of the identity $H(\cdot)$ of class l^1. Since $\{2\}$ and $\{(n-1)/n\}$ $(n = 2, 3, 4, \ldots)$ are open-and-closed subsets of $\sigma(S + F(\{1\}))$, we obtain from Proposition 5.27

$$G(\{2\}) = H(\{2\}), G(\{(n-1)/n\}) = H(\{(n-1)/n\}) \qquad (n = 2, 3, 4, \ldots).$$

Hence

$$H(\{2\}) = F(\{1\}), H(\{(n-1)/n\}) = E(\{(n-1)/n\}) \qquad (n = 2, 3, 4, \ldots).$$

Now let $f \in l^1, x \in l^\infty$ and let τ be the relatively open subset $\{\frac{1}{2}, \frac{2}{3}, \ldots\}$ of $\sigma(S + F(\{1\}))$. Then

$$\langle f, H(\tau)x \rangle = \sum_{n=2}^{\infty} \langle f, H(\{(n-1)/n\})x \rangle$$

$$= \sum_{n=2}^{\infty} \langle f, E(\{(n-1)/n\})x \rangle = \langle f, E(\tau)x \rangle,$$

since $H(\cdot)$ and $E(\cdot)$ are resolutions of the identity of class l^1. Since l^1 is a total subspace of $(l^\infty)^*$

$$H(\tau) = E(\tau) = I - E(\{1\}).$$

Now, since $H(\cdot)$ is a resolution of the identity for $S + F(\{1\})$, $H(\tau)H(\{2\}) = H(\{2\})H(\tau)$ and so

$$F(\{1\})(I - E(\{1\})) = (I - E(\{1\}))F(\{1\})$$

$$F(\{1\})E(\{1\}) = E(\{1\})F(\{1\}). \tag{4}$$

(3) and (4) give a contradiction. Hence $S + F(\{1\})$ is not a scalar-type operator

of class l^1. Since $F(\{1\})$ and S commute, we obtain

$$\begin{aligned}\exp[2\pi i(S + F(\{1\}))] &= \exp(2\pi iS)\exp[2\pi iF(\{1\})] \\ &= \exp(2\pi iS)\left[I + \frac{2\pi iF(\{1\})}{1!} + \frac{(2\pi i)^2[F(\{1\})]^2}{2!} + \dots\right] \\ &= \exp(2\pi iS)[I + (F(\{1\})(\exp 2\pi i - 1)]\end{aligned}$$

since $F(\{1\})$ is a projection. Hence

$$\exp[2\pi i(S + F(\{1\}))] = \exp(2\pi iS).$$

Now, by Proposition 5.8, $\exp(2\pi iS)$ is a scalar-type operator of class l^1. However, $2\pi i(S + F(\{1\}))$ is a logarithm for this operator which is not a scalar-type operator of class l^1.

Let $m \geqslant 2$ be a positive integer. We can deduce that there is an invertible scalar-type operator of class l^1 such that there exists an mth root of the operator which is not a scalar-type operator of class l^1.

Define $A = \exp(2\pi iS)$ and $A_0 = \exp[2\pi im^{-1}(S + F(\{1\}))]$. Clearly $A_0^m = A$. We claim that A_0 is not a scalar-type operator of class l^1. By the spectral mapping theorem, $\sigma(A_0)$ is a countable subset of the unit circle. Hence, there is g in $C(\sigma(A_0))$ such that

$$g(\exp(2\pi im^{-1}\lambda)) = 2\pi im^{-1}\lambda \qquad (\lambda \in \sigma(S + F(\{1\}))).$$

If A_0 were a scalar-type operator of class l^1, then $g(A_0)$ would be a scalar-type operator of class l^1, by Proposition 5.8. Hence $2\pi im^{-1}(S + F(\{1\}))$ would be a scalar-type operator of class l^1. This gives a contradiction.

We observe that one cannot replace "scalar-type spectral operators" by "scalar-type operators of class Γ" in the statement of Theorem 10.5. We have given an example of a scalar-type operator on l^∞ of class l^1 and a projection F such that $SF = FS$, F has one-dimensional range and $S + F$ is not a scalar-type operator of class l^1. However $S + F$ is a scalar-type operator of some other class. We have not been able to construct an example of a scalar-type operator S and a projection F with $SF = FS$ such that $S + F$ is not a scalar-type operator of any class. Such an example would also yield an example of a scalar-type operator, one of whose logarithms was not prespectral of any class.

The crucial step in the proof of Theorem 10.9 is the construction of a logarithm of an invertible spectral operator A which commutes with every operator commuting with A. Of course, not every logarithm of an invertible scalar-type spectral operator possesses this commutativity property, as the following example shows.

Example 10.12. Let $X = \mathbf{C}^2$. Let operators I, P, N respectively be defined on X by the matrices

$$\begin{bmatrix} 1 & 0 \\ 0 & 1 \end{bmatrix}, \begin{bmatrix} 1 & 0 \\ 0 & 0 \end{bmatrix} \begin{bmatrix} 0 & 1 \\ 0 & 0 \end{bmatrix}$$

relative to the basis $(1, 0), (0, 1)$. Then $NI = IN$ but $NP \neq PN$. However, since P is a projection, $\exp(2\pi i P) = I$.

The crucial step in the proof of Theorem 10.10 is the construction of an mth root of an invertible spectral operator A which commutes with every operator commuting with A. Of course, not every mth root (indeed not every square root) of an invertible scalar-type spectral operator possesses this commutativity property, as the following example shows.

Example 10.13. Let $X = \mathbf{C}^2$. Let operators I, A, N respectively be defined on X by the matrices

$$\begin{bmatrix} 1 & 0 \\ 0 & 1 \end{bmatrix}, \begin{bmatrix} -1 & 0 \\ 0 & 1 \end{bmatrix}, \begin{bmatrix} 0 & 1 \\ 0 & 0 \end{bmatrix}$$

relative to the basis $(1, 0), (0, 1)$. Then $NI = IN$ but $NA \neq AN$. However, $A^2 = I$.

We now prove theorems giving sufficient conditions on the spectrum to guarantee the existence of a logarithm and mth root of a prespectral operator possessing the commutativity property referred to above.

THEOREM 10.14. *Let A, in $L(X)$, be a prespectral operator of class Γ. Suppose that the point 0 lies in the unbounded component of $\rho(A)$. Then there is T, in $L(X)$, with the following properties.*

(i) *T is prespectral of class Γ and* $\exp T = A$.

(ii) *If B, in $L(X)$, commutes with A then B commutes with T.*

Proof. For each λ in $\sigma(A)$, there is an open disc $\Omega(\lambda)$ with $\lambda \in \Omega(\lambda)$ but $0 \notin \Omega(\lambda)$.

A finite family $\Omega(\lambda_1), \ldots, \Omega(\lambda_n)$ of these discs cover $\sigma(A)$. Let Ω be the open set formed by taking the union of $\bigcup_{r=1}^{n} \Omega(\lambda_r)$ and the bounded components of $\rho(A)$. Then Ω is simply connected and so by Theorem 13.18 (*g*) of [**14**; p. 262–3] there is f analytic in Ω such that

$$\exp f(\lambda) = \lambda \qquad (\lambda \in \Omega).$$

Since $\sigma(A) \subseteq \Omega$, $f \in C(\sigma(A))$. Let $E(\cdot)$ be the resolution of the identity of class Γ for T. Define

$$S = \int_{\sigma(A)} \lambda E(d\lambda),\ N = A - S,\ S_0 = \int_{\sigma(A)} f(\lambda)E(d\lambda),$$

$$Q = N \exp(-S_0)$$

$$N_0 = \frac{1}{2\pi i} \int_C (\lambda I - Q)^{-1} \log(1 + \lambda) d\lambda,$$

where C denotes the circle, centre the origin, radius $\frac{1}{2}$, described once counter-clockwise. Let $T = S_0 + N_0$. Then, as in the proof of Theorem 10.7, T is prespectral of class Γ and $\exp T = A$. Suppose now that $BA = AB$. By Theorem 5.12 we obtain $BS = SB$, $BS_0 = S_0 B$, and so $BN = NB$. Hence $BQ = QB$,

$$B(\lambda I - Q)^{-1} = (\lambda I - Q)^{-1} B \qquad (\lambda \in \rho(Q)),$$

and so $BN_0 = N_0 B$. Therefore $BT = TB$ and the proof is complete.

Clearly, the hypothesis of Theorem 10.14 could be weakened. Only the existence of a continuous logarithm on $\sigma(A)$ is required.

THEOREM 10.15. *Let $m \geqslant 2$ be a positive integer. Let A, in $L(X)$, be a prespectral operator of class Γ. Suppose that the point 0 lies in the unbounded component of $\rho(A)$. Then there is T, in $L(X)$, with the following properties.*

(i) *T is prespectral of class Γ and $T^m = A$.*

(ii) *If B, in $L(X)$, commutes with A then B commutes with T.*

Proof. By Theorem 10.14 there is T_0, in $L(X)$, with the following properties.

(a) T_0 is prespectral of class Γ and $\exp T_0 = A$.

(b) If B, in $L(X)$, commutes with A then B commutes with T_0.
Define $T = \exp(m^{-1}T_0)$. Then T has the required properties.

Clearly the hypothesis of Theorem 10.15 could be weakened. Only the existence of a continuous mth root on $\sigma(A)$ is required.

We now show that there is a scalar-type operator, none of whose logarithms possess the commutativity property of Theorem 10.14.

Example 10.16. Let $X = l^1(C)$, the space of complex functions g, defined on $C = \{z \in \mathbf{C}: |z| = 1\}$, vanishing off a countable set $G(g)$ and such that $\sum_{z \in \mathbf{C}} |g(z)| < \infty$. With the vector space operations

$$(\alpha f + \beta g)(z) = \alpha f(z) + \beta g(z) \qquad (f, g \in X; \alpha, \beta \in \mathbf{C}; z \in C)$$

and norm

$$\|g\| = \sum_{z \in \mathbf{C}} |g(z)| \qquad (g \in X),$$

X is a complex Banach space. We may and shall identify X^* with the space of bounded complex functions on C with the vector space operations defined as above and endowed with the supremum norm. Define S, in $L(X^*)$, by

$$(Sf)(z) = zf(z) \qquad (z \in C).$$

S is an invertible isometric operator. Hence, by Proposition 1.31, $\sigma(S) \subseteq C$. Also, for every λ in C, let χ_λ denote the characteristic function of the set $\{\lambda\}$. Then

$$(S\chi_\lambda)(z) = \lambda\chi_\lambda(z) \qquad (z \in C)$$

and so λ is an eigenvalue of S. Hence $\sigma(S) = C$. Observe that S is a scalar-type operator on X^* with resolution of the identity $E(\cdot)$ of class X given by

$$(E(\tau)f)(z) = \chi(\tau; z)f(z) \qquad (f \in X^*, \tau \in \Sigma_p, z \in C).$$

Note also that it follows easily from Lemma 5.30 that $Sx = \lambda x$ if and only if $E(\{\lambda\})x = x$. Thus the only eigenvectors of S are of the form $\alpha\chi_\lambda$ where $\alpha \in \mathbf{C}$ and $\lambda \in C$.

Now let T, in $L(X^*)$, satisfy $\exp T = S$. Clearly $TS = ST$ and so

$$ST\chi_\lambda = TS\chi_\lambda = T\lambda\chi_\lambda = \lambda T\chi_\lambda.$$

Therefore $T\chi_\lambda$ is an eigenvector of S and so there is a complex function h on C such that

$$T\chi_\lambda = h(\lambda)\chi_\lambda \qquad (\lambda \in C),$$

$$\exp h(\lambda) = \lambda \qquad (\lambda \in C).$$

h is bounded because T is bounded. Also h is discontinuous, so there is a point a in C and a sequence $\{\lambda_n\}$ of distinct points of C with $\lambda_n \neq a$, for all n, such that

$$\lambda_n \to a \text{ but } h(\lambda_n) \nrightarrow h(a) \text{ as } n \to \infty.$$

Since $|\lambda_n| = 1$, $\lambda_n h(\lambda_n) - \lambda_n h(a) \nrightarrow 0$ as $n \to \infty$. Define a function η on C by

$$\eta(z) = 0, \text{ if } z \neq \lambda_n \text{ for all } n; \eta(\lambda_n) = \lambda_n.$$

We now divide the proof into two cases.

First, suppose that $(T\eta)(\lambda) = h(\lambda)\eta(\lambda)$ $(\lambda \in C)$. Let M be the subspace of X^* consisting of functions continuous at a. If ξ is defined by

$$\xi(z) = 0, \text{ if } z \neq \lambda_n \text{ for all } n,$$

$$\xi(\lambda_n) = \lambda_n h(\lambda_n) - \lambda_n h(a),$$

then $\xi \notin M$. On the subspace of X^* generated by M and the vector ξ we can define a bounded linear functional L so that $L(\xi) = 1$ and $L(f) = 0$ $(f \in M)$. By the Hahn-Banach theorem we can extend L to a bounded linear functional on X^*. Denote the extended functional by L also. Define A, in $L(X^*)$, by

$$(Af)(z) = L(f)\chi(\{a\}; z) \qquad (z \in C).$$

Clearly

$$SAf = aAf \qquad (f \in X^*).$$

Also from the identity $zf(z) = (z - a)f(z) + af(z)$ $(z \in C)$ it follows that

$$ASf = aAf \qquad (f \in X^*).$$

Hence $AS = SA$. However, $(AT\eta)(a) - (TA\eta)(a) = L(\xi) \neq 0$. Therefore $AT \neq TA$.

Next, suppose that $(T\eta)(b) \neq h(b)\eta(b)$ for some b in C. Let M be the subspace of X^* consisting of functions vanishing at b. Observe that $T\eta - h(b)\eta \notin M$. On the subspace of X^* generated by M and the vector $T\eta - h(b)\eta$ we can define a bounded linear functional L so that $L(T\eta - h(b)\eta) \neq 0$ and $L(f) = 0$ $(f \in M)$. By the Hahn-Banach theorem we can extend L to a bounded linear functional on X^*. Denote the extended functional by L also. Define B, in $L(X^*)$, by

$$(Bf)(z) = L(f)\chi(\{b\}; z) \qquad (z \in C).$$

Clearly

$$SBf = bBf \qquad (f \in X^*).$$

Also from the identity $zf(z) = (z - b)f(z) + bf(z)$ $(z \in C)$ it follows that

$$BSf = bBf \qquad (f \in X^*).$$

Hence $BS = SB$. However $(BT\eta)(b) - (TB\eta)(b) = L(T\eta - h(b)\eta) \neq 0$. Therefore $TB \neq BT$.

We observe that for each positive integer m greater than one, none of the mth roots of S possess the commutativity property of Theorem 10.15. This may be proved by the argument given above for the corresponding property for logarithms together with the following observation. If h is a complex function defined on C and

$$(h(\lambda))^m = \lambda \qquad (\lambda \in C)$$

then h is discontinuous.

We now given an application of Theorem 10.7.

THEROEM 10.17. *Let T, in $L(X)$, be an invertible prespectral operator of class Γ.*

Suppose that T is power-bounded in the sense that there is a real number K such that

$$\|T^n\| \leqslant K \qquad (n \in \mathbf{Z}).$$

Then T is a scalar-type operator of class Γ.

Proof. By Proposition 1.31, $\sigma(T) \subseteq \{z: |z| = 1\}$. By Theorem 10.7 there is H, in $L(X)$, such that $\exp iH = T$ and H is prespectral of class Γ. By the spectral mapping theorem $\sigma(H)$ is real. Consider the group of operators $\{\exp i\alpha H: \alpha \in \mathbf{R}\}$. Since the map $\beta \rightarrow \exp i\beta H$ is a continuous map of $[0, 1]$ into $L(X)$

$$\|\exp i\alpha H\| \leqslant K \sup_{\beta \in [0, 1]} \|\exp i\beta H\| < \infty \qquad (\alpha \in \mathbf{R}).$$

Define

$$|||x||| = \sup\{\|\exp i\alpha H x\|: \alpha \in \mathbf{R}\} \qquad (x \in X).$$

Then $|||\ \ |||$ is a norm on X equivalent to $\|\ \ \|$; moreover

$$|||\exp i\alpha H||| = 1 \qquad (\alpha \in \mathbf{R}).$$

This means that under an equivalent renorming of X, H is hermitian. Let $E(\cdot)$ be the resolution of the identity of class Γ for H. Define

$$R = \int_{\sigma(H)} \lambda E(d\lambda) = \int_{\sigma(H)} \operatorname{Re} \lambda E(d\lambda),$$

$$Q = H - R.$$

By Theorem 5.40, there is an equivalent norm on X with respect to which R is hermitian. Since $HR = RH$ then, by Theorem 4.18, there is an equivalent norm on X with respect to which H and R are simultaneously hermitian. Assume that this renorming has been carried out. Then in the equation $H - R = Q$, the left-hand side is hermitian and the right-hand side quasi-nilpotent. Hence, by Sinclair's theorem (4.10), $Q = 0$ and $H = R$. Thus H is a scalar-type operator of class Γ and, by Proposition 5.8, so is $T = \exp iH$. This completes the proof.

The special case of this result in which $\Gamma = X^*$ was first proved independently by Foguel (Theorem 3 of [**2**; p. 62]) and Fixman (Theorem 4.2 of [**1**; p. 1041]).

We conclude this chapter with the following results.

THEOREM 10.18. *Let T, in $L(X)$, be an invertible operator with real spectrum such that, for every integer n, $\|T^n\| \leqslant M < \infty$. Then $T^2 = I$.*

Proof. Since T is power-bounded, $\sigma(T) \subseteq \{z : |z| = 1\}$. Hence, $\sigma(T) \subseteq \{-1, +1\}$ and so $\sigma(T^2 - I) = \{0\}$. Thus $T^2 = I + Q$, where Q is quasinilpotent. Since T^2 is a power-bounded spectral operator it is of scalar-type by Theorem 10.7. Hence $Q = 0$ and $T^2 = I$.

THEOREM 10.19. *Let $T \in L(X)$. Suppose that T is both a spectral operator and a hermitian operator. Then T is a scalar-type spectral operator.*

Proof. This result may be established by the argument used in proving Theorem 10.17.

11. Further Properties of Prespectral Operators

The main purpose of this chapter is to present some results of Foguel on the relationships between a spectral operator and its scalar part. Analogues of some of these results are valid for a prespectral operator. If this is the case we state and prove the more general version. Examples are given to show that not all of the results generalize to the case of a prespectral operator.

Let T be a prespectral operator on X with resolution of the identity $E(\cdot)$ of class Γ. Throughout this chapter

$$S = \int_{\sigma(T)} \lambda E(d\lambda), \; N = T - S.$$

LEMMA 11.1. *Let T be a prespectral operator on X with resolution of the identity $E(\cdot)$ of class Γ. Then S is in the closed subspace of $L(X)$ generated by $\{E(\tau): \tau \in \Sigma_p, 0 \notin \bar{\tau}\}$.*

Proof. Let $\varepsilon > 0$ be given. By the definition of the integral there is a partition $\tau_0, \tau_1, \ldots, \tau_n$ of $\sigma(S)$ into Borel sets with the point 0 in at most one of the closures $\bar{\tau}_i$ and with

$$\left\| S - \sum_{i=0}^{n} \lambda_i E(\tau_i) \right\| < \varepsilon$$

for any choice of the complex numbers λ_i in τ_i. If $0 \notin \sigma(S)$ this proves the lemma. If $0 \in \sigma(S)$ we may without loss of generality take $0 \in \tau_0$ and $\lambda_0 = 0$ in the inequality above, which proves the lemma in this case too.

THEOREM 11.2. *Let T be a prespectral operator on X with resolution of the identity $E(\cdot)$ of class Γ. Let T belong to the right (left) ideal J in $L(X)$. Then* *every projection $E(\tau)$ with $0 \notin \bar{\tau}$ belongs to J. If J is closed, then S and N also belong to J.*

Proof. Suppose that $\tau \in \Sigma_p$ and $0 \notin \bar{\tau}$. Let $T_\tau = T|E(\tau)X$. Since $\sigma(T|E(\tau)X) \subseteq \bar{\tau}$, it follows that $0 \in \rho(T_\tau)$, and hence T_τ^{-1} exists as a bounded linear operator on the space $E(\tau)X$. Let V_τ, on $L(X)$, be defined by

$$V_\tau x = T_\tau^{-1}E(\tau)x \qquad (x \in X).$$

Then $TV_\tau = E(\tau) = V_\tau T$, which proves that $E(\tau) \in J$. It follows from Lemma 11.1 that S and hence N also belong to J if J is closed.

COROLLARY 11.3. *Let T be a prespectral operator on X with resolution of the identity $E(\cdot)$ of class Γ. If T is compact, then so are S, N and every projection $E(\tau)$ with $0 \notin \bar{\tau}$.*

Proof. By Corollary 2.9 the compact operators on X form a closed two-sided ideal in $L(X)$.

COROLLARY 11.4. *Let T be a prespectral operator on X with resolution of the identity $E(\cdot)$ of class Γ. If T is weakly compact, then so are S,N and every projection $E(\tau)$ with $0 \notin \bar{\tau}$.*

Proof. By VI.4.6 of [**5**; p. 484] the weakly compact operators on X form a closed two-sided ideal in $L(X)$.

Observe that if Y is a closed subspace of X, the set $\{A \in L(X): AX \subseteq Y\}$ is a closed right ideal in $L(X)$. Hence we can deduce the following result from Theorem 11.2.

COROLLARY 11.5. *Let T be a prespectral operator on X with resolution of the identity $E(\cdot)$ of class Γ. Then the ranges of S,N and $E(\tau)$ with $0 \notin \bar{\tau}$ are contained in the closure of the range of T.*

Let A_0, in $L(X)$, be fixed. Then the sets $\{A \in L(X): A_0A = 0\}$, $\{A \in L(X): AA_0 = 0\}$ are respectively closed right and left ideals of $L(X)$. Hence our next result also follows from Theorem 11.2.

COROLLARY 11.6. *Let T be a prespectral operator on X with resolution of the identity $(E\cdot)$ of class. If $A_0T = 0$ (respectively $TA_0 = 0$) then $A_0S = A_0N = A_0E(\tau) = 0$ if $0 \notin \bar{\tau}$ (respectively $SA_0 = NA_0 = E(\tau)A_0$ if $0 \notin \bar{\tau}$).*

For the definition of a prespectral operator of finite type m the reader is referred to Definition 5.18.

COROLLARY 11.7. *Let T be a prespectral operator on X with radical part N. Then T is of finite type if and only if $N^pT = 0$ for some positive integer p.*

Proof. If T is of finite type, then $N^n = 0$ for some positive integer n and so $N^nT = TN^n = 0$. Conversely if some power of N annihilates T, say $N^pT = TN^p = 0$, then it follows from Corollary 11.6 that $N^{p+1} = 0$ and so T is of finite type.

COROLLARY 11.8. *Let T be a prespectral operator on X with resolution of the identity $E(\cdot)$ of class Γ. Let $x \in X$. If $Tx = 0$ and $0 \notin \bar{\tau}$ then $Sx = Nx = E(\tau)x = 0$.*

Proof. For a given x in X, the set $\{A \in L(X)\colon Ax = 0\}$ is a closed left ideal in $L(X)$.

COROLLARY 11.9. *Let T be a prespectral operator on X with resolution of the identity $E(\cdot)$ of class Γ. Let $0 \notin \bar{\tau}$ and let $\{x_n\}$ be a sequence in X for which $\{Tx_n\}$ is convergent (respectively convergent to 0). Then the sequence $\{E(\tau)x_n\}$ is convergent (respectively convergent to 0). If, in addition, $\{x_n\}$ is bounded, then the sequences $\{Sx_n\}$ and $\{Nx_n\}$ are also convergent (respectively convergent to 0).*

Proof. The set of all A, in $L(X)$, for which $\{Ax_n\}$ is convergent (respectively convergent to 0) is a left ideal, and this ideal is closed if $\{x_n\}$ is bounded. Thus the corollary follows directly from Theorem 11.2.

Let T be a prespectral operator on X with resolution of the identity $E(\cdot)$ of class Γ. Define $N_0 = NE(\{0\})$. Then $N_0 = TE(\{0\}) = E(\{0\})T$.

THEOREM 11.10. *Let T be a prespectral operator on X with resolution of the identity $E(\cdot)$ of class Γ. Let $A \in L(X)$. Then $TA = 0$ if and only if $A - E(\{0\})A = N_0A = 0$.*

Proof. If $N_0A = A - E(\{0\})A = 0$ then $E(\tau)A = E(\tau)E(\mathbf{C}\backslash\{0\})A = 0$ if $0 \notin \bar{\tau}$ and so, by Lemma 11.1, $SA = 0$. Now, since N and $E(\cdot)$ commute

$$NA = E(\{0\})NA + E(\mathbf{C}\backslash\{0\})NA = N_0A + N(A - E(\{0\})A) = 0.$$

Thus $TA = SA + NA = 0$. Conversely, if $TA = 0$ then $N_0 A = E(\{0\})TA = 0$ and, by Corollary 11.5, $E(\tau)A = 0$ if $0 \notin \bar{\tau}$. Now, for each x in X and y in Γ

$$\langle E(\mathbf{C}\backslash\{0\})Ax, y\rangle = \lim_{n\to\infty} \langle E(\tau_n)Ax, y\rangle,$$

where $\tau_n = \{z : |z| \geqslant n^{-1}\}$. Since Γ is total, $A - E(\{0\})A = 0$ and the proof is complete.

COROLLARY 11.11. *Let T be a prespectral operator on X with resolution of the identity $E(\cdot)$ of class Γ. If $E(\{0\}) = 0$ then $TA = 0$ if and only if $A = 0$.*

Proof. By Theorem 11.10, $TA = 0$ if and only if $A = 0$ in this case.

We now prove some results on the point spectrum of a prespectral operator.

THEOREM 11.12. *Let T be a prespectral operator on X with resolution of the identity $E(\cdot)$ of class Γ. Let $x \in X$ and let n be a positive integer. Then $(\lambda I - T)^n x = 0$ if and only if $E(\{\lambda\})x = x$ and $N^n x = 0$.*

Proof. Let $(\lambda I - T)^n x = 0$ and let τ be a closed subset of $\mathbf{C}$ such that $\lambda \notin \tau$. Let $T_\tau = T|E(\tau)X$ and $I_\tau = I|E(\tau)X$. Then $\lambda \in \rho(T_\tau)$ and so

$$E(\tau)x = (\lambda I_\tau - T_\tau)^{-n}(\lambda I - T)^n E(\tau)x,$$
$$E(\tau)x = (\lambda I_\tau - T_\tau)^{-n} E(\tau)(\lambda I - T)^n x = 0.$$

Let $\tau_n = \{z : |z - \lambda| \geqslant n^{-1}\}$ and let $y \in \Gamma$. Then

$$\langle E(\mathbf{C}\backslash\{\lambda\})x, y\rangle = \lim_{n\to\infty} \langle E(\tau_n)x, y\rangle = 0.$$

Since Γ is total we obtain

$$E(\mathbf{C}\backslash\{\lambda\})x = 0, \; E(\{\lambda\})x = x.$$

Hence

$$Sx = SE(\{\lambda\})x = \int_{\{\lambda\}} \mu E(d\mu)x = \lambda E(\{\lambda\})x = \lambda x,$$

which shows that

$$(\lambda I - T)x = -Nx$$

and hence that

$$0 = (\lambda I - T)^n x = (-1)^n N^n x.$$

This proves the necessity of the conditions. Now, conversely, suppose that $E(\{\lambda\})x = x$ and $N^n x = 0$. It follows as above that $(\lambda I - S)x = 0$ and hence that $(\lambda I - T)^n x = (-1)^n N^n x$. Therefore $(\lambda I - T)^n x = 0$.

COROLLARY 11.13. *Let T be a prespectral operator on X with scalar part S. Then $\sigma_p(T) \subseteq \sigma_p(S)$.*

Proof. This result follows easily from the first part of the proof of Theorem 11.12.

Next, we characterize the point spectrum of a scalar-type operator in terms of the values of one of its resolutions of the identity.

THEOREM 11.14. *Let S be a scalar-type operator on X with resolution of the identity $E(\cdot)$ of class Γ. Then $\lambda \in \sigma_p(S)$ if and only if $E(\{\lambda\}) \neq 0$. Moreover if for some x in X and some positive integer n we have $(\lambda I - S)^n x = 0$ then $Sx = \lambda x$. Thus if $\lambda \in \sigma_p(S)$ the ascent of the operator $\lambda I - S$ is one.*

Proof. This result follows easily from Theorem 11.12.

THEOREM 11.15. *Let X be separable and let T, in $L(X)$, be a prespectral operator with resolution of the identity $E(\cdot)$ of class Γ. Then $\sigma_p(T)$ is countable.*

Proof. By Theorem 11.12, $\sigma_p(T) \subseteq \{\lambda: E(\{\lambda\}) \neq 0\}$. There is $M > 0$ such that

$$\| E(\tau) \| \leqslant M < \infty \; (\tau \in \Sigma_p).$$

Let $x_\lambda, x_\mu \in X$, $\| x_\lambda \| = \| x_\mu \| = 1$, $E(\{\lambda\})x_\lambda = x_\lambda$ and $E(\{\mu\})x_\mu = x_\mu$. Then if μ and λ are distinct points of $\sigma_p(T)$ we have

$$\| x_\lambda - x_\mu \| \geqslant M^{-1} \| E(\{\lambda\})(x_\lambda - x_\mu) \| = M^{-1} \| x_\lambda \| = M^{-1}.$$

Since X is separable, it follows that $\sigma_p(T)$ is countable.

For the case of a spectral operator we can obtain further results of this type.

THEOREM 11.16. *Let T be a spectral operator on X with resolution of the identity $E(\cdot)$. Let $A \in L(X)$. Then $AT = 0$ if and only if $AN = 0$ and $AE(\mathbf{C}\backslash\{0\}) = 0$. Similarly $TA = 0$ if and only if $E(\mathbf{C}\backslash\{0\})A = NA = 0$.*

Proof. If $AT = 0$ then it follows from Corollary 11.6 that $AN = 0$ and, for each positive integer n, $AE(\tau_n) = 0$ where $\tau_n = \{\lambda : |\lambda| \geqslant n^{-1}\}$. Hence $AE(\mathbf{C}\backslash\{0\}) = 0$ from the strong countable additivity of $E(\cdot)$.

Now suppose that $AN = 0$ and $AE(\mathbf{C}\backslash\{0\}) = 0$. If $\tau \in \Sigma_p$ and $0 \notin \bar{\tau}$ then $\bar{\tau} \subseteq \mathbf{C}\backslash\{0\}$. Hence $E(\tau) = E(\tau)E(\mathbf{C}\backslash\{0\})$. Therefore

$$AE(\tau) = AE(\mathbf{C}\backslash\{0\})E(\tau) = 0.$$

Thus $E(\tau)$ belongs to the closed ideal $\{C \in L(X) : AC = 0\}$. It follows from Lemma 11.1 that $AS = 0$. Since $AN = 0$, we have $AT = 0$ also. The second part of the theorem may be proved in a similar fashion or alternatively it can be deduced from Theorem 11.10.

COROLLARY 11.17. *Let T be a spectral operator on X with resolution of the identity $E(\cdot)$. Let $E(\{0\}) = 0$. If either $AT = 0$ or $TA = 0$ and $A \in L(X)$ then $A = 0$.*

Proof. If either $AT = 0$ or $TA = 0$ then, by Theorem 11.16, either $A = AE(\{0\})$ or $A = E(\{0\})A$. Thus $A = 0$.

COROLLARY 11.18. *Let T be a spectral operator on X with resolution of the identity $E(\cdot)$. If, for some complex number λ, $E(\{\lambda\}) = 0$ then $(\lambda I - T)X$ is dense in X.*

Proof. First suppose that $\lambda = 0$. If TX is not dense in X then there is y in X^* with $y \neq 0$ and $\langle Tx, y\rangle = 0$ $(x \in X)$. Let x_1 be a unit vector in X. Define A, in $L(X)$, by $Ax = \langle x, y\rangle x_1$ so that $A \neq 0$. However $AT = 0$, which contradicts Corollary 11.17. Now for an arbitrary λ, Theorem 5.16 shows that $\lambda I - T$ is a spectral operator with resolution of the identity $F(\cdot)$ such that $F(\{\lambda\}) = E(\{0\})$. Thus it follows by what has been proved that $(\lambda I - T)X$ is dense in X.

THEOREM 11.19. *Let T, in $L(X)$, be a prespectral operator of class Γ. If T has a closed range then so does S, the scalar part of T.*

Proof. Let $E(\cdot)$ be the resolution of the identity of class Γ for T. The proof will be divided into two cases depending on whether the projection $E(\{0\})$ is 0 or not. First suppose that $E(\{0\}) = 0$. Since TX is closed, the image of $\{x \in X: \|x\| < 1\}$ under T contains a set of the form $\{y \in TX: \|y\| < \delta\}$, by the open mapping theorem. Thus if $y \in TX$, $y \neq 0$, the vector $\delta y/2\|y\|$ is the image under T of a vector z with $\|z\| \leqslant 1$. Hence, if $x = 2\|y\|z/\delta$, we have $Tx = y$ and $\|x\| \leqslant (2/\delta)\|y\|$. It follows that there is a positive real K such that $\|Tx\| \geqslant K\|x\|$, for all x in X. By Theorem 5.47, $0 \notin \sigma_a(T) = \sigma(T) = \sigma(S)$ and so $SX = X$.

Next suppose that $E(\{0\}) \neq 0$. We note first that for any Borel subset τ of $\mathbf{C}$ the restriction of T to the subspace $E(\tau)X$ is a prespectral operator with resolution of the identity $F(\cdot)$ of class Γ given by

$$F(\delta) = E(\delta \cap \tau) \qquad (\delta \in \Sigma_p).$$

This follows immediately from the definition of a prespectral operator. Let $\tau = \mathbf{C}\backslash\{0\}$, $V = T|E(\tau)X$ and $F(\cdot) = E(\cdot)|E(\tau)X$. Observe that $F(\{0\}) = 0$ and so we may apply to V the result already proved in the first part as soon as it is shown that the range of V is closed. Let $y \in \overline{VX}$. Then for some sequence $\{x_n\}$ in $E(\mathbf{C}\backslash\{0\})X$ we have $Vx_n \to y$ and, since the range of T is closed, there is an x in X with $Tx = y$. Hence

$$\begin{aligned} VE(\mathbf{C}\backslash\{0\})x &= TE(\mathbf{C}\backslash\{0\})x = E(\mathbf{C}\backslash\{0\})Tx \\ &= E(\mathbf{C}\backslash\{0\})y = y. \end{aligned}$$

Thus V satisfies the conditions assumed for T in the first part of the proof and so we conclude that the scalar part of V maps $E(\mathbf{C}\backslash\{0\})X$ onto itself. It follows from Theorem 5.13 that the scalar part of V is $S|E(\mathbf{C}\backslash\{0\})X$ and so

$$SE(\mathbf{C}\backslash\{0\})X = E(\mathbf{C}\backslash\{0\})X.$$

However $SE(\{0\}) = 0$ and so $SX = E(\mathbf{C}\backslash\{0\})X$, which shows that S has a closed range. The proof is complete.

THEOREM 11.20. *Let T be a prespectral operator on X with resolution of the* *identity* $E(\cdot)$ *of class* Γ. *Then* T *has a closed range if and only if the following two conditions hold.*

(i) *Either* $0 \in \rho(T)$ *or* 0 *is an isolated point of* $\sigma(T)$.

(ii) *The operator* $TE(\{0\})$ *has a closed range.*

Proof. Suppose that T has a closed range. It was shown in the course of proving Theorem 11.19 that $0 \in \rho(T \mid E(\mathbf{C}\backslash\{0\})X)$ in the case $E(\{0\}) \neq 0$. Hence, by Proposition 1.37,

$$\sigma(T) = \{0\} \cup \sigma(T \mid E(\mathbf{C}\backslash\{0\})X)$$

and so 0 is an isolated point of $\sigma(T)$. Also if $E(\{0\}) = 0$ then $0 \in \rho(T)$. This proves (i). To prove (ii) let y be in the closure of the range of $TE(\{0\})$ and let $\{x_n\}$ be a sequence in X such that $TE(\{0\})x_n \to y$. Since T has a closed range, there is an x in X with $Tx = y$ and so

$$TE(\{0\})x = E(\{0\})Tx = E(\{0\})y = y$$

which proves (ii).

Conversely we assume (i) and (ii). If $0 \in \rho(T)$, then $TX = X$, and so we may assume that $0 \in \sigma(T)$. Let $y \in \overline{TX}$ and let $\{x_n\}$ be a sequence in X such that $Tx_n \to y$. Then

$$TE(\{0\})x_n \to E(\{0\})y$$

and, since the range of $TE(\{0\})$ is closed, there is a vector w with $TE(\{0\})w = E(\{0\})y$. Since 0 is an isolated point of $\sigma(T)$ it is in $\rho(T \mid E(\mathbf{C}\backslash\{0\})X)$ and so for some z in $E(\mathbf{C}\backslash\{0\})X$ we have $Tz = E(\mathbf{C}\backslash\{0\})y$. Hence

$$T(z + E(\{0\})w) = E(\mathbf{C}\backslash\{0\})y + E(\{0\})y = y$$

which proves that the range of T is closed.

PROPOSITION 11.21. *Let* T *be a spectral operator of finite type on* X. *The residual spectrum of* T *is void. Also* λ *is in the point spectrum of* T *if and only if* $E(\{\lambda\}) \neq 0$, *where* $E(\cdot)$ *is the resolution of the identity of* T.

Proof. Let $\lambda \in \sigma(T)$. If $E(\{\lambda\}) \neq 0$ then $E(\{\lambda\})x = x$ for some non-zero x in X and $N^n = 0$ for some positive integer n. It follows from Theorem 11.12 that

$\lambda \in \sigma_p(T)$. If $E(\{\lambda\}) = 0$ then it follows from Theorem 11.12 that $\lambda I - T$ is one-to-one. By Corollary 11.18 the set $(\lambda I - T)X$ is dense in X and so $\lambda \in \sigma_c(T)$. This completes the proof.

COROLLARY 11.22. *Let T be a spectral operator on X and let S be the scalar part of T. Then*

(i) $\sigma(S) = \sigma_p(S) \cup \sigma_c(S)$,

(ii) $\sigma_c(S) \subseteq \sigma_c(T)$,

(iii) $\sigma_p(T) \cup \sigma_r(T) \subseteq \sigma_p(S)$.

Proof. Since S is a spectral operator of finite type, $\sigma_r(S)$ is empty and so the first statement follows. Next let $\lambda \in \sigma_c(S)$. Let $E(\cdot)$ be the resolution of the identity for T and S. It follows from Proposition 11.21 that $E(\{\lambda\}) = 0$. From this fact and Theorem 11.12 we deduce that $\lambda \notin \sigma_p(T)$. By Corollary 11.18, $(\lambda I - T)X$ is dense in X and hence, since $\lambda \in \sigma(T)$, it follows that $\lambda \in \sigma_c(T)$. This proves the second statement, and the final assertion follows from this by taking complements in $\sigma(T) = \sigma(S)$ and using the fact that $\sigma_r(S) = \varnothing$.

COROLLARY 11.23. *Let T be a spectral operator on X with resolution of the identity $E(\cdot)$. Let $\tau \in \Sigma_p$ and $\lambda \in \mathbf{C} \backslash \tau$. Then λ is either in the resolvent set of $T | E(\tau)X$ or in the continuous spectrum of $T | E(\tau)X$.*

Proof. It was pointed out in the course of proving Theorem 11.19 that $S | E(\tau)X$ is the scalar part of $T | E(\tau)X$. Hence, by Corollary 11.22, $\sigma_c(S | E(\tau)X) \subseteq \sigma_c(T | E(\tau)X)$ and so to complete the proof it suffices to show that if $\lambda \in \sigma(S | E(\tau)X)$ then $\lambda \in \sigma_c(S | E(\tau)X)$. Let $(S - \lambda I)x = 0$, where $x \in E(\tau)X$. Since $E(\cdot)$ is the resolution of the identity for S, it follows from Theorem 11.14 that $E(\{\lambda\})x = x$. Since $\{\lambda\}$ and τ are disjoint we have $x = E(\tau)x = 0$. Thus $S - \lambda I$ is one-to-one on $E(\tau)X$, and so λ is either in the resolvent set of $S | E(\tau)X$, and hence of $T | E(\tau)X$, or else in the continuous spectrum of $S | E(\tau)X$, since $\sigma_r(S | E(\tau)X) = \varnothing$. The proof is complete.

THEOREM 11.24. *Let T be a spectral operator on X with resolution of the identity $E(\cdot)$. Let $\lambda \in \sigma(T)$. If $E(\{\lambda\}) = 0$ then $\lambda \in \sigma_c(T)$. If $E(\{\lambda\}) \neq 0$ then*

(i) *$\lambda \in \sigma_p(T)$ if and only if $0 \in \sigma_p(N | E(\{\lambda\})X)$;*

(ii) *$\lambda \in \sigma_r(T)$ if and only if $0 \in \sigma_r(N | E(\{\lambda\})X)$;*

(iii) *$\lambda \in \sigma_c(T)$ if and only if $0 \in \sigma_c(N | E(\{\lambda\})X)$.*

Proof. If $E(\{\lambda\}) = 0$ it follows from Theorem 11.12 that $\lambda \notin \sigma_p(T)$. We then see from Corollary 11.18 that $\lambda \in \sigma_c(T)$. Now suppose that $E(\{\lambda\}) \neq 0$. Since $SE(\{\lambda\}) = \lambda E(\{\lambda\})$ we have

$$(T - \lambda I)E(\{\lambda\}) = NE(\{\lambda\}). \tag{1}$$

Now

$$(T - \lambda I)X = (T - \lambda I)E(\{\lambda\})X \oplus (T - \lambda I)E(\mathbf{C}\backslash\{\lambda\})X. \tag{2}$$

Since Corollary 11.23 shows that $(T - \lambda I)E(\mathbf{C}\backslash\{\lambda\})X$ is dense in $E(\mathbf{C}\backslash\{\lambda\})X$ it follows from (2) that $(T - \lambda I)X$ is dense in X if and only if $(T - \lambda I)E(\{\lambda\})X$ is dense in $E(\{\lambda\})X$. Since, by Theorem 11.12, any eigenvector for T corresponding to λ must be in the space $E(\{\lambda\})X$, the three statements (i), (ii) and (iii) now follow from (1).

THEOREM 11.25. *Let X be separable. Let T be a spectral operator on X with resolution of the identity $E(\cdot)$. Then the point and residual spectra of T are countable.*

Proof. This result may be established by the argument used to prove Theorem 11.15 together with the following observation. By Theorem 11.24

$$\sigma_p(T) \cup \sigma_r(T) \subseteq \{\lambda : E(\{\lambda\}) \neq 0\}.$$

We now give some examples which throw light on the results proved in this chapter.

Example 11.26. Let $X = l^1$. Define N, in $L(X)$, by

$$N\{x_1, x_2, x_3, \ldots\} = \{x_2, 0, x_4, \ldots\}$$

and let $S = 0$. Observe that $N^2 = 0$. Define $T = S + N$. Then T is a spectral operator. S is compact while T is not even weakly compact.

Example 11.27. Let X be the closed subspace of $C[0, 1]$ consisting of functions vanishing at the point 0. For each f in X define $Nf = g$, where

$$g(x) = \int_0^x f(s)ds \qquad (0 \leqslant x \leqslant 1).$$

Note first that if $g = 0$ then $f = 0$ so that N is one-to-one. We show that N is quasinilpotent. Observe that

$$|(Nf)(t)| = |g(t)| \leqslant t\|f\| \qquad (0 \leqslant t \leqslant 1)$$

and, inductively,

$$|(N^n f)(t)| \leqslant \frac{t^n}{n!}\|f\| \qquad (0 \leqslant t \leqslant 1)$$

which shows that

$$\|N^n\| \leqslant \frac{1}{n!} \qquad (n \in \mathbf{N}).$$

It follows that N is quasinilpotent and so a spectral operator. Also from the Weierstrass polynomial theorem it follows that the range of N is dense in X. Let $S = 0$. Then S has a closed range while $T = S + N$ does not. $0 \in \sigma_p(S)$ but $0 \in \sigma_c(T)$.

Example 11.28. Let N be defined as in Example 11.27 and let $S = I$. Since $\sigma(T) = \sigma(S) = \{1\}$ then T and S have closed ranges but the range of N is not closed.

Example 11.29. Let $X = l^\infty$ and let S, in $L(X)$, be defined by

$$S\{\xi_n\} = \{n^{-1}\xi_n\}.$$

Observe that $\sigma(S) = \{0, 1, \frac{1}{2}, \frac{1}{3}, \ldots\}$. S is a scalar-type operator with resolution of the identity $E(\cdot)$ of class l^1 given by

$$E(\{0\}) = 0$$
$$E(\{k^{-1}\})\{\xi_n\} = \{\delta_{nk}\xi_n\} \qquad (k = 1, 2, 3, \ldots).$$

We show first that SX is not dense in X. Define

$$y = \{1, 1, 1, \ldots\}$$

and let $x = \{\xi_n\} \in X$. We can choose a positive integer n so large that

$$\|x\| = \sup_m |\xi_m| < \frac{n}{2}.$$

Then

$$\|Sx - y\| \geqslant 1 - n^{-1}|\xi_n| > \tfrac{1}{2}.$$

This shows that Corollary 11.18 fails in general for prespectral operators. Also $0 \in \sigma_r(S)$. By Proposition 11.21, a scalar-type spectral operator has empty residual spectrum, so this result does not extend to scalar-type operators. Next, we note that $\overline{SX}$ is a proper closed subspace of X and $y \notin \overline{SX}$. On the subspace of X generated by $\overline{SX}$ and y we can define a linear functional x^* such that

$$x^*(\overline{SX}) = 0, \qquad x^*(y) = 1.$$

By the Hahn-Banach theorem we can extend x^* to a bounded linear functional on l^∞ of norm 1. Denote the extension by x^* also. Define

$$Px = x^*(x)y \qquad (x \in X).$$

Note that $\|Px\| \leqslant \|x^*\| \|x\| \|y\| = \|x\|$ $(x \in X)$ and so $\|P\| \leqslant 1$. Also P is linear and so $P \in L(X)$. We have

$$PSx = 0 \qquad (x \in X)$$

but $P \neq 0$ since $Py = y \neq 0$. This shows that Corollary 11.17 fails to generalize to prespectral operators. (Compare this result with Corollary 11.11.)

12. Restrictions of Spectral Operators

The main purpose of this chapter is to discuss the following problem: given a spectral operator T on X and a closed subspace Y of X invariant under T, when is the restriction of T to Y also a spectral operator? It is shown that this is a generalization of the following problem: given a normal operator T on a complex Hilbert space H and a closed subspace Y of H invariant under T, when is Y a reducing subspace for T in the Hilbert space sense?

We begin with the following necessary condition for a restriction of a spectral operator to be spectral.

PROPOSITION 12.1. *Let T be a spectral operator on X and let Y be a closed subspace of X invariant under T. If $T|Y$ is a spectral operator, then $\sigma(T|Y) \subseteq \sigma(T)$.*

Proof. By Theorem 5.47, $\sigma_a(T) = \sigma(T)$ and $\sigma_a(T|Y) = \sigma(T|Y)$. Also, by Theorem 1.16 (i), $\sigma_a(T|Y) \subseteq \sigma_a(T)$ and so the desired conclusion follows.

The next two theorems, giving necessary and sufficient conditions for a restriction of a spectral operator to be spectral were first proved by Fixman [**1**; p. 1032–4]. The short proof of the second theorem given here is due to the author [**1**; p. 442].

THEOREM 12.2. *Let T be a spectral operator on X and let Y be a closed subspace of X invariant under T such that $T|Y$ is spectral. Let $E(\cdot)$ and $F(\cdot)$ be the resolutions of the identity of T and $T|Y$ respectively. Then for each τ in Σ_p, $E(\tau)Y \subseteq Y$ and $F(\tau) = E(\tau)|Y$.*

Proof. Let $y \in Y$. Since T and $T|Y$ are spectral operators they have the single-valued extension property by Theorem 5.31. Let y_X and y_Y be the maximal

X-valued and Y-valued analytic functions such that

$$(\zeta I - T)y_X(\zeta) = y, \qquad (\zeta I - T)y_Y(\zeta) = y.$$

Let the domains of definition of y_X and y_Y be $\rho_X(y)$ and $\rho_Y(y)$ respectively. Let

$$\sigma_X(y) = \mathbf{C}\backslash\rho_X(y), \qquad \sigma_Y(y) = \mathbf{C}\backslash\rho_Y(y).$$

Since y_Y may be regarded as X-valued, $\rho_Y(y) \subseteq \rho_X(y)$ and so

$$\sigma_X(y) \subseteq \sigma_Y(y) \qquad (y \in Y). \tag{1}$$

If τ is a closed subset of $\mathbf{C}$ then $F(\tau)Y = \{y \in Y: \sigma_Y(y) \subseteq \tau\}$ by Theorem 5.33 and so for every y in Y we have $\sigma_Y(F(\tau)y) \subseteq \tau$. Hence by (1)

$$\sigma_X(F(\tau)y) \subseteq \tau \qquad (y \in Y).$$

Therefore, by Theorem 5.33, $F(\tau)y \in E(\tau)X$ and so

$$E(\tau)F(\tau)y = F(\tau)y \qquad (y \in Y). \tag{2}$$

Let δ be closed and disjoint from τ. We have similarly

$$E(\delta)F(\delta)y = F(\delta)y \qquad (y \in Y). \tag{3}$$

Operating on (3) with $E(\tau)$ gives $E(\tau)F(\delta)y = 0$. Since $T|Y$ is a spectral operator, $F(\cdot)$ is countably additive in the strong operator topology of $L(Y)$. There is an increasing sequence $\{\delta_n\}$ of compact subsets of $\mathbf{C}\backslash\tau$ with union $\mathbf{C}\backslash\tau$. We have $E(\tau)F(\delta_n)y = 0$ for each n and every y in Y. Now

$$\lim_n F(\delta_n)y = F(\mathbf{C}\backslash\tau)y$$

in the norm of X and, since $E(\tau) \in L(X)$, it follows that

$$E(\tau)F(\mathbf{C}\backslash\tau)y = 0 \qquad (y \in Y). \tag{4}$$

From (2) and (4) we deduce that if τ is a closed subset of $\mathbf{C}$ then

$$E(\tau)y = F(\tau)y \qquad (y \in Y).$$

Let $y \in Y, z \in X^*$ and consider the measure μ defined by

$$\mu(\tau) = \langle E(\tau)y - F(\tau)y, z \rangle \qquad (\tau \in \Sigma_p).$$

The class of sets in Σ_p on which μ vanishes forms a σ-algebra containing all closed sets. Hence μ vanishes identically. Since z is arbitrary

$$E(\tau)y = F(\tau)y \qquad (y \in Y, \ \tau \in \Sigma_p)$$

and this suffices to complete the proof.

THEOREM 12.3. *Let T be a spectral operator on X and let Y be a closed subspace of X invariant under T. Let $E(\cdot)$ be the resolution of the identity of T. If $E(\tau)Y \subseteq Y$ $(\tau \in \Sigma_p)$ then $T|Y$ is spectral. Moreover, if T is of scalar type (respectively finite type m) then $T|Y$ is of scalar type (finite type m).*

Proof. Define

$$S = \int_{\sigma(T)} \lambda E(d\lambda), \ N = T - S. \tag{5}$$

Observe that it follows from Theorem 5.17 that

$$(\zeta I - T)^{-1} = \sum_{n=0}^{\infty} N^n \int_{\sigma(T)} \frac{E(d\lambda)}{(\zeta - \lambda)^{n+1}} \qquad (\zeta \in \rho(T)). \tag{6}$$

It is easily verified that $T|Y$ satisfies the axioms for a spectral operator with the exception of $\sigma(T|E(\tau)Y) \subseteq \tau$ $(\bar{\tau} \in \Sigma_p)$. In the case $Y = E(\delta)X$ we have

$$\sigma(T|E(\delta)E(\tau)X) = \sigma(T|E(\delta \cap \tau)X) \subseteq \overline{\tau \cap \delta} \cap \bar{\tau} \qquad (\tau \in \Sigma_p).$$

Hence $T|E(\delta)X$ is spectral. Now observe that, by (5) and (6) the hypothesis $E(\tau)Y \subseteq Y$ $(\tau \in \Sigma_p)$ implies that S,N and $(\zeta I - T)^{-1}$ $(\zeta \in \rho(T))$ leave Y invariant. Hence $\sigma(T|Y) \subseteq \sigma(T)$. This argument may be applied to the spectral operator $T|E(\tau)X$ and its restriction to the closed invariant subspace $E(\tau)Y$. Therefore

$$\sigma(T|E(\tau)Y) \subseteq \sigma(T|E(\tau)X) \subseteq \bar{\tau} \qquad (\tau \in \Sigma_p).$$

Since the last statement of the theorem is trivial, the proof is complete.

Restrictions of normal operators were first studied by Halmos [**1**]. We have the following result.

PROPOSITION 12.4. *Let $H \neq \{0\}$ be a complex Hilbert space and let A be a normal operator on H. Let Y be a closed subspace of H invariant under A. The following three statements are equivalent.*

(i) *$A|Y$ is normal.*

(ii) *$A|Y$ is spectral.*

(iii) *$A^*Y \subseteq Y$, i.e. Y is reducing.*

Proof. Trivially (iii) implies (i) and (i) implies (ii). Finally (ii) implies (iii) by Theorem 12.2 and Note 7.19.

PROPOSITION 12.5. *Let $H \neq \{0\}$ be a complex Hilbert space and let A be a normal operator on H. Let Y be a closed subspace of H invariant under A. If $A|Y$ is normal then $\sigma(A|Y) \subseteq \sigma(A)$.*

Proof. This follows at once from Propositions 12.1 and 12.4.

The restriction of a spectral operator to a closed invariant subspace need not be spectral, as the following example shows.

Example 12.6. Let H be a separable complex Hilbert space and let

$$\{\phi_n : n = 0, \pm 1, \pm 2, \ldots\}$$

be an orthonormal basis for H. Define a unitary operator U on H by

$$U\phi_n = \phi_{n+1} \qquad (n = 0, \pm 1, \pm 2, \ldots).$$

Let $Y = \text{clm}\{\phi_r : r = 1, 2, 3, \ldots\}$. Then $UY \subseteq Y$. However $U^*\phi_1 = U^{-1}\phi_1 = \phi_0 \notin Y$. It follows from Proposition 12.4 that $U|Y$ is not a spectral operator.

The next theorem characterizes the ranges of projections among the closed invariant subspaces of the spectral operator. Two preliminary lemmas are required.

LEMMA 12.7. *Let $T \in L(X)$ and let Y be a closed subspace of X invariant under T. If T has the single-valued extension property then so does $T|Y$.*

Proof. Let $y \in Y$, and let f and g be analytic Y-valued functions, defined on open sets $D(f)$ and $D(g)$ respectively, such that

$$(\zeta I - T)f(\zeta) = y \qquad (\zeta \in D(f)),$$
$$(\zeta I - T)g(\zeta) = y \qquad (\zeta \in D(g)).$$

Since T has the single-valued extension property,

$$f(\zeta) = g(\zeta) \qquad (\zeta \in D(f) \cap D(g)).$$

Therefore $T \mid Y$ has the single-valued extension property.

LEMMA 12.8. *Let T be a spectral operator on X and let $E(\cdot)$ be the resolution of the identity of T. Then for each τ in Σ_p,*

$$E(\tau)X = \operatorname{clm}\{\cup E(\delta)X : \delta \subseteq \tau, \delta \textit{ compact}\}.$$

Proof. Let Y denote the closed subspace on the right. Trivially, $Y \subseteq E(\tau)X$. Let $x \in E(\tau)X$ and $y \in Y^{\perp}$. Then $x = E(\tau)x$, and for each compact $\delta \subseteq \tau$, $\langle E(\delta)x, y\rangle = 0$. Now $\langle E(\cdot)x, y\rangle$ is a finite measure on the Borel subsets of $\mathbf{C}$ and so is regular. Therefore $\langle x, y\rangle = \langle E(\tau)x, y\rangle = 0$. By the standard properties of the annihilator it follows that $E(\tau)X \subseteq Y$ completing the proof.

THEOREM 12.9. *Let T be a spectral operator on X and let $E(\cdot)$ be the resolution of the identity of T. Define, for each τ in Σ_p,*

$$\begin{aligned} G(\tau) = \operatorname{clm}\{&\textit{the union of all closed subspaces } Y \textit{ of } X \\ &\textit{such that } TY \subseteq Y, \sigma(T \mid Y) \subseteq \tau, \textit{ and} \\ &T \mid Y \textit{ is spectral}\}; \\ H(\tau) = \operatorname{clm}\{&\textit{the union of all closed subspaces } Y \textit{ of } X \\ &\textit{such that } TY \subseteq Y \textit{ and } \sigma(T \mid Y) \subseteq \tau\}. \end{aligned}$$

Then

$$E(\tau)X = G(\tau) = H(\tau) \qquad (\tau \in \Sigma_p).$$

Proof. Trivially

$$G(\tau) \subseteq H(\tau) \qquad (\tau \in \Sigma_p). \tag{7}$$

For closed δ we have $\sigma(T|E(\delta)X) \subseteq \delta$. As in the proof of Theorem 12.3, $T|E(\delta)X$ is spectral and so $E(\delta)X \subseteq G(\delta)$. Now let $\tau \in \Sigma_p$. Since the spectrum of each restriction is a compact set,

$$G(\tau) = \text{clm}\{\cup G(\delta) \colon \delta \subseteq \tau, \delta \text{ compact}\}.$$

By Lemma 12.8, $E(\tau)X = \text{clm}\{\cup E(\delta)X \colon \delta \subseteq \tau, \delta \text{ compact}\}$. Therefore

$$E(\tau)X \subseteq G(\tau) \qquad (\tau \in \Sigma_p). \tag{8}$$

Let Y be a closed subspace of X invariant under T with $\sigma(T|Y) \subseteq \tau$ and let $y \in Y$. By Lemma 12.7 and Theorem 5.31, $T|Y$ has the single-valued extension property. Let y_X and y_Y be the maximal X-valued and Y-valued analytic functions which satisfy

$$(\zeta I - T)y_X(\zeta) = y \qquad (\zeta \in \rho_X(y)),$$
$$(\zeta I - T)y_Y(\zeta) = y \qquad (\zeta \in \rho_Y(y)),$$

where $\rho_X(y)$ and $\rho_Y(y)$ are the domains of definition of these functions. Let $\sigma_X(y)$ and $\sigma_Y(y)$ be the complements of these sets. y_Y may be regarded as X-valued, and so, by the maximality of $\rho_X(y)$,

$$\rho_Y(y) \subseteq \rho_X(y),$$
$$\sigma_X(y) \subseteq \sigma_Y(y) \subseteq \sigma(T|Y) \subseteq \tau.$$

Now $\sigma_X(y)$ is a closed set and so, by Theorem 5.33, $y \in E(\sigma_X(y))X \subseteq E(\tau)X$. y is an arbitrary element of Y, and Y is an arbitrary closed invariant subspace with $\sigma(T|Y) \subseteq \tau$. Therefore

$$H(\tau) \subseteq E(\tau)X. \tag{9}$$

(7), (8) and (9) suffice to prove the theorem.

Definition 12.10. Let T be a spectral operator on X with resolution of the identity $E(\cdot)$. Let Y be a closed subspace of X invariant under T. Let $\{Y_\alpha \colon \alpha \in A\}$ be the class of all closed subspaces of X which satisfy

$$TY_\alpha \subseteq Y_\alpha,$$
$$E(\tau)Y_\alpha \subseteq Y_\alpha \qquad (\tau \in \Sigma_p),$$
$$Y \subseteq Y_\alpha.$$

The set $Y_0 = \bigcap_{\alpha \in A} Y_\alpha$ is closed and a subspace of X (possibly equal to X or Y). It is invariant under T, and under $\{E(\tau): \tau \in \Sigma_p\}$. Therefore $T \mid Y_0$ can be defined and $T \mid Y_0$ is spectral. $T \mid Y_0$ is called the *minimal spectral extension* of the restriction $T \mid Y$.

PROPOSITION 12.11. *Let T be a spectral operator on X with resolution of the identity $E(\cdot)$. Let Y be a closed subspace of X invariant under T, and let $T \mid Y_0$ be the minimal spectral extension of $T \mid Y$. Then $\sigma(T \mid Y_0) \subseteq \sigma(T \mid Y)$.*

Proof. Let $\sigma(T \mid Y) = \delta$. Then by Theorem 12.9, $Y \subseteq E(\delta)X$. By the definition of Y_0, $Y_0 \subseteq E(\delta)X$ and $T \mid Y_0$ is spectral. By applying Proposition 12.1 to the spectral operator $T \mid E(\delta)X$ and its restriction $T \mid Y_0$, we obtain

$$\sigma(T \mid Y_0) \subseteq \sigma(T \mid E(\delta)X) \subseteq \delta = \sigma(T \mid Y).$$

Note 12.12. Observe that the inclusion relation here is in the opposite direction to that of Proposition 12.1. In this case the operator on the smaller space has the larger spectrum.

Results for normal operators, analogous to Theorem 12.9 and Proposition 12.11 were obtained by Halmos [**2**], using different methods.

We now discuss restrictions of scalar-type spectral operators. The operational calculus for such an operator together with results on uniform approximation on compact subsets of the complex plane described in Chapter 5 (see Theorem 5.37 et seq.) enable us to simplify Fixman's conditions in the case in which the spectrum of the operator is suitably thin. In order to prove the main theorems we require the following preliminary result.

PROPOSITION 12.13. *Let S be a scalar-type spectral operator on X with resolution of the identity $E(\cdot)$. Define*

$$\psi(f) = \int_{\sigma(S)} f(\lambda)E(d\lambda) \qquad (f \in C(\sigma(S))).$$

Let Y be a closed subspace of X invariant under $\{\psi(f): f \in C(\sigma(S))\}$. Then $E(\tau)Y \subseteq Y\ (\tau \in \Sigma_p)$.

Proof. Let $x \in Y$, $y \in Y^\perp$ and $\mu(\tau) = \langle E(\tau)x, y\rangle\ (\tau \in \Sigma_p)$. Since, by Note 5.7, $\sigma(S)$ is the support of $E(\cdot)$ we may regard μ as being defined on the σ-algebra of Borel subsets of $\sigma(S)$. μ is a regular measure. By Proposition 5.8

$$0 = \langle \psi(f)x, y \rangle = \int_{\sigma(S)} f(\lambda)\mu(d\lambda) \qquad (f \in C(\sigma(S)))$$

and so the Riesz representation theorem shows that $\mu = 0$. Since the support of $E(\cdot)$ is $\sigma(S)$

$$\langle E(\tau)x, y \rangle = 0 \qquad (\tau \in \Sigma_p, x \in Y, y \in Y^\perp).$$

By the standard properties of the annihilator we have $E(\tau)Y \subseteq Y$ $(\tau \in \Sigma_p)$.

THEOREM 12.14. *Let S be a scalar-type spectral operator on X with resolution of the identity $E(\cdot)$. Suppose that $\sigma(S)$ is nowhere dense and $\rho(S)$ is connected. Let Y be a closed subspace of X invariant under S. Then $S|Y$ is spectral.*

Proof. If p is a polynomial, then $p(S)Y \subseteq Y$. By Lavrentieff's theorem and Proposition 5.9 (ii)

$$\int_{\sigma(S)} f(\lambda)E(d\lambda)Y \subseteq Y \qquad (f \in C(\sigma(S))).$$

It follows from Proposition 12.13 that $E(\tau)Y \subseteq Y$ $(\tau \in \Sigma_p)$ and so $S|Y$ is spectral by Theorem 12.3.

THEOREM 12.15. *Let S, in $L(X)$, be a scalar-type spectral operator whose spectrum is in R-set. Let Y be a closed subspace of X invariant under S. Then $S|Y$ is spectral if and only if $\sigma(S|Y) \subseteq \sigma(S)$.*

Proof. The necessity of the condition follows immediately from Proposition 12.1. To prove sufficiency, note that for any polynomial p, $p(S)Y \subseteq Y$. For each ζ in $\rho(S)$, $\zeta \in \rho(S|Y)$ and so, by Lemma 1.28, we have $(\zeta I - S)^{-1}Y \subseteq Y$. Therefore, for each rational function r, with poles lying in $\rho(S)$, we have $r(S)Y \subseteq Y$. Since $\sigma(S)$ is an R-set it follows from Proposition 5.9 (ii) that

$$\int_{\sigma(S)} f(\lambda)E(d\lambda)Y \subseteq Y \qquad (f \in C(\sigma(S)))$$

where $E(\cdot)$ is the resolution of the identity of S. It follows from Proposition 12.13 that $E(\tau)Y \subseteq Y$ $(\tau \in \Sigma_p)$ and so $S|Y$ is spectral by Theorem 12.3.

The next result gives a stronger sufficient condition than Theorem 12.15 for a restriction of a scalar-type spectral operator to be spectral, since it is an immediate consequence of the Tietze extension theorem that a closed subset of an R-set is an R-set.

THEOREM 12.16. *Let S be a scalar-type spectral operator on X. Let Y be a closed subspace of X invariant under S and such that $\sigma(S|Y)$ is an R-set. Then $S|Y$ is spectral.*

Proof. Let $S|Y_0$ be the minimal spectral extension of $S|Y$. By Proposition 12.11, $\sigma(S|Y_0) \subseteq \sigma(S|Y)$. Since $\sigma(S|Y)$ is an R-set it is nowhere dense and so, by Theorem 1.29, $\sigma(S|Y) \subseteq \sigma(S|Y_0)$. Hence $\sigma(S|Y) = \sigma(S|Y_0)$. However, $S|Y_0$ is spectral and of scalar-type by Theorem 12.3 so that, by Theorem 12.15, $S|Y$ is spectral. (Hence $Y = Y_0$.)

PROPOSITION 12.17. *Let S, in $L(X)$, be a scalar-type spectral operator whose spectrum is a simple closed curve Γ. Let Y be a closed subspace of X invariant under S. Then $S|Y$ fails to be spectral if and only if*

$$\sigma(S|Y) = \Gamma \cup (\Gamma')_1,$$

where $(\Gamma')_1$ denotes the bounded component of $\Gamma' = \mathbf{C}\backslash\Gamma$.

Proof. By the Jordan curve theorem, Γ' has precisely two components, $(\Gamma')_1$ and $(\Gamma')_\infty$ say, the latter being unbounded. Also Γ is the boundary of each component. The set Γ is an R-set and so, by Theorem 12.15, $S|Y$ fails to be spectral if and only if $\sigma(S|Y) \not\subseteq \sigma(S)$. By Theorem 1.29 this is equivalent to

$$(\Gamma')_1 \subseteq \sigma(S|Y),$$
$$(\Gamma')_\infty \cap \sigma(S|Y) = \varnothing.$$

Since $\sigma(S|Y)$ is closed and Γ is the boundary of $(\Gamma')_1$ this is equivalent to $\sigma(S|Y) = \Gamma \cup (\Gamma')_1$. This concludes the proof.

It will be shown in the next chapter, when we discuss restrictions of normal operators, that Theorem 12.14 is 'best possible'. For this purpose it is convenient to state the special cases of Theorems 12.14, 12.15, 12.16 and Proposition 12.17 in which the scalar-type spectral operator is a normal operator. In the next four results H is a non-zero complex Hilbert space.

THEOREM 12.18. *Let S be a normal operator on H such that $\sigma(S)$ is nowhere dense and $\rho(S)$ is connected. Let Y be a closed subspace of H invariant under S. Then $S \mid Y$ is a normal operator.*

THEOREM 12.19. *Let S be a normal operator on H such that $\sigma(S)$ is an R-set. Let Y be a closed subspace of H invariant under S. Then $S \mid Y$ is normal if and only if $\sigma(S \mid Y) \subseteq \sigma(S)$.*

THEOREM 12.20. *Let S be a normal operator on H and let Y be a closed subspace of H invariant under S such that $\sigma(S \mid Y)$ is an R-set. Then $S \mid Y$ is normal.*

PROPOSITION 12.21. *Let S be a normal operator on H and let $\sigma(S)$ be a simple closed curve Γ. Let Y be a closed subspace of H invariant under S. Then $S \mid Y$ fails to be normal if and only if*

$$\sigma(S \mid Y) = \Gamma \cup (\Gamma')_1,$$

where $(\Gamma')_1$ denotes the bounded component of $\Gamma' = \mathbf{C} \backslash \Gamma$.

We now show that Theorem 12.19 and hence also Theorem 12.15 may fail if the hypothesis that the spectrum of the operator is an R-set is omitted.

Example 12.22. Let H be the complex Hilbert space $L^2(\Omega, \mu)$, where $\mu(\cdot)$ denotes plane Lebesgue measure on the closed unit disc Ω in $\mathbf{C}$. Define

$$\begin{aligned} f_0(z) &= z \qquad && (z \in \Omega), \\ (Sf)(z) &= zf(z) \qquad && (z \in \Omega, f \in H), \\ Y &= \text{clm}\{f_0, Sf_0, S^2f_0, \ldots\}. \end{aligned}$$

Then Y is a closed subspace of H invariant under S. However, $S^*f_0 \notin Y$ since if n is a positive integer

$$\begin{aligned} \langle S^*f_0, S^{n-1}f_0 \rangle &= \int_\Omega (z\bar{z} \,.\, \bar{z}^n)\mu(dz) \\ &= \int_0^1 r^{n+2}\left(\int_0^{2\pi} e^{-ni\theta}\, d\theta\right) r\, dr = 0. \end{aligned}$$

Observe that $S^*f_0 \neq 0$. Hence Y is not a reducing subspace for S. By Proposition 12.4, $S|Y$ is not a normal operator. However, $\sigma(S) = \Omega$ and, since $\rho(S)$ is connected, it follows from Theorem 1.29 that $\sigma(S|Y) \subseteq \sigma(S)$.

To conclude this chapter we indicate the extent to which the preceding results generalize to non-scalar-type spectral operators.

THEOREM 12.23. *Let T, in $L(X)$, be a spectral operator whose spectrum is totally disconnected. Let Y be a closed subspace of X invariant under T. Then $T|Y$ is spectral.*

Proof. Let $E(\cdot)$ be the resolution of the identity for T and let τ be an open-and-closed subset of $\sigma(T)$. By Proposition 5.27

$$E(\tau) = \frac{1}{2\pi i}\int_\Gamma (\lambda I - T)^{-1} d\lambda,$$

where Γ is a suitable finite family of rectifiable Jordan curves enclosing τ but no other point of $\sigma(T)$. Now $\rho(T)$ is connected by [**11**; p. 123] and so, by Lemma 1.28 and Theorem 1.29,

$$(\lambda I - T)^{-1} Y \subseteq Y \qquad (\lambda \in \rho(T)).$$

It follows by norm continuity that $E(\tau)Y \subseteq Y$ and so

$$\langle E(\tau)x, y\rangle = 0 \qquad (x \in Y, y \in Y^\perp)$$

for each open-and-closed subset τ of $\sigma(T)$. Now $\sigma(T)$ is a separable metric space. Its topology has a base of open-and-closed subsets. By Lindelöf's theorem a countable base of open-and-closed subsets can be constructed. Hence if τ is an open subset of $\mathbf{C}$,

$$\langle E(\tau)x, y\rangle = \langle E(\tau \cap \sigma(T))x, y\rangle = 0 \qquad (x \in Y, y \in Y^\perp).$$

Since the class of sets on which a measure vanishes forms a σ-algebra

$$\langle E(\tau)x, y\rangle = 0 \qquad (x \in Y, y \in Y^\perp, \tau \in \Sigma_p).$$

From the standard properties of the annihilator $E(\tau)Y \subseteq Y$ $(\tau \in \Sigma_p)$ and so $T|Y$ is spectral by Theorem 12.3.

THEOREM 12.24. *Let T be a spectral operator on X. Let Y be a closed subspace of X invariant under T and such that $\sigma(T|Y)$ is totally disconnected. Then $T|Y$ is spectral.*

Proof. Let $T|Y_0$ be the minimal spectral extension of $T|Y$. By Proposition 12.11, $\sigma(T|Y_0) \subseteq \sigma(T|Y)$. Since $\sigma(T|Y)$ is totally disconnected it is nowhere dense and so, by Theorem 1.29, $\sigma(T|Y) \subseteq \sigma(T|Y_0)$. Hence $\sigma(T|Y) = \sigma(T|Y_0)$. However, $T|Y_0$ is spectral and so, by Theorem 12.23, $T|Y$ is spectral. (Hence $Y = Y_0$.)

The following example shows that Theorems 12.14 and 12.15 fail to generalize to the case of non-scalar-type spectral operators.

Example 12.25. Let X be the complex Hilbert space $L^2[0, 1] \oplus L^2[0, 1]$ where the measure is Lebesgue linear measure on $[0, 1]$ and $\oplus$ denotes orthogonal direct sum. Let S and N, in $L(X)$, be defined by

$$\begin{aligned} S&: (f_1(t), f_2(t)) \to (tf_1(t), tf_2(t)) \qquad (0 \leqslant t \leqslant 1), \\ N&: (f_1(t), f_2(t)) \to (0, f_1(t)) \qquad (0 \leqslant t \leqslant 1). \end{aligned}$$

Note that $SN = NS$ and $N^2 = 0$. Define $T = S + N$. Then T is a spectral operator with resolution of the identity $E(\cdot)$ given by

$$E(\tau): (f_1(t), f_2(t)) \to (\chi(\tau; t) f_1(t), \chi(\tau; t) f_2(t))$$

for $0 \leqslant t \leqslant 1$ and $\tau \in \Sigma_p$. Also $\sigma(T) = \sigma(S) = [0, 1]$. Let x be the element $(1, 0)$ of H. Consider the closed subspace Y of H defined by

$$\begin{aligned} Y &= \operatorname{clm}\{x, Tx, T^2x, \ldots\} \\ &= \operatorname{clm}\{(1, 0), (t, 1), (t^2, 2t), \ldots\}. \end{aligned}$$

Observe that $TY \subseteq Y$.

We show that we may regard Y as consisting of elements of H of the form (f, f'), where f is absolutely continuous on $[0, 1]$ and $f' \in L^2[0, 1]$. Also if $\{(f_n, f_n')\}$ is a sequence in Y converging to (f, f') (where f_n, f are absolutely continuous on $[0, 1]$ then $\{f_n\}$ converges to f uniformly on $[0, 1]$).

For each positive integer n let p_n be a polynomial. Suppose that the sequence $\{(p_n, p_n')\}$ converges to (f_1, f_2) in H. By Hölder's inequality,

$f_2 \in L^1[0, 1]$ and $\{p'_n\}$ converges to f_2 in $L^1[0, 1]$. On integrating we obtain

$$\left| \int_0^x p'_n(t)dt - \int_0^x f_2(t)dt \right| \leqslant \int_0^1 |p'_n - f_2| \qquad (0 \leqslant x \leqslant 1; n = 1, 2, 3, \ldots).$$

It follows that $\{p_n(x) - p_n(0)\}$ converges uniformly on $[0, 1]$ and hence this sequence of functions is a Cauchy sequence in $L^2[0, 1]$. However, $\{p_n\}$ is also a Cauchy sequence in $L^2[0, 1]$ and so the sequence $\{p_n(0)\}$ converges. Hence $\{p_n\}$ converges uniformly on $[0, 1]$ to some indefinite integral of f_2. However $\{p_n\}$ converges to f_1 in $L^2[0, 1]$. It follows that f_1 is a.e. equal to an absolutely continuous function, whose derivative is a.e. equal to f_2. Conversely, if f is absolutely continuous on $[0, 1]$ and $f' \in L^2[0, 1]$, then we can find a sequence of polynomials converging to f' in $L^2[0, 1]$. By integrating and adding a suitable constant we obtain a sequence $\{p_n\}$ of polynomials such that $\{(p_n, p'_n)\}$ converges to (f, f') in H. The correspondence between elements of Y and ordered pairs (f, f') with f absolutely continuous on $[0, 1]$ and $f' \in L^2[0, 1]$ is one-to-one. Henceforth we shall assume that each element of Y has been expressed in this way. The argument given previously for the special case of a sequence $\{(p_n, p'_n)\}$ in Y, where p_n is a polynomial, suffices to prove the following result. If $\{(f_n, f'_n)\}$ is a sequence in Y converging to (f, f') then $\{f_n\}$ converges uniformly to f on $[0, 1]$.

Observe that

$$T|Y: (f(t), f'(t)) \rightarrow (tf(t), f(t) + tf'(t)) \qquad (0 \leqslant t \leqslant 1).$$

Let $\chi_{[0, \frac{1}{2}]}$ be the characteristic function of the interval $[0, \frac{1}{2}]$. Since

$$(\chi_{[0, \frac{1}{2}]}, 0) = E([0, \tfrac{1}{2}])x \notin Y$$

it follows from Theorem 12.2 that $T|Y$ is not spectral.

13. Property (P) for Normal Operators

In the last chapter we initiated the study of the following problem: when is a closed invariant subspace of a normal operator reducing in the Hilbert space sense? The purpose of the present chapter is to study this question in more detail. Throughout this chapter we use the term spectral measure to mean a spectral measure whose values are self-adjoint projections on a Hilbert space.

We begin with the following definition.

Definition 13.1. Let T be a normal operator on H. Then T is said to have *property* (P) if every closed subspace of H invariant under T is reducing for T.

The following result, first proved by Wermer in 1952, follows immediately from Theorem 12.18 and Proposition 12.4.

THEOREM 13.2. *Let T be a normal operator on H such that $\sigma(T)$ is nowhere dense and $\rho(T)$ is connected. Then T has property (P).*

We show that this result is 'best possible'.

THEOREM 13.3. *Let K be a compact subset of* $\mathbf{C}$, *which has either non-empty interior or a disconnected complement. Then there exist a Hilbert space H, and a normal operator S in $L(H)$ such that $\sigma(S) = K$ and property (P) fails for S.*

Proof. Let $P(K)$ denote the closed subspace of $C(K)$ generated by the polynomials. Since $P(K) \neq C(K)$, there is a continuous linear functional on the Banach space $C(K)$ which annihilates $P(K)$ but is not identically zero. By the Riesz representation theorem, this linear functional determines a unique complex measure μ_1 defined and countably additive on Σ_K, the σ-algebra of Borel subsets of K. Define, for each Borel subset τ of $\mathbf{C}$, $\mu_1'(\tau) = \mu_1(\tau \cap K)$. Introduce the total variation $\mu(\tau)$ of μ_1' on τ for each τ in Σ_p. μ is a finite Borel measure. Also, the support of μ (K_0 say) is equal to the support of μ'_1, and so $K_0 \subseteq K$.

Let H_1 be the Hilbert space $L^2(\mu)$, and let S_1, in $L(H_1)$, be the normal operator defined by

$$(S_1 f)(t) = tf(t) \qquad (t \in K_0, f \in H_1).$$

Then $\sigma(S_1) = K_0 \subseteq K$. If $K_0 \neq K$, there exists an at most countable dense subset $\{\lambda_n : n = 1, 2, 3, \ldots\}$ of $K \backslash K_0$. Let $\{\phi_n : n = 1, 2, 3, \ldots\}$ be a complete orthonormal system in a Hilbert space H_2. Define S_2, in $L(H_2)$, by

$$S_2 \phi_n = \lambda_n \phi_n \qquad (n = 1, 2, 3, \ldots).$$

Clearly $K \backslash K_0 \subseteq \sigma(S_2) \subseteq K$. Define S, in $L(H_1 \oplus H_2)$, by $S = S_1 \oplus S_2$, where $\oplus$ denotes orthogonal direct sum. Let $\lambda \in \mathbf{C} \backslash K$. Then

$$\lambda \in \rho(S_1) \cap \rho(S_2).$$

Moreover, $(\lambda I | H_1 - S_1)^{-1} \oplus (\lambda I | H_2 - S_2)^{-1}$ is in $L(H_1 \oplus H_2)$ and is inverse to $\lambda I - S$. Therefore $\mathbf{C} \backslash K \subseteq \rho(S)$ and $\sigma(S) \subseteq K$. Since S_1 and S_2, restrictions of the normal operator S, are also normal we have by Proposition 12.5

$$K_0 = \sigma(S_1) \subseteq \sigma(S) \quad \text{and} \quad K \backslash K_0 \subseteq \sigma(S_2) \subseteq \sigma(S).$$

If follows that $K = \sigma(S)$. If $K_0 = K$, take $S = S_1$. Again $K = \sigma(S)$. We shall show that property (P) fails for S. Since μ'_1 is absolutely continuous with respect to μ, it follows from the Radon-Nikodym theorem that there is f in $L^1(\mu)$ such that

$$\mu'_1(\tau) = \int_\tau f(\lambda)\mu(d\lambda) \qquad (\tau \in \Sigma_p).$$

Define

$$f_1(t) = |f(t)|^{\frac{1}{2}} \qquad (t \in K_0),$$

$$f_2(t) = \frac{f(t)}{|f(t)|^{\frac{1}{2}}} \qquad (f(t) \neq 0),$$

$$f_2(t) = 0 \qquad (f(t) = 0).$$

Then $f_1, f_2 \in L^2(\mu)$ and $f_1 f_2 = f$. Also if χ_τ denotes the characteristic function of τ then

$$\mu_1'(\tau) = \int_{K_0} (\chi_\tau f)(\lambda)\mu(d\lambda) = \int_{K_0} (\chi_\tau f_1)(f_2)(\lambda)\mu(d\lambda) = \langle E(\tau)f_1, \bar{f}_2\rangle \qquad (\tau \in \Sigma_p)$$

where $E(\cdot)$ is the resolution of the identity of S_1. Also

$$\int_{K_0} \lambda^n \mu_1'(d\lambda) = 0 \qquad (n = 0, 1, 2, \ldots).$$

This is equivalent to

$$\langle S_1^n f_1, \bar{f}_2\rangle = 0 \qquad (n = 0, 1, 2, \ldots).$$

However, for some τ in Σ_p, $\langle E(\tau)f_1, \bar{f}_2\rangle \neq 0$. Let

$$Y = \mathrm{clm}\{S_1^n f_1 : n = 0, 1, 2, \ldots\}.$$

By the properties of the orthogonal complement and Theorem 12.2, $S_1 | Y$ is not spectral. Hence $S | Y$ is not spectral and so, by Proposition 12.4, property (P) fails for S.

Wermer observed that possession of property (P) cannot be characterized in terms of the spectrum of the normal operator. The following two examples illustrate this point.

Example 13.4. Let H be a separable Hilbert space and let

$$\{\phi_n : n = 0, \pm 1, \pm 2, \ldots\}$$

be an orthonormal basis for H. Define U, in $L(H)$, by

$$U\phi_n = \phi_{n+1} \qquad (n = 0, \pm 1, \pm 2, \ldots).$$

Let

$$Y = \mathrm{clm}\{\phi_r : r = 1, 2, 3, \ldots\}.$$

Then $UY \subseteq Y$. However, $U^*\phi_1 = U^{-1}\phi_1 = \phi_0 \notin Y$. Hence property (P)

fails for the unitary operator U. By Theorem 13.2

$$\sigma(U) = \{z : |z| = 1\}.$$

Example 13.5. Let H be a separable Hilbert space and let $\{\phi_n : n \in \mathbf{N}\}$ be an orthonormal basis for H. Let $\{r_n : n \in \mathbf{N}\}$ be an enumeration of the rationals in $[0, 1]$. Let A, in $L(H)$, be the unitary operator defined by

$$A\phi_n = \alpha_n \phi_n \qquad (n \in \mathbf{N})$$

where

$$\alpha_n = \exp(2\pi i r_n) \qquad (n \in \mathbf{N}).$$

Suppose that, for each integer n, $r_n = p_n/q_n$, where $q_n \geqslant 1$ and p_n and q_n are mutually prime. Let $\Gamma = \{z \in \mathbf{C} : |z| = 1\}$. Then $\sigma(A) \subseteq \Gamma$. Also $\sigma(A)$ is closed and contains a dense subset of Γ. Hence $\sigma(A) = \Gamma$. We show that A has property (P). Let Y be a closed subspace of H with $AY \subseteq Y$, and let $x \in Y$. Then x can be expressed in the form $\sum_{i=1}^{\infty} \beta_i \phi_i$, where $\sum_{i=1}^{\infty} |\beta_i|^2 < \infty$. Let $\varepsilon > 0$ be given. Choose k such that

$\sum_{i=k+1}^{\infty} |\beta_i|^2 < (\frac{1}{2}\varepsilon)^2$. Let $x = y + z$, where $y = \sum_{i=1}^{k} \beta_i \phi_i$ and $z = \sum_{k+1}^{\infty} \beta_i \phi_i$.

Now

$$A^{-1} \sum_{i=1}^{k} \beta_i \phi_i = A^{q_1 q_2 \cdots q_k - 1} \sum_{i=1}^{k} \beta_i \phi_i.$$

Denoting $q_1 q_2 \cdots q_k - 1$ by n_k, we can write the preceding equation as

$$A^{-1} y = A^{n_k} y.$$

Now $\| A^{n_k} z \| = \| z \| < \frac{1}{2}\varepsilon$ and $\| A^{-1} z \| = \| z \| < \frac{1}{2}\varepsilon$, whence

$$\| A^{n_k} x - A^{-1} x \| \leqslant \| A^{n_k} y - A^{-1} y \| + \| A^{n_k} z - A^{-1} z \| < \varepsilon.$$

Since Y is closed, $A^{-1} x = \lim_{k \to \infty} A^{n_k} x \in Y$, and so $A^* Y \subseteq Y$. Therefore A has property (P).

Observe that we have in fact shown that A^* is the limit in the strong operator topology of a sequence of polynomials in A.

We shall next give a necessary and sufficient condition on the spectral measure of a unitary operator for it to possess property (P). Some preliminary definitions and results are required.

Definition 13.6. Let $E(\cdot)$ be a spectral measure and $\mu(\cdot)$ a measure defined on Σ_p. Then $\mu(\cdot)$ is called *absolutely continuous with respect* to $E(\cdot)$, and we write $\mu(\cdot) \ll E(\cdot)$ if $\tau \in \Sigma_p$, $E(\tau) = 0$ together imply $\mu(\tau) = 0$.

Definition 13.7. Let $E(\cdot)$ be a spectral measure defined on Σ_p and with values in $L(H)$. The *cyclic subspace* $M(x)$ corresponding to x in H is defined by

$$M(x) = \operatorname{clm}\{E(\tau)x : \tau \in \Sigma_p\}.$$

PROPOSITION 13.8. *Let $E(\cdot)$ be a spectral measure defined on Σ_p and with values in $L(H)$. If $x \in H$, and if μ is defined on Σ_p by*

$$\mu(\tau) = \langle E(\tau)x, x\rangle \qquad (\tau \in \Sigma_p)$$

then there is an isometric isomorphism U from $L^2(\mu)$ onto the subspace $M(x)$ such that

$$U^{-1}E(\tau)Uf = \chi_\tau f \qquad (f \in L^2(\mu), \tau \in \Sigma_p),$$

where χ_τ denotes the characteristic function of τ.

Proof. We write $U\chi_\tau = E(\tau)x$ for every τ in Σ_p. If the definition of U is extended from characteristic functions to simple functions by the requirement of linearity, then, in view of the definition of $M(x)$, U becomes a linear transformation from a dense subset of $L^2(\mu)$ onto a dense subset of $M(x)$. The additivity of $E(\cdot)$ guarantees that U is well-defined. Since the equations

$$\|\chi_\tau\|^2 = \mu(\tau) = \langle E(\tau)x, x\rangle = \|E(\tau)x\|^2 = \|U\chi_\tau\|^2$$

show that U is norm-preserving, U may be extended to an isomorphism. If $\tau, \delta \in \Sigma_p$, then

$$\begin{aligned} U(\chi_\tau \cdot \chi_\delta) &= U(\chi_{\tau \cap \delta}) = E(\tau \cap \delta)x \\ &= E(\tau)E(\delta)x = E(\tau)U\chi_\delta. \end{aligned}$$

This means that $U(\chi_\tau f) = E(\tau)Uf$ whenever $f = \chi_\delta$. Since the simple functions are norm dense in $L^2(\mu)$ the desired conclusion follows.

COROLLARY 13.9. *$M(x)$ is separable, for every x in H.*

PROPOSITION 13.10. *Let $E(\cdot)$ be a spectral measure defined on Σ_p and with values in $L(H)$. Suppose that v is a finite measure on Σ_p, and x is a vector in H such that $v \ll \langle E(\cdot)x, x\rangle$. Then there is y in $M(x)$ such that*

$$v(\tau) = \langle E(\tau)y, y\rangle \qquad (\tau \in \Sigma_p).$$

Proof. If $\mu(\tau) = \langle E(\tau)x, x\rangle$ $(\tau \in \Sigma_p)$, then, by the Radon-Nikodym theorem, there exists a non-negative function g in $L^1(\mu)$ such that

$$v(\tau) = \int_\tau g(\lambda)\mu(d\lambda) \qquad (\tau \in \Sigma_p).$$

If f is the non-negative square root of g, then $f \in L^2(\mu)$. If $y = Uf$, where U is the isomorphism described in Proposition 13.8, then

$$\begin{aligned} v(\tau) &= \int_\tau |f(\lambda)|^2\mu(d\lambda) = \|\chi_\tau f\|^2 = \|U(\chi_\tau f)\|^2 \\ &= \|E(\tau)Uf\|^2 = \|E(\tau)y\|^2 = \langle E(\tau)y, y\rangle \qquad (\tau \in \Sigma_p). \end{aligned}$$

Definition 13.11. Let $E(\cdot)$ be a spectral measure defined on Σ_p and with values in $L(H)$. A vector x in H is called a *separating vector* for $E(\cdot)$ if $\tau \in \Sigma_p$, $E(\tau)x = 0$ together imply that $E(\tau) = 0$.

PROPOSITION 13.12. *Let H be separable and let $E(\cdot)$ be a spectral measure defined on Σ_p and with values in $L(H)$. Then there is a separating vector x for $E(\cdot)$.*

Proof. We are assuming that $H \neq \{0\}$. Choose a maximal family $\{x_\alpha\}$ of unit vectors in H such that $M(x_\alpha) \perp M(x_\beta)$, if $\alpha \neq \beta$. Since H is separable, this family is countable. Call it $\{x_r : r = 1, 2, 3, \ldots\}$. We claim that if

$$Y = M(x_1) \oplus M(x_2) \oplus \ldots \oplus M(x_r) \oplus \ldots$$

then $Y = H$. Suppose the contrary, and let $y \in Y^\perp$ with $\|y\| = 1$. Then for each r, and every δ, τ in Σ.

$$0 = \langle E(\tau \cap \delta)x_r, y\rangle = \langle E(\tau)E(\delta)x_r, y\rangle = \langle E(\delta)x_r, E(\tau)y\rangle$$

and so $M(x_r) \perp M(y)$, for every r. This contradicts the maximality of the family $\{x_r : r = 1, 2, 3, \ldots\}$. Now let $x = \sum_{r=1}^{\infty} 2^{-r}x^r$. If for some τ in Σ_p we have $\langle E(\tau)x, x\rangle = 0$, then since

$$0 = \langle E(\tau)x, x\rangle = \| E(\tau)x \|^2 = \sum_{r=1}^{\infty} \frac{\| E(\tau)x_r \|^2}{2^{2r}}$$

we obtain $\| E(\tau)x_r \| = 0$, for every r. Hence $E(\tau)x_r = 0$ and $E(\tau)|M(x_r) = 0$, for every r. Therefore $E(\tau) = 0$, completing the proof.

PROPOSITION 13.13. *Let H be separable. Let $E(\cdot)$ be a spectral measure defined on Σ_p with values in $L(H)$. If μ is a finite measure on Σ_p with $\mu(\cdot) \ll E(\cdot)$, then there is a vector x in H such that*

$$\mu(\tau) = \langle E(\tau)x, x\rangle \qquad (\tau \in \Sigma_p).$$

Proof. By Proposition 13.12 there is a separating vector y for $E(\cdot)$. Hence

$$\mu \ll \langle E(\cdot)y, y\rangle = \| E(\cdot)y \|^2.$$

Now by Proposition 13.10 there is a vector x in $M(y)$ such that

$$\mu(\tau) = \langle E(\tau)x, x\rangle \qquad (\tau \in \Sigma_p).$$

The next two results, giving necessary and sufficient conditions for a unitary operator to possess property (P), are due to Wermer.

THEOREM 13.14. *Let U be a unitary operator on H. Then U has property (P) if and only if there does not exist an orthonormal sequence $\{\phi_k : -\infty < k < \infty\}$ with $U\phi_k = \phi_{k+1}$, for all k.*

Proof. Suppose there exists such a set $\{\phi_k\}$. Then if $Y = \text{clm}\{\phi_n : n = 1, 2, 3, \ldots\}$ we have $UY \subseteq Y$ but $U^*\phi_1 = U^{-1}\phi_1 = \phi_0 \notin Y$. Hence property (P) fails for U. Conversely, assume there exists a closed subspace Y of H invariant under U but not under U^*. Hence there is ϕ in Y such that $U^*\phi = U^{-1}\phi \notin Y$. Then $U^{-1}\phi = \psi + \phi_1$, where $\phi \in Y$, $\phi_1 \neq 0$ and $\phi_1 \in Y^\perp$. Now

$$U^n\phi_1 = U^{n-1}\phi - U^n\psi \in Y \qquad (n = 1, 2, 3, \ldots)$$

and so $\phi_1 \perp U^n\phi_1$, for all $n \geqslant 1$. Hence if we define

$$\phi_k = U^{k-1}\phi_1 \qquad (k = 0, \pm 1, \pm 2, \ldots)$$

then $\langle \phi_k, \phi_m \rangle = \langle U^{k-1}\phi_1, U^{m-1}\phi_1 \rangle = \langle \phi_1, U^{m-k}\phi_1 \rangle = 0\,(m \neq k)$. In other words, the $\{\phi_k\}$ form an orthonormal set and $U\phi_k = \phi_{k+1}$, for every k.

THEOREM 13.15. *Let H be separable and let U be a unitary operator on H. Let $E(\cdot)$ be the resolution of the identity for U. Let $m(\cdot)$ denote Lebesgue lineal measure on $T = \{z: |z| = 1\}$. Then property (P) fails for U if and only if $m(\cdot) \ll E(\cdot)$.*

Proof. Suppose that property (P) fails for U. Then by Theorem 13.14 we can find x in H such that $x \neq 0$ and $\langle U^k x, x \rangle = 0$ $(k \neq 0)$. Let

$$\mu(\tau) = \langle E(\tau)x, x \rangle \qquad (\tau \in \Sigma_p).$$

Then

$$\int_T \xi^k \, d\mu(\xi) = 0 \qquad (k \neq 0).$$

Observe that

$$\int_T \xi^k dm(\xi) = \int_0^{2\pi} e^{ik\theta} d\theta = 0 \qquad (k \neq 0).$$

The set of finite linear combinations of functions of the form $\{\xi^k : k \in \mathbf{Z}\}$ is, by the Stone-Weierstrass theorem, dense in $C(T)$. Hence there is $\lambda > 0$ such that

$$\int_T \lambda f(\xi) dm(\xi) = \int_T f(\xi) d\mu(\xi) \qquad (f \in C(T)),$$

and so, by the Riesz representation theorem, $\mu(\cdot) = \lambda m(\cdot)$. Hence, since $\mu(\cdot) \ll E(\cdot)$, we have also $m(\cdot) \ll E(\cdot)$. Conversely, suppose that $m(\cdot) \ll E(\cdot)$. Then by Proposition 13.13, there is an x in H with

$$m(\tau) = \langle E(\tau)x, x\rangle \qquad (\tau \in \Sigma_p).$$

Hence

$$\langle U^k x, x\rangle = \int_T \xi^k dm(\xi) = \int_0^{2\pi} e^{ik\theta} d\theta = 0 \qquad (k \neq 0)$$

and $\langle x, x\rangle = 2\pi$. Therefore $x \neq 0$ and property (P) fails for U by Theorem 13.14. The proof is complete.

Note that the proof of the preceding theorem shows that, in a general complex Hilbert space, if property (P) fails for U then $m(\cdot) \ll E(\cdot)$. It will be shown later in this chapter that in general the converse is false.

Using conformal mapping techniques, Wermer established the following result.

THEOREM 13.16. *Let H be separable. Let T be a normal operator on H with resolution of the identity $E(\cdot)$. Suppose that the spectrum of T lies in a closed rectifiable curve Γ which bounds a simply connected region D. Let μ be the unique measure defined on Σ_p which is zero on subsets of $\mathbf{C}\backslash\Gamma$ and which assigns to each arc of Γ its length. Then property (P) fails for T if and only if $\mu \ll E(\cdot)$.*

For a proof of this generalization of Theorem 13.15 see Theorem 9 of Wermer [**1**; p. 276–7].

Consider the following classical result due to F. and M. Riesz. See for example [**14**; p. 335].

If μ is a complex Borel measure on the unit circle T and if

$$\int_T e^{-nit} d\mu(e^{it}) = 0 \qquad (n = -1, -2, -3, \ldots)$$

then $\mu(\cdot) \ll m(\cdot)$, where $m(\cdot)$ denotes Lebesgue lineal measure on T.

We deduce the following improved version of this theorem, which has been proved by different methods. See [**10**; p. 51–2].

THEOREM 13.17. *If μ is a complex Borel measure on the unit circle T, $\mu \neq 0$, and if*

$$\int_T e^{-nit} d\mu(e^{it}) = 0 \qquad (n = -1, -2, -3, \ldots)$$

then $m(\cdot) \ll |\mu|$ and so $m(\cdot)$ and $|\mu|$ are equivalent.

Proof. $|\mu|$ denotes the total variation of μ. (See for example [**14**; p. 117–21]. Then $\mu \ll |\mu|$. Define a unitary operator U on $L^2(|\mu|)$ by

$$(Uf)(t) = tf(t) \qquad (t \in T, f \in L^2(|\mu|)).$$

Let $E(\cdot)$ be the resolution of the identity of U. The spectral measure $E(\cdot)$ is clearly equivalent to $|\mu|$. By the Radon-Nikodym theorem there is f in $L^1(|\mu|)$ such that

$$\mu(\tau) = \int_\tau f d|\mu| \qquad (\tau \in \Sigma_T),$$

where Σ_T denotes the class of Borel subsets of T. Define

$$f_1(t) = |f(t)|^{1/2} \qquad (t \in T),$$

$$f_2(t) = \frac{f(t)}{|f(t)|^{1/2}} \qquad (f(t) \neq 0),$$

$$f_2(t) = 0 \qquad (f(t) = 0).$$

Then $f_1, f_2 \in L^2(|\mu|)$ and $f_1 f_2 = f$. Also

$$\mu(\tau) = \int_T \chi_\tau f d|\mu| = \int_T (\chi_\tau f_1)(f_2) d|\mu| = \langle E(\tau) f_1, \bar{f}_2 \rangle \qquad (\tau \in \Sigma_T).$$

Observe that

$$\langle U^m f_1, \bar{f}_2 \rangle = \int_T e^{mi\theta} d\mu(e^{i\theta}) \qquad (m = 0, \pm 1, \pm 2, \ldots)$$

and that if m is a positive integer the right-hand side of the equation is 0. If this were so for every integer m, then by the Stone-Weierstrass theorem we should have

$$\int_T f(e^{i\theta})d\mu(e^{i\theta}) = 0 \qquad (f \in C(T))$$

and hence by the Riesz representation theorem $\mu = 0$. Since $\mu \neq 0$, we have $\langle U^m f_1, \bar{f}_2 \rangle \neq 0$, for some m. Hence $\text{clm}\{U^n f_1 : n = 1, 2, 3, \ldots\}$ is an invariant but not a reducing subspace for U. Therefore property (P) fails for U. Hence by Theorem 13.15

$$m(\cdot) \ll E(\cdot) \ll |\mu|(\cdot).$$

Also $|\mu| \ll m(\cdot)$ follows from the F. and M. Riesz theorem and Proposition 6.8 (e) of [**14**; p. 121].

We next show Proposition 13.13 and Theorem 13.15 may both fail if the hypothesis of separability on the Hilbert space is omitted.

Example 13.18. Let $T = \{z : |z| = 1\}$. Consider the set H consisting of all complex functions f, defined on T, vanishing off some countable set $G(f) \subseteq T$, and such that

$$\sum_{t \in T} |f(t)|^2 < \infty.$$

With the usual vector space operations, defined by setting

$$(\alpha f + \beta g)(t) = \alpha f(t) + \beta g(t) \qquad (t \in T;\ f, g \in H)$$

inner-product and norm defined respectively by

$$\langle f, g \rangle = \sum_{t \in T} f(t)\overline{g(t)} \qquad (f, g \in H),$$

$$\| f \| = \left(\sum_{t \in T} |f(t)|^2 \right)^{1/2} \quad (f \in H),$$

H has the structure of a Hilbert space. (For a proof of the completeness of H, see for example Theorem 3.13 of [**15**; p. 100].) Define a unitary operator U on H by

$$(Uf)(t) = tf(t) \qquad (t \in T,\ f \in H).$$

The resolution of the identity $E(\cdot)$ for U is given by

$$(E(\tau)f)(t) = \chi_\tau(t)\, f(t) \qquad (t \in T,\, f \in H,\, \tau \in \Sigma_T).$$

Next we show that if $\tau \in \Sigma_T$ and $\tau \neq \varnothing$ then $E(\tau) \neq 0$. Suppose that $\lambda \in \tau$. Then $E(\tau)\chi_{\{\lambda\}} = \chi_{\{\lambda\}}$. It follows immediately that if $m(\cdot)$ denotes Lebesgue lineal measure on T then $m(\cdot) \ll E(\cdot)$. Also if $x \in H$, then $\langle E(\cdot)x, x\rangle$ vanishes off some countable subset of T. Therefore, for all x in H, $m(\cdot) \neq \langle E(\cdot)x, x\rangle$. This shows that Proposition 13.13 may fail if the hypothesis of separability is omitted. Suppose now that property (P) fails for U. The argument of Theorem 13.15 shows that there is a vector y in H such that $m(\cdot) = \langle E(\cdot)y, y\rangle$. As explained above this is impossible. Hence although we have $m(\cdot) \ll E(\cdot)$, property (P) holds in this case and so Theorem 13.15 may also fail if the hypothesis of separability is dropped.

The theory of $*$-algebras of operators on a Hilbert space has been extensively developed by Murray, von Neumann and many others. We present only a single result from the theory due to von Neumann. This is the double commutant theorem, which is of fundamental importance in the theory of $*$-algebras of operators.

Definition 13.19. A *$*$-algebra* of operators on H is a subalgebra $\mathscr{A}$ of $L(H)$ containing I with the property: if $T \in \mathscr{A}$ then $T^* \in \mathscr{A}$.

Definition 13.20. If $\mathscr{A} \subseteq L(H)$, the *commutant* $\mathscr{A}'$ of $\mathscr{A}$ is the set of all operators on H which commute with every operator in $\mathscr{A}$.

LEMMA 13.21. *The commutant $\mathscr{A}'$ of a $*$-algebra $\mathscr{A}$ of operators on H is also a $*$-algebra. Moreover $\mathscr{A}'$ is closed in the weak operator topology of $L(H)$.*

Proof. Let $A \in \mathscr{A}$, $B \in \mathscr{A}'$. Then $A^* \in \mathscr{A}$ and $BA^* = A^*B$. Hence $AB^* = B^*A$, for every A in $\mathscr{A}$, and so $B^* \in \mathscr{A}'$. If $\{B_\alpha\}$ is a generalized sequence in $\mathscr{A}'$ converging in the weak operator topology to an operator B, then for every A in $\mathscr{A}$ and x, y in H we have

$$\langle BAx, y\rangle = \lim_\alpha \langle B_\alpha Ax, y\rangle = \lim_\alpha \langle AB_\alpha x, y\rangle = \langle ABx, y\rangle,$$

and so $B \in \mathscr{A}'$.

We now prove von Neumann's double commutant theorem.

THEOREM 13.22. *A $*$-algebra of operators on H is equal to the commutant of its commutant if and only if it is closed in the weak operator topology of $L(H)$.*

Proof. It is clear from the definition of the commutant that $\mathscr{A} \subseteq (\mathscr{A}')'$ and from Lemma 13.21 that $\mathscr{A}$ is closed in the weak operator topology if $\mathscr{A} = (\mathscr{A}')'$. To prove the theorem it suffices to show that if $\mathscr{A}$ is closed in the weak operator topology, then every neighbourhood of a point B in $(\mathscr{A}')'$ contains an element of $\mathscr{A}$. In order to illustrate the idea of the proof this statement will be demonstrated first for a neighbourhood of the form

$$N_1 = \{A \in L(H): |\langle x_1, (A - B)y_1\rangle < \varepsilon\},$$

where $\varepsilon > 0$ and x_1, y_1 are fixed non-zero elements of H. Let E be the orthogonal projection of H onto $M = \text{clm}\{Ay_1 : A \in \mathscr{A}\}$. For every A in $\mathscr{A}$ we have $AM \subseteq M$ and $A^*M \subseteq M$; hence $EAE = AE$ and $EA^*E = A^*E$. Taking adjoints in this last equation, it follows that $EA = EAE = AE$ and so $E \in \mathscr{A}'$. Hence $By_1 = BEy_1 = EBy_1 \in M$ and so there is an A in $\mathscr{A}$ with $|Ay_1 - By_1| < \varepsilon/\|x_1\|$, which implies that A is in N_1, as required.

We now consider an arbitrary neighbourhood

$$N = \{A \in L(H): |\langle x_i, (A - B)y_i\rangle| < \varepsilon, i = 1, \ldots, n\}$$

of B, and we may assume that $\|x_i\| \leqslant 1$. Let $H^{(n)}$ be the orthogonal direct sum of H with itself n times, and for each operator A defined on H let $A^{(n)}$ be the operator defined on $H^{(n)} = H \oplus \ldots \oplus H$ by the equation

$$A^{(n)}[z_1, \ldots, z_n] = [Az_1, \ldots, Az_n].$$

The set of $A^{(n)}$ with A in $\mathscr{A}$ will be denoted by $\mathscr{A}^{(n)}$. Now the general bounded linear operator C on $H^{(n)}$ has the form

$$C[z_1, \ldots, z_n] = \left[\sum_{i=1}^{n} C_{1i}z_i, \ldots, \sum_{i=1}^{n} C_{ni}z_i\right],$$

and therefore $(\mathscr{A}^{(n)})'$ consists of those C for which all the operators C_{ij} are in $\mathscr{A}'$. From this we see that $((\mathscr{A}^{(n)})')'$ consists of those operators of the form $\mathscr{A}^{(n)}$ with A in $(\mathscr{A}')'$. In view of this representation for the general element in $((\mathscr{A}^{(n)})')'$, the argument used in the case $n = 1$ may be applied to obtain an

$A^{(n)}$ in $\mathscr{A}^{(n)}$ with

$$\|(A^{(n)} - B^{(n)})[y_1, \ldots, y_n]\| = \sum_{i=1}^{n} \|(A - B)y_i\| < \varepsilon$$

from which it follows that A is in N.

This theorem is of importance since it characterizes $*$-algebras, closed in the weak operator topology, in terms of an algebraic property and suggests that they are of particular interest. This leads to the following definition.

Definition 13.23. A $*$-algebra of operators on H which is closed in the weak operator topology of $L(H)$ is called a *von Neumann algebra.*

Sarason [**1**] has proved an analogue of von Neumann's double commutant theorem valid for certain non self-adjoint algebras of operators on a Hilbert space. In order to motivate Sarason's theorem we begin by proving two corollaries to the double commutant theorem.

PROPOSITION 13.24. *Let A be a normal operator on H. Suppose that B, in $L(H)$, commutes with every self-adjoint projection that commutes with A. Then B is in the von Neumann algebra $\mathscr{A}$ generated by I and A.*

Proof. By Lemma 13.21, $\mathscr{A}'$ is a $*$-subalgebra of $L(H)$. Let $T \in \mathscr{A}'$. Then

$$T = \tfrac{1}{2}(T + T^*) + i\left\{\frac{1}{2i}(T - T^*)\right\}. \tag{1}$$

If now S is an arbitrary self-adjoint operator in $\mathscr{A}'$ and $A \in \mathscr{A}$, then by Fuglede's theorem A commutes with all the spectral projections of S, and hence these all lie in $\mathscr{A}'$. By hypothesis, B commutes with these spectral projections and so, by the spectral theorem, $BS = SB$. Since an arbitrary operator T in $\mathscr{A}'$ can, by (1), be expressed as a linear combination of self-adjoint operators in $\mathscr{A}'$, B commutes with every operator in $\mathscr{A}'$. Hence, by von Neumann's double commutant theorem, $B \in \mathscr{A}$.

The last theorem may be restated in the following form.

COROLLARY 13.25. *Let A be a normal operator on H. Suppose that B, in $L(H)$, is reduced by every closed subspace of H reducing A. Then B is in the closed $*$-subalgebra of $L(H)$, generated by I and A in the weak operator topology.*

It is in this form that Sarason's analogue of von Neumann's double commutant theorem is obtained. Observe that a stronger hypothesis leads to a stronger conclusion.

THEOREM 13.26. *Let A be a normal operator on H. Suppose that B, in $L(H)$, leaves invariant every closed subspace of H left invariant by A. Then B is in the closed subalgebra of $L(H)$, generated by I and A in the weak operator topology.*

In order to prove this result we require some preliminary notation and a lemma. Let H and A be as in Theorem 13.26 and let $E(\cdot)$ be the resolution of the identity of A. For m a natural number, we let H_m denote the orthogonal direct sum of m copies of H and A_m denotes the direct sum of A with itself m times (regarded as an operator on H_m). Then $E_m(\cdot)$, the direct sum of $E(\cdot)$ with itself m times, is the resolution of the identity of A_m.

LEMMA 13.27. *With the notations above, let B be an operator on H which leaves invariant every closed subspace of H invariant under A. Then B_m leaves invariant every closed subspace of H_m invariant under A_m ($m = 1, 2, 3, \ldots$).*

Proof. Let $x \in H_m$ and let Y be the smallest closed reducing subspace of A_m containing x. Suppose that $x = (x_1, \ldots, x_m)$. Then

$$\langle E_m(\cdot)x, x\rangle = \sum_{r=1}^{m} \langle E(\cdot)x_r, x_r\rangle.$$

Define

$$\tilde{H} = \operatorname{clm}\{E(\tau)x_r : \tau \in \Sigma_p, r = 1, \ldots, m\}.$$

By Corollary 13.9, each $M(x_r)$ is separable. Hence, since a countable union of countable sets is countable, $\tilde{H}$ is a closed separable subspace of H. Also H is invariant under $\{E(\tau): \tau \in \Sigma_p\}$. Let $\tilde{E}(\cdot) = E(\cdot)|\tilde{H}$. By Proposition 13.12, there is a separating vector $\tilde{x}$ for $\tilde{E}(\cdot)$ with the property

$$\begin{aligned}
\tau \in \Sigma_p, \langle E(\tau)\tilde{x}, \tilde{x}\rangle = 0 &\Rightarrow \tilde{E}(\tau) = 0 \\
&\Rightarrow \langle E(\tau)x_r, x_r\rangle = 0 \qquad (r = 1, \ldots, m) \\
&\Rightarrow \langle E_m(\tau)x, x\rangle = 0.
\end{aligned}$$

Hence $\langle E_m(\cdot)x, x\rangle \ll \langle E(\cdot)\tilde{x}, \tilde{x}\rangle$. Therefore, by Proposition 13.10, there is y in $M(\tilde{x}) \subseteq H$ such that $\langle E(\cdot)y, y\rangle = \langle E_m(\cdot)x, x\rangle$. Let N be the smallest closed reducing subspace for A containing y; that is $N = M(y)$. Since $\langle E_m(\cdot)x, x\rangle = \langle E(\cdot)y, y\rangle$ the operators $A_m|Y$ and $A|N$ are unitarily equivalent, using Proposition 13.8. Hence there is an isometry V of N onto Y such that $A_m|Y = VAV^{-1}$. It follows that if q is any complex polynomial in two variables then

$$q(A_m, A_m^*)|Y = Vq(A, A^*)V^{-1}.$$

However, by Proposition 13.24, there is a net $\{q_\alpha\}$ of such polynomials with $q_\alpha(A, A^*) \to B$ in the weak operator topology. Therefore also $q_\alpha(A_m, A_m^*) \to B_m$ in the weak operator topology of $L(H_m)$. It follows that $B_m|Y = VBV^{-1}$. Hence V maps closed invariant subspaces for B onto closed invariant subspaces for $B_m|Y$. Let L be the smallest closed subspace of H_m invariant under A_m and containing x. Then $V^{-1}L$ is invariant under A, and therefore also under B. Hence L is invariant under B_m. If now Y_0 is an arbitrary closed subspace of H_m, containing x and invariant under A_m, then $B_m x \in L \subseteq Y_0$. Hence $B_m Y_0 \subseteq Y_0$ and the proof is complete.

Proof of Theorem 13.26. Let $x_1, \ldots, x_m, y_1, \ldots, y_m$ be unit vectors in H. Let $\varepsilon > 0$ be given. Define U to be the set of all operators T in $L(H)$ satisfying

$$|\langle Tx_j, y_j\rangle - \langle Bx_j, y_j\rangle| < \varepsilon \qquad (j = 1, \ldots, m).$$

Then U is a neighbourhood of B, and the family of all such sets U is a base of neighbourhoods of B in the weak operator topology. Hence it will suffice to show that U contains a polynomial in A. To do this form the vector

$$x = (x_1, \ldots, x_m) \in H_m.$$

By Lemma 13.27, $B_m x \in \mathrm{clm}\{x, A_m x, A_m^2 x, \ldots\}$. Hence there is a polynomial p such that $\| p(A_m)x - B_m x \| < \varepsilon$. This implies that

$$\| p(A)x_j - Bx_j \| < \varepsilon \qquad (j = 1, \ldots, m)$$

and therefore for $j = 1, \ldots, m$

$$|\langle p(A)x_j, y_j\rangle - \langle Bx_j, y_j\rangle| \leqslant \| p(A)x_j - Bx_j\| \, \|y_j\| < \varepsilon.$$

Thus $p(A) \in U$, and the proof is complete.

THEOREM 13.28. *Let A be a normal operator on H and let $B \in L(H)$. The following two properties are equivalent.*

(i) *B leaves invariant every closed subspace of H left invariant by A.*

(ii) *B is in the closed subalgebra of $L(H)$ generated by I and A in the weak operator topology.*

Proof. By Theorem 13.26, (i) implies (ii). Now suppose that (ii) holds. Then there is a net $\{p_\alpha\}$ of polynomials in A converging to B in the weak operator topology. Let Y be a closed subspace of H invariant under A. Let $y \in Y$ and $z \in Y^\perp$. Then

$$\begin{aligned} \langle A^n y, z\rangle &= 0 \qquad (n = 0, 1, 2, \ldots). \\ \therefore \langle p_\alpha(A)y, z\rangle &= 0 \qquad (\text{for every } \alpha). \\ \therefore \langle By, z\rangle &= 0 \qquad (y \in Y, z \in Y^\perp). \end{aligned}$$

Hence $BY \subseteq Y$, and the proof is complete.

Finally, if we take $B = A^*$ in Theorem 13.28 we obtain the following result.

THEOREM 13.29. *Let A be a normal operator on H. Then A has property (P) if and only if A^* is in the closed subalgebra of $L(H)$ generated by I and A in the weak operator topology.*

In Example 13.5, we exhibited a unitary operator A whose spectrum is the whole unit circle, such that property (P) holds for A and, moreover A^* is the limit in the strong operator topology of a sequence of polynomials in A. The purpose of the next example is to exhibit a normal operator A which possesses property (P) but is such that no sequence of polynomials in A converges to A^* in the weak operator topology. Observe that by Theorem 13.29 there exists a net of polynomials in A converging to A^* in the weak operator topology. However we show that no such net is norm-bounded.

Example 13.30. Let $\{r_n\}$ be an enumeration of the rationals in $(0, 1)$. Define

$$\alpha_n = \exp(2\pi i r_n) \qquad (n = 1, 2, 3, \ldots).$$

(a) *If $\{w_i\}$ is a complex sequence such that*

$$\sum_{i=1}^{\infty} |w_i| < \infty \text{ and } \sum_{i=1}^{\infty} \alpha_i^n w_i = 0 \qquad (n = 0, 1, 2, \ldots)$$

then $w_i = 0$ $(i = 1, 2, \ldots)$.

Proof. Suppose that for each n, $r_n = p_n/q_n$, where p_n, q_n are integers, $q_n \geqslant 1$, and p_n, q_n are mutually prime. Let l be a non-negative integer and let $n_k = (q_1 \ldots q_k) - l$. Then

$$\alpha_i^{n_k} = a_i^{-l} = \bar{\alpha}_i^l \qquad (i = 1, 2, \ldots, k).$$

Now let $k \to \infty$. Then by hypothesis and the dominated convergence theorem

$$\lim_{k\to\infty} \sum_{i=1}^{\infty} \alpha_i^{n_k} w_i = \sum_{i=1}^{\infty} \bar{\alpha}_i^l w_i = 0 \qquad (l = 0, 1, 2, \ldots).$$

Hence, if $p(z, \bar{z})$ is any polynomial in z and $\bar{z}$

$$\sum_{i=1}^{\infty} p(\alpha_i, \bar{\alpha}_i) w_i = 0.$$

By the Stone-Weierstrass theorem, if f is continuous on $\{z: |z| = 1\}$,

$$\sum_{i=1}^{\infty} f(\alpha_i) w_i = 0. \tag{2}$$

Fix i. Define a sequence $\{f_n\}$ of continuous functions on $\{z: |z| = 1\}$ by

$$f_n(z) = 1 - \inf\{1, n|z - \alpha_i|\}.$$

Observe that $|f_n(z)| \leqslant 1$ on the domain of f_n, and $\{f_n\}$ converges pointwise to the characteristic function of the set $\{\alpha_i\}$. Hence, by (2) and the dominated convergence theorem, we obtain $w_i = 0$ $(i = 1, 2, \ldots)$.

(b) *If* $\{v_i\}$ *and* $\{w_i\}$ *are complex sequences such that* $\sum_{i=1}^{\infty} (|v_i| + |w_i|) < \infty$ *and*

$$\sum_{i=1}^{\infty} \frac{\alpha_i^n}{2^n} v_i + \sum_{i=1}^{\infty} \alpha_i^n w_i = 0 \qquad (n = 0, 1, 2, \ldots) \tag{3}$$

then $v_i = 0$, $w_i = 0$ $(i = 1, 2, \ldots)$.

Proof. Let l be a non-negative integer. If in (3) we take $n = (q_1 \ldots q_k) + l$ and let $k \to \infty$, then by hypothesis and the dominated convergence theorem

$$\sum_{i=1}^{\infty} |w_i| < \infty; \sum_{i=1}^{\infty} \alpha_i^l w_i = 0 \qquad (l = 0, 1, 2, \ldots).$$

By (a), $w_i = 0$ $(i = 1, 2, 3, \ldots)$. Consequently

$$\sum_{i=1}^{\infty} |v_i| < \infty; \sum_{i=1}^{\infty} \alpha_i^n v_i = 0 \qquad (n = 0, 1, 2, \ldots)$$

and another application of (a) gives the desired result.

Let H be a separable complex Hilbert space, and $\{\phi_n : n = 1, 2, \ldots\}$ an orthonormal basis for H. Define a unitary operator U on H by

$$U\left(\sum_{i=1}^{\infty} \beta_i \phi_i\right) = \sum_{i=1}^{\infty} \alpha_i \beta_i \phi_i \qquad \left(\sum_{i=1}^{\infty} |\beta_i|^2 < \infty\right).$$

Let $\oplus$ denote orthogonal direct sum. If n is a positive integer, let H_n denote the orthogonal direct sum of n copies of H.

(c) *If $c_1, \ldots, c_n$ are either $\frac{1}{2}$ or 1, and Y is a closed invariant subspace for the operator $T = c_1 U \oplus \ldots \oplus c_n U$ on H_n, then Y is reducing.*

Proof. Let $x \in Y$, and let $\{\lambda_i\}$ be an enumeration of the eigenvalues of T. We can write $x = \sum_{i=1}^{\infty} \chi_i$, where $T\chi_i = \lambda_i \chi_i$. Let $y \in Y^\perp$. We can write $y = \sum_{i=1}^{\infty} e_i$, where $Te_i = \lambda_i e_i$. Since $T^n x \in Y$ $(n = 0, 1, 2, \ldots)$,

$$\langle T^n x, y \rangle = \sum_{i=1}^{\infty} \lambda_i^n \langle \chi_i, e_i \rangle = 0.$$

Let $\langle \chi_i, e_i \rangle = w_i$. Then

$$\sum_{i=1}^{\infty} \lambda_i^n w_i = 0 \qquad (n = 0, 1, 2, \ldots).$$

Now eigenvectors, corresponding to distinct eigenvalues, are orthogonal.

Hence

$$\sum_{i=1}^{\infty} \|\chi_i\|^2 < \infty \text{ and } \sum_{i=1}^{\infty} \|e_i\|^2 < \infty.$$

Therefore we also have $\sum_{i=1}^{\infty} |w_i| < \infty$. Now each λ_i is of the form α_m or $\alpha_m/2$, for some m, and so by (b)

$$\langle \chi_i, y \rangle = w_i = 0 \qquad (i = 1, 2, \ldots).$$

It follows that each χ_i is in Y. The equation $T\chi_i = \lambda\chi_i$ implies that $T^*\chi_i = \bar{\lambda}_i\chi_i$ and so $T^*x \in Y$, since Y is closed. This completes the proof.

Now define the operator $A = U \oplus \frac{1}{2}U$ on $H \oplus H$.

(d) *A^* is in the closed algebra of operators generated by I and A in the weak operator topology.*

Proof. By (c), A has property (P). It follows from this and Theorem 13.29 that (d) holds.

(e) *There is no sequence of polynomials in A converging to A^* in the weak operator topology.*

Proof. Let $\{\lambda_i\}$ be an enumeration of the eigenvalues of A. Let e_i be the unit eigenvector of A corresponding to the eigenvalue λ_i. Then $\{e_i : i = 1, 2, \ldots\}$ is an orthonormal basis for $H \oplus H$. Then if $\langle P_n(A)x, y\rangle \to \langle A^*x, y\rangle$ for all x, y in $H \oplus H$ we have

$$\sum_{i=1}^{\infty} P_n(\lambda_i)\xi_i\bar{\eta}_i \to \sum_{i=1}^{\infty} \bar{\lambda}_i\xi_i\bar{\eta}_i$$

on putting

$$x = \sum_{i=1}^{\infty} \xi_i e_i, \qquad y = \sum_{i=1}^{\infty} \eta_i e_i.$$

Let $B_i^n = P_n(\lambda_i) - \bar{\lambda}_i$. It follows that

$$\sum_{i=1}^{\infty} B_i^n \omega_i \to 0, \text{ for all } \{\omega_i\} \in l^1 \left(\text{i.e. with } \sum_{i=1}^{\infty} |\omega_i| < \infty\right).$$

Fix n. Since $\sigma(A)$ is compact, $|B_i^n| < K_n$, for all i, where K_n is real. Hence the $\{B_i^n\}_{n=1}^{\infty}$ form a pointwise converging sequence of bounded linear functionals on l^1, and so by the uniform boundedness principle $|B_i^n|$ is bounded for all i, n. Therefore, for some real K,

$$|P_n(\lambda_i)| \leqslant K.$$

As $n \to \infty$, $P_n(\lambda_i) \to \bar{\lambda}_i$, for all i. Hence it follows that

$$\begin{aligned} P_n(\alpha_i) &\to \bar{\alpha}_i \qquad (i = 1, 2, \ldots), \\ P_n(\tfrac{1}{2}\alpha_i) &\to \tfrac{1}{2}\bar{\alpha}_i \qquad (i = 1, 2, \ldots), \\ |P_n(\alpha_i)| \leqslant K; |P_n(\tfrac{1}{2}\alpha_i)| &\leqslant K \qquad (i, n = 1, 2, \ldots). \end{aligned}$$

Since the $\{\alpha_i\}$ are dense in $\{z : |z| = 1\}$ it follows that $|P_n(z)| \leqslant K$ ($|z| = 1$), and so by the maximum modulus theorem

$$|P_n(z)| \leqslant K \qquad (|z| \leqslant 1; n = 1, 2, \ldots).$$

Hence by Montel's theorem (see, for example Theorem 14.6 of [**14**; p. 272]) the $\{P_n\}$ constitute a normal family of analytic functions in $\{z : |z| < 1\}$. Hence some subsequence $\{P_{n_k}\}$ converges uniformly in $\{z : |z| < \frac{3}{4}\}$ to f analytic in this region. Since $P_n(\frac{1}{2}\alpha_i) \to \frac{1}{2}\bar{\alpha}_i$ and $\alpha_i\bar{\alpha}_i = 1$, we obtain

$$f(\tfrac{1}{2}\alpha_i) = \tfrac{1}{2}\bar{\alpha}_i = \frac{1}{2\alpha_i}, \text{ for all } i.$$

Consider the functions whose values are $f(z)$ and $1/(4z)$ at each point of $\{z : 0 < |z| < \frac{3}{4}\}$. The last equation shows that these functions coincide on a dense subset of $\{z : |z| = \frac{1}{2}\}$ and by the identity theorem they coincide in $\{z : 0 < |z| < \frac{3}{4}\}$. This is absurd since one function is bounded in this region and the other is unbounded there. This contradiction suffices to complete the proof.

(f) *There is no norm-bounded net of polynomials in A converging to A^* in the weak operator topology.*

Proof. To deduce (f) from (e) it is sufficient to observe that since $H \oplus H$ is separable, the weak operator topology on any positive real multiple of the unit ball of $L(H \oplus H)$ is metrizable.

Let $\mathscr{A}(I, A)$ denote the algebra of polynomials in A. Let $\mathscr{A}^w(I, A, A^*)$ denote the commutative von Neumann algebra generated by I, A, A^*. Observe that we have shown that in the weak operator topology $\mathscr{A}(I, A)$ is dense in $\mathscr{A}^w(I, A, A^*)$, but no positive real multiple of the unit ball of $\mathscr{A}(I, A)$ is dense in the unit ball of $\mathscr{A}^w(I, A, A^*)$. Hence no analogue of the Kaplansky density theorem holds in this situation. For completeness we state the Kaplansky density theorem.

Let $\mathscr{A}$ and $\mathscr{B}$ be $*$-algebras of operators on H with $\mathscr{A} \subseteq \mathscr{B}$. Let $\mathscr{A}_1$ and $\mathscr{B}_1$ respectively denote the unit balls of $\mathscr{A}$ and $\mathscr{B}$. Then in the weak operator topology of L(H), $\mathscr{A}_1$ is dense in $\mathscr{B}_1$.

For a proof of this result, see [**13**; p. 24].

The last main result in this chapter is Theorem 13.33, which states that if the spectrum of a normal operator A has non-empty interior then property (P) fails for A. This result was first proved by Scroggs [**1**] in 1959. In fact Scroggs proved a slightly more general result, which we now state.

Let A be a normal operator on H. Suppose that the spectrum of A contains an infinite sequence $\{\Gamma_n\}_{n=1}^{\infty}$ of rectifiable Jordan curves with the property that Γ_n lies inside of Γ_{n+1}, for every n. Then property (P) fails for A.

For a proof of this result see Theorem 3 of Scroggs [**1**; p. 99].

The proof of Theorem 13.33 which we shall give is due to Kelly [**1**]. However, see also earlier papers of Sarason [**1**], [**2**]. Two preliminary results are required.

PROPOSITION 13.31. *Let T be a normal operator on H. Then there is a closed separable subspace Y of H such that Y reduces T and $\sigma(T \mid Y) = \sigma(T)$.*

Proof. Let $E(\cdot)$ be the resolution of the identity for T. Let $\{\lambda_r : r \in \mathbf{N}\}$ be a dense countable subset of $\mathbf{C}$, containing a dense subset of $\sigma(T)$. Let λ_k be a particular element of the set. With λ_k as centre construct a sequence of open discs, say $\{S_m : m \in \mathbf{N}\}$, so that, if r_m is the radius of S_m, $\lim_{m \to \infty} r_m = 0$. For each disc, choose a vector x_{km} belonging to the range of the projection $E(S_m)$. If $S_m \cap \sigma(T) = \emptyset$. then $x_{km} = 0$; otherwise we choose $x_{km} \neq 0$. In this way we obtain an infinite sequence of vectors $\{x_{km} : m \in \mathbf{N}\}$ associated with λ_k. Repeating this process for each $\lambda_k (k = 1, 2, \ldots)$, we obtain a doubly indexed sequence of vectors $\{x_{km} : k \in \mathbf{N}, m \in \mathbf{N}\}$. Now, by Corollary 13.9, $M(x_{km})$, the cyclic subspace corresponding to the vector x_{km}, is separable. Hence since a countable union of countable sets is countable it follows that Y, the closed subspace generated by $\cup \{M(x_{km}) : k \in \mathbf{N}, m \in \mathbf{N}\}$, is separable. Clearly Y

reduces T. Finally we show that $\sigma(T|Y) = \sigma(T)$. By Propositions 12.4 and 12.5 we have $\sigma(T|Y) \subseteq \sigma(T)$ and so it is enough to prove that $\sigma(T) \subseteq \sigma(T|Y)$. Suppose that $\lambda_k \in \sigma(T)$. Then, given any neighbourhood $N(\lambda_k)$ of λ_k, some $S_i(\lambda_k)$ has the following properties:

$$\text{(i) } S_i(\lambda_k) \subseteq N(\lambda_k) \qquad \text{(ii) } S_i(\lambda_k) \cap \sigma(T) \neq \varnothing.$$

By construction, there is an $x_{ki} \neq 0$ such that $E(S_i)x_{ki} = x_{ki}$. However, this means that $S_i(\lambda_k)$ and hence $N(\lambda_k)$ contains a point of the set $\sigma(T|Y)$. If this point is not λ_k, then this shows that λ_k is a cluster point of $\sigma(T|Y)$ and hence $\lambda_k \in \sigma(T|Y)$, since this set is closed. Hence $\sigma(T) \subseteq \sigma(T|Y)$, since $\sigma(T)$ has a dense subset consisting of the points λ_k, and the proof is complete.

PROPOSITION 13.32. *Let X be a separable Banach space. A convex subset A of X^* is weak*-closed if and only if A is weak*-sequentially closed.*

Proof. It is enough to show that $y_n \in A$ and $\lim_{n\to\infty} \langle x, y_n\rangle = \langle x, y\rangle$ for all x in X and some y in X^* together imply that $y \in A$.

By Theorem V.5.1 of [**5**; p. 426], the closed unit ball B of X^* under the weak*-topology is a metric space. Hence $B \cap A$, being sequentially closed, is closed in the weak*-topology. A similar statement holds if B is replaced by any positive multiple of B. Now, by a result of Krein and Šmulian (Theorem V.5.7 of [**5**; p. 429]), a convex set in X^* is weak*-closed if and only if its intersection with every positive multiple of the closed unit ball of X^* is weak*-closed. The required result follows.

THEOREM 13.33. *Let T be a normal operator on H. If the spectrum of T has non-empty interior, property (P) fails for T.*

Proof. Suppose that T has property (P). By Proposition 13.31 there exists a separable closed subspace Y of H which is reducing for T and such that $\sigma(T|Y) = \sigma(T)$. By Proposition 13.12 there is a separating vector $\tilde{x}$ for the spectral measure $E(\cdot)|Y$. Now $\sigma(T|M(\tilde{x}))$ is the support of the spectral measure $E(\cdot)|M(\tilde{x})$ and this is the same as the support of $E(\cdot)|Y$. Hence $\sigma(T|M(\tilde{x})) = \sigma(T)$. Suppose that $\sigma(T)$ has non-empty interior.

Let $S = T|M(\tilde{x})$. Then S is a normal operator and S has property (P). By Theorem 13.29, S^* lies in the closure in the weak operator topology of $L(M(\tilde{x}))$ of the set of polynomials in S. Hence there is a net $\{p_\alpha\}$ of polynomials such that

$$\lim_{\alpha} \langle p_\alpha(S)x, y\rangle = \langle S^* x, y\rangle, \text{ for all } x, y \text{ in } M(\tilde{x}).$$

Define $\mu(\tau) = \langle E(\tau)\tilde{x}, \tilde{x}\rangle$ $(\tau \in \Sigma_p)$. Then μ is a finite measure with compact support supp $\mu = \sigma(T | M(\tilde{x})) = \sigma(T)$. By Proposition 13.8, there exists an isometric isomorphism U of $L^2(\mu)$ onto $M(\tilde{x})$ with the property that

$$U^{-1}E(\tau)Uf = \chi_\tau f \qquad (\tau \in \Sigma_p, f \in L^2(\mu)).$$

We have

$$\begin{aligned}\langle p_\alpha(S)x, y\rangle &= \langle p_\alpha(S)Uf, Ug\rangle, \text{ for some } f, g \text{ in } L^2(\mu),\\ &= \int_{\sigma(T)} p_\alpha(\lambda)\langle E(d\lambda)Uf, Ug\rangle.\end{aligned}$$

Now for all τ in Σ_p we have

$$\begin{aligned}\langle E(\tau)Uf, Ug\rangle &= \langle UU^{-1}E(\tau)Uf, Ug\rangle = \langle U^{-1}E(\tau)Uf, g\rangle\\ &= \int_\tau f(\lambda)\overline{g(\lambda)}\mu(d\lambda).\end{aligned}$$

Hence for each α,

$$\langle p_\alpha(S)x, y\rangle = \int_{\sigma(T)} p_\alpha(\lambda) f(\lambda)\overline{g(\lambda)}\mu(d\lambda).$$

Therefore

$$\lim_{\alpha} \int_{\sigma(T)} p_\alpha(\lambda) f(\lambda)\overline{g(\lambda)}\mu(d\lambda) = \int_{\sigma(T)} \bar{\lambda} f(\lambda)\overline{g(\lambda)}\mu(d\lambda) \qquad (f, g \in L^2(\mu))$$

and so

$$\lim_{\alpha} \int_{\sigma(T)} p_\alpha(\lambda) h(\lambda)\mu(d\lambda) = \int_{\sigma(T)} \bar{\lambda} h(\lambda)\mu(d\lambda) \qquad (h \in L^1(\mu)).$$

Define $\eta(\lambda) = \bar{\lambda}\ (\lambda \in \sigma(T))$. Observe that we have shown that η lies in the weak*-closure of the polynomials in $L^\infty(\mu)$.

Consider the subspace M of functions in $L^\infty(\mu)$ which are analytic in G, the interior of $\sigma(T)$. We show that M is closed in the weak*-topology of $L^\infty(\mu)$. By Proposition 13.32, it is enough to show that M is weak*-sequentially closed in $L^\infty(\mu)$. By considering a sequence $\{f_n\}$ converging weak* in $L^\infty(\mu)$ to f, we see from the uniform boundedness theorem that $\{f_n\}$ is a bounded sequence in $L^\infty(\mu)$. Hence this sequence of functions is uniformly bounded in G. By Montel's theorem (see, for example, Theorem 14.6 of [**14**; p. 272]), $\{f_n\}$ has a subsequence which converges uniformly on compact subsets of G to the function g, say, where g is analytic in G. Hence $f = g$ (μ – a.e.) in G. Thus $f \in M$. Therefore M is closed in the weak*-topology of $L^\infty(\mu)$ and this of course shows that η does not belong to the weak*-closure of the polynomials in $L^\infty(\mu)$, since $\eta \notin M$. This contradiction suffices to complete the proof.

14. Restrictions of Prespectral Operators

In Chapter 12 we presented Fixman's necessary and sufficient conditions for a restriction of a spectral operator to be spectral. It was shown that in certain cases, when the spectrum of the operator is suitably thin, these conditions simplified considerably. In the present chapter we consider the analogous problem for prespectral operators. A complete analogue of Fixman's conditions is obtained. A study of restrictions of adjoints of spectral operators leads to complete analogues of results in Chapter 12 for the following problem: when is the operator, induced on a quotient space by a spectral operator, also spectral? An example is given to show that no analogue of the author's results on restrictions of spectral operators with suitably thin spectra holds in the case of prespectral operators.

We begin with the following necessary condition for a restriction of a prespectral operator to be prespectral.

PROPOSITION 14.1. *Let T be a prespectral operator on X and let Y be a closed subspace of X invariant under T. If $T|Y$ is a prespectral operator then $\sigma(T|Y) \subseteq \sigma(T)$.*

Proof. By Theorem 5.47, $\sigma_a(T) = \sigma(T)$ and $\sigma_a(T|Y) = \sigma(T|Y)$. Also by Theorem 1.16 (i), $\sigma_a(T|Y) \subseteq \sigma_a(T)$ and so the desired conclusion follows.

The next result gives a sufficient condition for a restriction of a prespectral operator to be prespectral.

THEOREM 14.2. *Let T, in $L(X)$, be a prespectral operator with resolution of the identity $E(\cdot)$ of class Γ. Let Y be a closed subspace of X invariant under T and $\{E(\tau): \tau \in \Sigma_p\}$. Then $T|Y$ is prespectral with resolution of the identity $E(\cdot)|Y$ of class Γ.*

Proof. It is easily verified that $T|Y$ satisfies the axioms for a prespectral operator with the exception of $\sigma(T|E(\tau)Y) \subseteq \bar{\tau}$, for all τ in Σ_p. In the case $Y = E(\delta)X$ we have for all τ in Σ_p

$$\sigma(T|E(\delta)E(\tau)X) = \sigma(T|E(\delta \cap \tau)X) \subseteq \overline{\tau \cap \delta} \subseteq \bar{\tau}.$$

Hence $T|E(\delta)X$ is prespectral of class Γ. Now define

$$S = \int_{\sigma(T)} \lambda E(d\lambda), N = T - S.$$

It follows from Theorem 5.17 that for all ζ in $\rho(T)$

$$(\zeta I - T)^{-1} = \sum_{n=0}^{\infty} N^n \int_{\sigma(T)} \frac{E(d\lambda)}{(\zeta - \lambda)^{n+1}},$$

where the series converges in the norm topology of $L(X)$. The hypotheses of the theorem therefore imply that S,N and $\{(\zeta I - T)^{-1}: \zeta \in \rho(T)\}$ all leave Y invariant. Hence $\sigma(T|Y) \subseteq \sigma(T)$. This argument may be applied to the prespectral operator $T|E(\tau)X$ and its restriction to the closed invariant subspace $E(\tau)Y$. Therefore for all τ in Σ_p,

$$\sigma(T|E(\tau)Y) \subseteq \sigma(T|E(\tau)X) \subseteq \bar{\tau}.$$

The next theorem will be used frequently throughout this chapter. A preliminary lemma is required.

LEMMA 14.3. *Let T, in $L(X)$, be a prespectral operator with resolution of the identity $E(\cdot)$ of class Γ and let δ be a closed subset of* **C**. *Then $E(\delta)X$ is the union of all closed subspaces Y of X with $TY \subseteq Y$ and $\sigma(T|Y) \subseteq \delta$.*

Proof. Let Y be a closed subspace of X with $TY \subseteq Y$ and $\sigma(T|Y) \subseteq \delta$. Lemma 12.7 shows that $T|Y$ has the single-valued extension property. Let y_X and y_Y be respectively the maximal X-valued and Y-valued analytic functions which satisfy

$$(\zeta I - T)y_X(\zeta) = y, \text{ for all } \zeta \text{ in } \rho_X(y),$$
$$(\zeta I - T)y_Y(\zeta) = y, \text{ for all } \zeta \text{ in } \rho_Y(y),$$

where $\rho_X(y)$ and $\rho_Y(y)$ are the domains of definition of these functions. Let $\sigma_X(y)$ and $\sigma_Y(y)$ be the complements of these sets. y_Y may be regarded as X-valued and so by the maximality of $\rho_X(y)$ we have $\rho_Y(y) \subseteq \rho_X(y)$. Therefore

$$\sigma_X(y) \subseteq \sigma\ (y) \subseteq \sigma(T \mid Y) \subseteq \delta.$$

By Theorem 5.33, $E(\delta)X = \{x \in X : \sigma_X(x) \subseteq \delta\}$ and so $Y \subseteq E(\delta)X$. Finally $\sigma(T \mid E(\delta)X) \subseteq \delta$ and the lemma follows.

THEOREM 14.4. *Let T, in $L(X)$, be a prespectral operator with resolution of the identity $E(\cdot)$ of class Γ. Let Y be a closed subspace of X with $TY \subseteq Y$. Define Y_0 to be the intersection of all closed subspaces of X that contain Y and are invariant under T and $\{E(\tau): \tau \in \Sigma_p\}$. Then $\sigma(T \mid Y) \supseteq \sigma(T \mid Y_0)$.*

Proof. Let $\sigma(T \mid Y) = \delta$. By the lemma, $Y \subseteq E(\delta)X$. Also $Y_0 \subseteq E(\delta)X$ by the definition of Y_0. By Theorem 14.2, $T \mid Y_0$ and $T \mid E(\delta)X$ are both prespectral operators. Apply Proposition 14.1 to $T \mid E(\delta)X$ and its restriction $T \mid Y_0$ to obtain

$$\sigma(T \mid Y_0) \subseteq \sigma(T \mid E(\delta)X) \subseteq \delta = \sigma(T \mid Y).$$

We now show that the sufficient condition of Theorem 14.2 is also necessary.

THEOREM 14.5. *Let T be a prespectral operator on X with resolution of the identity $E(\cdot)$ of class Γ. Let Y be a closed subspace of X invariant under T. If $T \mid Y$ is prespectral of class Γ, then $E(\tau)Y \subseteq Y$ $(\tau \in \Sigma_p)$.*

Proof. We consider the complex Banach space $X \oplus Y$ with norm defined by

$$\|(x, y)\| = \|x\| + \|y\| \qquad (x \in X, y \in Y).$$

Let P, Q be the projections onto the first and second co-ordinate spaces respectively. Then, by Lemma 10.1, the operator $T_1 = PTP + QTQ$ is prespectral on $X \oplus Y$ with resolution of the identity $G(\cdot)$ of class $\Gamma \oplus \Gamma$ given by

$$G(\tau) = PE(\tau)P + QF(\tau)Q \qquad (\tau \in \Sigma_p),$$

where $F(\cdot)$ is the resolution of the identity of class Γ for $T \mid Y$. Also, by

Proposition 14.1, $\sigma(T \mid Y) \subseteq \sigma(T)$ and so $\sigma(T_1) = \sigma(T)$. Let B be the bounded linear map of Y into X defined by $By = y\,(y \in Y)$. A trivial computation shows that T_1 commutes with the operator $A = PBQ$. By Theorem 5.12,

$$\int_{\sigma(T)} f(\lambda)G(d\lambda)A = A \int_{\sigma(T)} f(\lambda)G(d\lambda) \qquad (f \in C(\sigma(T))).$$

Another easy computation yields

$$\int_{\sigma(T)} f(\lambda)E(d\lambda)y = \int_{\sigma(T)} f(\lambda)F(d\lambda)y \qquad (f \in C(\sigma(T))).$$

Let Y_0 be the intersection of all closed subspaces of X that contain Y and are invariant under $\{E(\tau): \tau \in \Sigma_p\}$. By Theorem 14.4, $\sigma(T \mid Y) \supseteq \sigma(T \mid Y_0)$. It follows from Theorem 14.2 that $T \mid Y_0$ is a prespectral operator with resolution of the identity $H(\cdot)$ of class Γ where $H(\tau) = E(\tau) \mid Y_0\ (\tau \in \Sigma_p)$. By applying Proposition 14.1 to $T \mid Y_0$ and its restriction $T \mid Y$ we get $\sigma(T \mid Y) \subseteq \sigma(T \mid Y_0)$ and so $\sigma(T \mid Y) = \sigma(T \mid Y_0)$. The first part of the theorem applied to $T \mid Y_0$ and its restriction $T \mid Y$ shows that

$$\int_{\sigma(T \mid Y)} f(\lambda)H(d\lambda)y = \int_{\sigma(T \mid Y)} f(\lambda)F(d\lambda)y \qquad (y \in Y,\ f \in C(\sigma(T \mid Y))).$$

Let $z \in \Gamma$, $\mu_1(\cdot) = \langle H(\cdot)y, z\rangle$ and $\mu_2(\cdot) = \langle F(\cdot)y, z\rangle$. It follows that for all f in $C(\sigma(T \mid Y))$

$$\int_{\sigma(T \mid Y)} f(\lambda)\mu_1(d\lambda) = \int_{\sigma(T \mid Y)} f(\lambda)\mu_2(d\lambda).$$

The Riesz representation theorem shows that $\mu_1 = \mu_2$. Since Γ is total it follows that for all y in Y and τ in Σ_p

$$F(\tau)y = H(\tau)y = E(\tau)y.$$

Therefore $E(\tau)Y \subseteq Y$, for all τ in Σ_p, and the proof is complete.

We now give an example to show that the analogues of Theorems 12.14, 12.15 and 12.25 fail.

Example 14.6. Let $X = l^\infty$ and let Y be the closed subspace of X consisting of convergent sequences. Define S, in $L(X)$, by

$$S\{\xi_n\} = \left\{\frac{1}{n}\xi_n\right\}.$$

Observe that

(a) S is prespectral of class l^1 and $\sigma(S)$ is totally disconnected;

(b) S is a scalar-type operator of class l^1;

(c) S is the adjoint of a spectral operator;

(d) S^* is spectral, since $(l^\infty)^*$ is weakly complete. (See Theorem 6.11 in conjunction with IV.8.16 and IV.9.9 of [**5**].)

Suppose that $S|Y$ is prespectral with a resolution of the identity $F(\cdot)$ of some class Γ. Define disjoint sets τ and δ in Σ_p by

$$\tau = \{0, 1, \tfrac{1}{3}, \tfrac{1}{5}, \ldots\}, \qquad \delta = \{\tfrac{1}{2}, \tfrac{1}{4}, \tfrac{1}{6}, \ldots\}.$$

Observe that $F(\delta) + F(\tau) = I|Y$, $\sigma(S|F(\delta)Y) \subseteq \bar{\delta}$ and $\sigma(S|F(\tau)Y) \subseteq \tau$. By Lemma 14.3

$$F(\delta)Y \subseteq E(\bar{\delta})X \text{ and } F(\tau)Y \subseteq E(\tau)X,$$

where $E(\cdot)$ is the unique resolution of the identity of class l^1 for S. It follows that every element of Y can be expressed as the sum of two convergent sequences of the forms

$$\{a_1, 0, a_3, 0, \ldots\} \text{ and } \{0, a_2, 0, a_4, \ldots\}$$

and so every element of Y has limit 0. This is absurd. It follows that $S|Y$ is not prespectral of any class.

Let T be a prespectral operator with resolution of the identity $E(\cdot)$ of class Γ. Let Y be a closed invariant subspace for T. The argument of the proof of Theorem 14.5 shows that if $T|Y$ is prespectral then

$$\int_{\sigma(T)} f(\lambda)E(d\lambda)Y \subseteq Y \qquad (f \in C(\sigma(T))).$$

The last example shows that this necessary condition is not sufficient for $T|Y$ to be prespectral.

The results obtained on restrictions of adjoints of spectral operators can be applied to the study of operators induced on quotient spaces by spectral operators. For the appropriate definition, notation and properties which we shall use see Propositions 1.33, 1.34, Lemma 1.35 and Theorem 1.36 together with the discussion prior to Proposition 1.33.

THEOREM 14.7. *Let T be a spectral operator on X with resolution of the identity $E(\cdot)$. Let Y be a closed subspace of X with $TY \subseteq Y$. If T_Y is spectral with resolution of the identity $F(\cdot)$ then $E(\tau)Y \subseteq Y$ for all τ in Σ_p. Also*

$$F(\tau) = E_Y(\tau) \qquad (\tau \in \Sigma_p).$$

Proof. T^* is prespectral on X^* with resolution of the identity $E^*(\cdot)$ of class X. In view of Proposition 1.33 et seq. we may and shall identify $T_Y{}^*$ and $T^*|Y^\perp$. Hence $T^*|Y^\perp$ is prespectral on $Y^\perp$ with resolution of the identity $F^*(\cdot)$ of class X. By Theorem 14.5,

$$\langle x, E^*(\tau)y\rangle = \langle x, F^*(\tau)y\rangle \qquad (x \in X, y \in Y^\perp, \tau \in \Sigma_p).$$

Therefore

$$\begin{aligned}\langle E(\tau)z, y\rangle &= \langle z, E^*(\tau)y\rangle = \langle z, F^*(\tau)y\rangle \\ &= 0 \qquad\qquad (z \in Y, y \in Y^\perp, \tau \in \Sigma_p).\end{aligned}$$

It follows that $E(\tau)Y \subseteq Y$ and $F(\tau) = E_Y(\tau)$ for all τ in Σ_p.

THEOREM 14.8. *Let T be a spectral operator on X with resolution of the identity $E(\cdot)$. Let Y be a closed subspace of X invariant under T and $\{E(\tau): \tau \in \Sigma_p\}$. Then T_Y is a spectral operator on X/Y with resolution of the identity $E_Y(\cdot)$. If T is of finite type m or of scalar type then so is T_Y.*

Proof. We show first that $E_Y(\cdot)$ is a spectral measure. The axioms (i)–(v) are easily verified. (See after Theorem 5.4). Let $[x] \in X/Y$ and $\bar{x} \in (X/Y)^*$. The countable additivity of the set function $\langle E_Y(\cdot)[x], \bar{x}\rangle$ follows at once from the countable additivity of $E(\cdot)$ in the strong operator topology of $L(X)$ and the identification of $(X/Y)^*$ and $Y^\perp$. Hence $E_Y(\cdot)$ is a spectral measure of class $(X/Y)^*$. Clearly T_Y commutes with $E_Y(\tau)$ $(\tau \in \Sigma_p)$ and it remains only to show that

$$\sigma(T_Y | E_Y(\tau)X/Y) \subseteq \bar{\tau} \qquad (\tau \in \Sigma_p).$$

Define

$$S = \int_{\sigma(T)} \lambda E(d\lambda), \qquad T = S + N.$$

It follows from Theorem 5.17 that for all ζ in $\rho(T)$

$$(\zeta I - T)^{-1} = \sum_{n=0}^{\infty} N^n \int_{\sigma(T)} \frac{E(d\lambda)}{(\zeta - \lambda)^{n+1}},$$

where the series converges in the norm topology of $L(X)$. The hypotheses of the theorem therefore imply that S,N and $\{(\zeta I - T)^{-1} : \zeta \in \rho(T)\}$ all leave Y invariant. Hence, by Theorem 1.36, $\sigma(T_Y) \subseteq \sigma(T)$.

Now let $\tau \in \Sigma_p$; then $T|E(\tau)X$ is spectral. Also $E(\tau)Y$ is a closed subspace of $E(\tau)X$ invariant under $T|E(\tau)X$ and $\{E(\delta)|E(\tau)X : \delta \in \Sigma_p\}$. Applying the preceding argument to $T|E(\tau)X$ we obtain

$$\sigma(T_Y|E_Y(\tau)X/Y) = \sigma(T \| E(\tau)Y) \subseteq \sigma(T|E(\tau)X) \subseteq \bar{\tau} \qquad (\tau \in \Sigma_p)$$

where $T \| E(\tau)Y$ denotes the operator induced by T on the quotient space $E(\tau)X/E(\tau)Y$.

Therefore, by Theorem 6.5, T_Y is a spectral operator with resolution of the identity $E_Y(\cdot)$.

THEOREM 14.9. *Let T be a spectral operator on X and let Y be a closed subspace of X invariant under T. Then T_Y is a spectral operator if and only if $T|Y$ is a spectral operator.*

Proof. This follows from Theorems 12.2, 12.3, 14.8 and 14.7.

PROPOSITION 14.10. *Let T be a spectral operator on X. Let Y be a closed subspace of X, invariant under T and such that T_Y is a spectral operator. Then $\sigma(T_Y) \subseteq \sigma(T)$.*

Proof. By Theorem 14.9 and hypothesis, $T|Y$ is a spectral operator. Therefore, by Proposition 14.1, $\sigma(T|Y) \subseteq \sigma(T)$. Hence $\sigma(T|Y) \cap \rho(T) = \varnothing$. It now follows from Theorem 1.36 (ii) that $\sigma(T_Y) \cap \rho(T) = \varnothing$ and so $\sigma(T_Y) \subseteq \sigma(T)$.

THEOREM 14.11. *Let S be a scalar-type spectral operator on X. Suppose that the spectrum of S is nowhere dense and the resolvent set of S is connected. Let Y be a closed subspace of X invariant under S. Then S_Y is spectral.*

Proof. This follows from Theorem 12.14, the corresponding result for restrictions, and Theorem 14.9.

THEOREM 14.12. *Let S be a scalar-type spectral operator on X. Suppose that $\sigma(S)$ is an R-set. Let Y be a closed subspace of X invariant under S. Then S_Y is a spectral operator if and only if $\sigma(S_Y) \subseteq \sigma(S)$.*

Proof. For the definition of R-set the reader is referred to 5.38. The necessity of the condition follows from Proposition 14.10. To show sufficiency, note that $\sigma(S_Y) \subseteq \sigma(S)$ implies that $\sigma(S_Y) \cap \rho(S) = \emptyset$. By Theorem 1.36 (ii), $\sigma(S \mid Y) \cap \rho(S) = \emptyset$ and so $\sigma(S \mid Y) \subseteq \sigma(S)$. Therefore, by the corresponding result for restrictions, Theorem 12.15, $S \mid Y$ is spectral. An application of Theorem 14.9 completes the proof.

THEOREM 14.13. *Let S be a scalar-type spectral operator on X with resolution of the identity $E(\cdot)$. Let Y be a closed subspace of X invariant under S and such that $\sigma(S_Y)$ is an R-set. Then S_Y is a scalar-type spectral operator.*

Proof. We identify the dual space of X/Y with $Y^\perp$. Therefore $\sigma(S^* \mid Y^\perp) = \sigma(S_Y)$ is an R-set. Let M be the smallest closed subspace of X^* containing $Y^\perp$ and invariant under $\{E^*(\tau): \tau \in \Sigma_p\}$. By Theorem 14.4, $\sigma(S^* \mid M) \subseteq \sigma(S^* \mid Y^\perp)$. However, since $\sigma(S^* \mid M)$ and $\sigma(S^* \mid Y^\perp)$ are both nowhere dense, it follows from Theorem 1.29 that $\sigma(S^* \mid Y^\perp) \subseteq \sigma(S^* \mid M)$. Hence $\sigma(S^* \mid M) = \sigma(S^* \mid Y^\perp)$ is an R-set. Now $S^* \mid M$ is a scalar-type operator with resolution of the identity $F(\cdot)$ of class X, where $F(\cdot) = E^*(\cdot) \mid M$. Clearly

$$(\lambda I^* \mid M - S^* \mid M)^{-1} Y^\perp \subseteq Y^\perp, \text{ for all } \lambda \text{ in } \rho(S_Y).$$

If r is any rational function with poles in $\rho(S_Y)$ then

$$r(S^* \mid M) Y^\perp \subseteq Y^\perp.$$

Let $y \in Y$, $y^* \in Y^\perp$ and $\mu(\tau) = \langle y, F(\tau) y^* \rangle$, for all τ in Σ_p. Since $\sigma(S_Y)$ is an

R-set

$$\int_{\sigma(S_Y)} f(\lambda)\mu(d\lambda) = 0, \text{ for all } f \text{ in } C(\sigma(S_Y)).$$

It follows from the Riesz representation theorem that μ vanishes identically. Therefore for all τ in Σ_p, y in Y and y^* in $Y^\perp$

$$\langle y, F(\tau)y^* \rangle = 0.$$

Hence $F(\tau)Y^\perp \subseteq Y^\perp$ and so $E^*(\tau)Y^\perp \subseteq Y^\perp$, for all τ in Σ_p. It follows that $E(\tau)Y \subseteq Y$, for all τ in Σ_p, and so, by Theorem 14.8, S_Y is a scalar-type spectral operator.

THEOREM 14.14. *Let T be a spectral operator on X with totally disconnected spectrum. Let Y be a closed subspace of X invariant under T. Then T_Y is a spectral operator.*

Proof. This follows from Theorem 12.24, the corresponding result for restrictions, and Theorem 14.9.

THEOREM 14.15. *Let T be a spectral operator on X. Let Y be a closed subspace of X invariant under T and such that $\sigma(T_Y)$ has totally disconnected spectrum. Then T_Y is a spectral operator.*

Proof. Again we identify the dual space of X/Y with $Y^\perp$. Therefore $\sigma(T^* | Y^\perp) = \sigma(T_Y)$ is totally disconnected. Let M be the smallest closed subspace of X^* containing $Y^\perp$ and invariant under T^* and $\{E^*(\tau): \tau \in \Sigma_p\}$. By Theorem 14.4, $\sigma(T^*|M) \subseteq \sigma(T^*|Y^\perp)$. However, since both $\sigma(T^*|M)$ and $\sigma(T^*|Y^\perp)$ are nowhere dense it follows from Theorem 1.29 that $\sigma(T^*|Y^\perp) \subseteq \sigma(T^*|M)$. Hence $\sigma(T^*|M) = \sigma(T^*|Y^\perp)$ is totally disconnected. Now by Theorem 14.2, $T^*|M$ is prespectral with resolution of the identity $F(\cdot)$ of class X where $F(\cdot) = E^*(\cdot)|M$. By Proposition 5.27 the spectral projection corresponding to an open-and-closed subset δ of $\sigma(T_Y)$ is given by

$$F(\delta) = \frac{1}{2\pi i} \int_\Gamma (\lambda I^*|M - T^*|M)^{-1} d\lambda$$

where Γ is a suitable finite family of rectifiable Jordan curves enclosing δ but excluding $\sigma(T_Y)\backslash\delta$. Since $(\lambda I^*|M - T^*|M)^{-1}Y^\perp \subseteq Y^\perp$ for all λ in $\rho(T^*|M)$ it follows that for all y in Y and y^* in $Y^\perp$ we have $\langle y, F(\delta)y^*\rangle = 0$. If τ is any open subset of $\mathbf{C}$ then $\tau \cap \sigma(T_Y)$ can be expressed as a countable union of disjoint open-and-closed subsets of $\sigma(T_Y)$. Since $F(\cdot)$ is a spectral measure of class X with support $\sigma(T_Y)$, it follows that for every open set τ, all y in Y and y^* in $Y^\perp$ we have $\langle y, F(\tau)y^*\rangle = 0$.

Therefore all the measures $\langle y, F(\cdot)y^*\rangle$ vanish identically. Hence $F(\tau)Y^\perp \subseteq Y^\perp$ and so $E^*(\tau)Y^\perp \subseteq Y^\perp$, for all τ in Σ_p. It follows that $E(\tau)Y \subseteq Y$, for all τ in Σ_p, and so, by Theorem 14.8, T_Y is spectral.

Notes and Comments on Part Four

The theory of spectral operators was initiated by Dunford [**1**], [**2**], [**3**]. For a general survey of the theory of spectral operators up to 1958, the reader is referred to the article of Dunford [**4**], which gives the historical background and physical motivation for the theory. An extremely comprehensive account of the theory of spectral operators and the many important applications of the theory is given in Part III of [**5**]. In these "Notes and Comments" we restrict attention to the literature on the more general class of prespectral operators and those topics in the theory of spectral operators which are not treated in detail in [**5**].

The neat proof of Proposition 5.3 is due to P. G. Spain and has not been published elsewhere. Theorem 5.4 is due to Berkson [**1**] but the proof given here is due to P. G. Spain. The theory of prespectral operators was initiated by Dunford [**3**], who obtained Propositions 5.8, 5.9 and Theorem 5.10. Fixman [**1**] showed that the commutativity theorem (6.6), valid for spectral operators, may fail in general for a prespectral operator and that a prespectral operator may have distinct resolutions of the identity but of different classes. See Example 5.35. Berkson and Dowson [**1**] showed that for four important special classes of prespectral operators, such an operator has a unique resolution of the identity of a particular class and a unique Jordan decomposition for resolutions of the identity of all classes. Berkson, Dowson and Elliott [**1**] showed that if a prespectral operator S possessed a resolution of the identity $E(\cdot)$ of class Γ such that

$$S = \int_{\sigma(S)} \lambda E(d\lambda)$$

then S has a unique resolution of the identity of class Γ and $S + 0$ is the unique Jordan decomposition for S. Finally, Dowson [**9**] settled the question of uniqueness of resolution of the identity of a prespectral operator with full generality. The simpler approach presented in Chapter 5 is due to Dowson, Gillespie and Spain [**1**]. Dowson [**9**] proved Theorems 5.12, 5.13, 5.15, 5.23, 5.24, 5.25. In fact, Dowson [**9**] has proved a stronger version of Theorem 5.12 than is required in this book.

Let S be a scalar-type operator on X with resolution of the identity $E(\cdot)$ of class Γ. Let $A \in L(X)$. Then

$$A \int_{\sigma(S)} f(\lambda)E(d\lambda) = \int_{\sigma(S)} f(\lambda)E(d\lambda) \qquad (f \in C(\sigma(S)))$$

if and only if for every closed subset δ of $\mathbf{C}$ *we have $AE(\delta)X \subseteq E(\delta)X$.*

Using the theory of automatic continuity of linear transformations developed by Johnson [**1**], and Johnson and Sinclair [**1**], Turner [**1**] has also obtained this result.

Theorem 5.17, Propositions 5.19, 5.20 are due to Dunford [**3**]. Theorems 5.21, 5.22 are due to Berkson and Dowson [**1**]. Theorem 5.26 is due to Bade and Curtis [**1**]. Observe that in the statement of this theorem we only require A/R to be algebraically isomorphic to $C(\Omega)$. This follows from the uniqueness of the norm topology of a commutative semi-simple Banach algebra. See Corollary 2.5.18 of [**12**; p. 75]. That this last result remains true if the hypothesis of commutativity is dropped was proved by Johnson [**2**] in 1967. Proposition 5.27 is due to Dunford [**3**] as is Theorem 5.31. The example of an operator which fails to have the single-valued extension property (Example 5.29) is due to Dowson [**3**]. An earlier example of this was given by Kakutani. Dowson's example is of interest since it is an operator induced on a quotient space by a spectral operator, which has the single-valued extension property. Theorem 5.33 is due to Dunford [**3**].

Theorems 5.36 and 5.39 are due to Berkson and Dowson [**1**]. Theorem 5.40 was proved by Berkson [**1**]. Theorem 5.41 is also due to Berkson [**1**], but his original proof used the commutative version of the Vidav-Berkson-Glickfeld-Palmer theorem. The neat proof of Theorem 5.41 that we gave is due to T. A. Gillespie. Theorem 5.46 is due to Lumer [**2**]. Theorem 5.47 was proved for spectral operators by Foguel [**2**]. The result and proof of the theorem given here for the more general class of prespectral operators is due to Dowson [**6**].

Theorem 6.4 is due to Barry [**1**]. Theorems 6.5, 6.6, 6.7, 6.8, 6.9, 6.10, 6.11, 6.12, 6.13 are all due to Dunford [**3**], [**4**]. Theorem 6.14 was obtained by Berkson [**3**]. See also Bade [**1**]. Theorem 6.17 is due to Berkson [**1**]. Theorem

6.22 for the case of a reflexive Banach space is due to Lumer [**2**]. Theorem 6.24 is due to Spain [**1**].

The spectral theorem for bounded self-adjoint operators in Hilbert space is due to Hilbert [**1**; IV]. Reference should also be made to the proofs of F. Riesz [**1**], [**3**] which are quite modern in spirit. Many other proofs of the spectral theorem for self-adjoint, unitary, or normal operators have been given, both in the bounded and unbounded cases. We refer the reader to the books of Ahiezer and Glazman [**1**], Loomis [**1**], Riesz and Sz-Nagy [**1**], Stone [**1**], Sz-Nagy [**1**] and Wintner [**1**]. Our Chapter 7 is perhaps closest to [**7**], but the proof of the spectral theorem for normal operators presented is due to Whitley [**1**]. Theorem 7.21 is due to Fuglede [**1**]. We have given the neat proof of Fuglede's theorem due to Rosenblum [**1**], although Theorem 6.21 is a special case of Theorem 5.12. Theorem 7.24 and Example 7.25 are due to Berkson [**3**]. We observe that if in Chapter 4 we had defined hermitian operator in terms of the spatial numerical range then Theorem 7.23 would be trivial.

The fact that a finite number of commuting scalar-type spectral operators on a Hilbert space can be simultaneously transformed into normal operators by passing to an equivalent inner product is due to Wermer [**2**]. He based his argument on a result for spectral measures due to Lorch [**1**] and Mackey [**1**]. The proof given here depends on the result, due to Sz-Nagy [**2**], that a bounded multiplicative abelian group of operators on a Hilbert space is equivalent to a unitary group. See also Day [**1**] and Dixmier [**1**].

One of the most important applications of Wermer's theorem is to prove that the sum and product of commuting spectral operators in Hilbert space are also spectral operators. Kakutani [**1**] gave an example to show that this result may fail in general in a Banach space. By modifying Kakutani's construction, McCarthy [**1**] showed that neither the sum nor the product of two commuting scalar-type spectral operators on a separable reflexive Banach space need be a spectral operator. McCarthy's example was described in Chapter 9.

Kurepa [**1**] made a study of logarithms of spectral operators. The corresponding problem for prespectral operators was studied by Dowson [**10**]. Stampfli [**1**] made a study of *m*th roots of spectral operators. Dowson [**12**] investigated the corresponding question for prespectral operators. It was first proved independently by Foguel [**2**] and Fixman [**1**] that an invertible power-bounded spectral operator is of scalar type. Dowson [**13**] proved the corresponding result for prespectral operators by the method given in Chapter 10. Theorem 10.5 is due to Foguel [**1**].

The results of Chapter 11 pertaining to spectral operators are due to Foguel [**2**]. The remaining results in that chapter (on prespectral operators) are due to Dowson [**11**] with the exception of Theorems 11.19 and 11.20 which are due to Nagy [**1**].

Fixman [**1**] was the first to make a study of restrictions of spectral operators. He proved Theorems 12.2 and 12.3. The remainder of the results in Chapter 12 on restrictions of spectral operators are due to Dowson [**1**].

The results in Chapter 14 on restrictions of prespectral operators and operators induced on quotient spaces by spectral operators are due to Dowson [**3**] and [**6**] respectively.

Halmos [**1**], [**2**] initiated the study of restrictions of normal operators. He proved Proposition 12.5 and Theorem 12.9 for the case of a normal operator. The investigation was continued by Wermer [**1**], to whom Theorem 12.18 is due. Proposition 12.21 is due to Scroggs [**1**]. Theorem 13.3, which shows that Wermer's result is 'best possible' is due to Dowson [**1**]. Theorem 13.15 is due to Wermer [**1**]. Theorem 13.26 is due to Sarason [**1**]. See also Dowson and Moeti [**1**]. Example 13.30 is due to Dowson [**7**]. Theorem 13.33 was first proved by Scroggs [**1**] by another method. The proof given in Chapter 13 is due to Kelly [**1**] and is based on ideas in papers of Sarason [**1**], [**2**].

We shall now rewrite Sarason's result (Theorem 13.26) in modern terminology.

Definition. An algebra $\mathscr{A}$ of operators on X is said to be *reflexive* if and only if the algebra of operators on X that leave invariant every closed subspace of X invariant under $\mathscr{A}$ is precisely $\mathscr{A}$.

Definition. An operator A on X is said to be *reflexive* if and only if the closed subalgebra of $L(X)$ generated by I and A in the weak operator topology is reflexive.

Sarason's result may therefore be expressed as follows: every normal operator is reflexive. An extensive literature on reflexive algebras has grown out of Sarason's work. For an account of this the reader is referred to Radjavi and Rosenthal [**1**].

Before leaving Chapter 13, we mention the curious result of Dyer and Porcelli [**1**] on the invariant subspace conjecture. This states that for a separable Hilbert space H, every operator on H has a proper closed invariant subspace if and only if the following statement is true: the only operators on H, all of whose closed invariant subspaces are reducing, are normal.

Peter Rosenthal has asked the following question: is every scalar-type spectral operator reflexive? Berkson and Dowson [3] investigated this problem and showed that this is certainly the case if the spectrum is nowhere dense and the resolvent set connected. However the general question remains open. Berkson and Dowson [3] showed that if μ is any measure and $h \in L^{\infty}(\mu)$, then the operator of multiplication by h on $L^{p}(\mu)$ $(1 \leqslant p < \infty)$ is a reflexive scalar-type spectral operator. The proof uses a result of Dowson [2] on certain algebras of operators generated by a scalar-type spectral operator together with Bade's work [1], [2], [3] on Boolean algebras of projections. Gillespie [4] has obtained a different proof of this result, proving the stronger result that if S is a scalar-type spectral operator on X with resolution of the identity $E(\cdot)$ and if there exists x_0 in X such that $\text{clm}\{E(\tau)x_0 : \tau \in \Sigma_p\} = X$ then S is reflexive.

Bade [3] developed a multiplicity theory for complete Boolean algebras of projections on a Banach space.

Let **B** be a complete Boolean algebra of projections and let **B**′ be the commutant of **B**. Let $A(\mathbf{B})$ be the algebra of operators generated by **B** in the uniform operator topology. Bade [2; p. 356] proved that $A(\mathbf{B})$ is precisely the algebra of operators leaving invariant every closed invariant subspace for **B**. Dowson [4] introduced the following four properties pertaining to a complete Boolean algebra of projections:

I. **B** has uniform multiplicity one;
II. **B**′ is equal to $A(\mathbf{B})$;
III. there is no non-zero quasinilpotent in **B**′;
IV. **B**′ is commutative.

Dowson [4] proved that if **B** acts on a Hilbert space the four properties are equivalent. In the case of **B** acting on a Banach space, Dowson [4], showed that I implies II, II implies III, and, under the additional hypothesis of uniform finite multiplicity, III implies IV. This last result was proved using work of Tzafriri [1]. Two examples, constructed by Dieudonné, show that even for a complete Boolean algebra of projections of uniform finite multiplicity II does not necessarily imply I, and III does not necessarily imply II. In Dieudonné[2] there is an example of a complete Boolean algebra **B** of projections with uniform multiplicity two such that $\mathbf{B}' = A(\mathbf{B})$. The underlying Banach space may be chosen to be separable and reflexive. See also Dowson [5] for a simpler proof. In Dieudonné [1] it is shown that there is a complete Boolean algebra **B** of projections with uniform multiplicity two such that $A(\mathbf{B})$ is a proper subalgebra of **B**′ and such that there is no non-zero nilpotent in **B**′. A result of Foguel [3; p. 687] shows that in this case there is no non-zero quasinilpotent in $\mathbf{B}'$. It would seem to be an unsolved problem whether for a complete Boolean algebra of projections of uniform finite multiplicity IV implies III.

Foguel [**2**] observed that in any space the sum (or product) of two commuting spectral operators is spectral if and only if the sum (or product) of their scalar parts is spectral. In addition, Dunford [**3**] and Foguel [**1**] proved that if X is weakly complete, then the sum and product of two commuting spectral operators on X is spectral provided that the Boolean algebra generated by the values of their resolutions of the identity is bounded. McCarthy [**1**] showed that if the complete Boolean algebra of projections generated by the values of one of the resolutions of the identity has uniform finite multiplicity, then the sum of two commuting spectral operators is spectral; moreover, for certain separable reflexive Banach spaces this condition is necessary. Subsequently, McCarthy [**2**] proved that if $\mathscr{E}$ and $\mathscr{F}$ are commuting bounded Boolean algebras of projections in L^p with $1 < p < \infty$, then the Boolean algebra generated by $\mathscr{E}$ and $\mathscr{F}$ is bounded. As a consequence of this remarkable result it follows that the sum and product of commuting spectral operators in L^p $(1 < p < \infty)$ are spectral operators. There is no restriction on the operators or the separability of the space.

It follows from some deep work of Lindenstrauss and Pelczynski [**1**] that if X is a complemented subspace of an L^1-space and $\mathscr{E}$ is a bounded Boolean algebra of projections on X, then there exists a real constant M such that for every finite family $\{E_k: k = 1, \ldots, n\}$ of disjoint projections in $\mathscr{E}$ we have

$$\sum_{k=1}^{n} \|E_k x\| \leqslant M \left\| \left(\sum_{k=1}^{n} E_k \right) x \right\| \qquad (x \in X).$$

Similarly, if X^* is a complemented subspace of an L^1-space (in particular, if $X = C(K)$, where K is a compact Hausdorff space), then the analogous conditions imply that

$$\left\| \left(\sum_{k=1}^{n} E_k \right) x \right\| \leqslant M \sup_k \|E_k x\| \qquad (x \in X).$$

From this McCarthy and Tzafriri [**1**] showed that if X is a complemented subspace of an L^p-space with $1 \leqslant p < \infty$, then the Boolean algebra generated by two commuting bounded Boolean algebras of projections is bounded. Thus the sum and product of two commuting spectral operators in a complemented subspace X of an L^p-space with $1 \leqslant p \leqslant \infty$ are also spectral operators.

On the other hand it is proved that scalar-type spectral operators on an L^∞-space are of a very special form. If S is a scalar-type spectral operator on such a space then $S = \sum_{i=1}^{n} \lambda_i E_i$. Moreover, every scalar-type spectral operator on $L^\infty[0, 1]$ is similar to an operator of the form $Tf = gf$, where g is a simple function. This was proved by McCarthy and Tzafriri [**1**]. (See also Dean [**1**].) In the space $C[0, 1]$ every scalar-type spectral operator has a representation of the form

$$Sx = \sum_{n=1}^{\infty} \lambda_n E(\lambda_n) x,$$

where $\{\lambda_n\}$ is the sequence of eigenvalues of S. See McCarthy and Tzafriri [**1**; p. 539].

Part 5

WELL-BOUNDED OPERATORS

15. General Theory of Well-bounded Operators

The last section of this book is primarily concerned with the structure theory of well-bounded operators and the relationships between various classes of well-bounded operators and prespectral operators.

In this chapter, we follow Ringrose [3] to discuss well-bounded operators on a general Banach space X. It turns out that the well-boundedness of an operator T on X is equivalent to the existence of a family of projections $\{E(t): t \in \mathbf{R}\}$ on X^*, called a decomposition of the identity for T, satisfying certain natural properties and such that

$$\langle Tx, y\rangle = b\langle x, y\rangle - \int_a^b \langle x, E(t)y\rangle \, dt \qquad (x \in X, y \in X^*).$$

In this case, the family $\{E(t): t \in \mathbf{R}\}$ is not necessarily unique. A necessary and sufficient condition for its uniqueness is given.

In Chapters 16 and 17, we discuss three subclasses of well-bounded operators. These are well-bounded operators decomposable in X and well-bounded operators of types (A) and (B). The main results proved are that if T is a well-bounded operator decomposable in X then it is uniquely decomposable and that if T is a well-bounded operator of type (A) the algebra homomorphism from $AC(J)$ into $L(X)$ can be extended to an algebra homomorphism from $NBV(J)$ into $L(X)$. Several examples of well-bounded operators are given in Chapter 16. In Chapter 17, we follow Spain [2] to use an elementary integration theory to establish directly the characterization of well-bounded operators of type (B). We also show that if T is a well-bounded operator of type (B) and $\{F^*(t): t \in \mathbf{R}\}$ is the unique decomposition of the identity for T then we have

$$f(T) = \int_{a-}^b f(t)dF(t) \qquad (f \in AC(J)),$$

K*

where the integral exists as a strong limit of Riemann sums. Moreover, $F(s) - F(s-)$ is a projection of X onto $\{x \in X: Tx = sx\}$ and the residual spectrum of T is empty.

In Chapter 18 we describe some work of Gillespie. The main result proved is that the well-bounded operators on l^2 are a strictly larger class than the scalar-type spectral operators on l^2 with real spectrum. Also an example is given to show that the sum and product of two commuting well-bounded operators on a Hilbert space need not be well-bounded.

Some results on relationships between various classes of well-bounded operators and prespectral operators are proved in Chapter 19.

Finally in Chapter 20 we describe some more work of Gillespie. It is shown that if G is a locally compact abelian group then the translation operators on $L^p(G)$ $(1 < p < \infty)$ possess logarithms which are scalar multiples of well-bounded operators. Necessary and sufficient conditions for these operators to be spectral are given.

Let $J = [a, b]$ be a compact interval of the real line. Let $BV(J)$ be the Banach algebra of complex-valued functions of bounded variation on J with norm $|||\ |||$ defined by

$$|||f||| = |f(b)| + \operatorname{var}(f, J) \qquad (f \in BV(J)),$$

where $\operatorname{var}(f, J)$ is the total variation of f over J.

Let $AC(J)$ be the Banach subalgebra of $BV(J)$ consisting of absolutely continuous functions on J. For f in $AC(J)$

$$|||f||| = |f(b)| + \int_a^b |f'(t)|\,dt.$$

Let $NBV(J)$ be the Banach subalgebra of $BV(J)$ consisting of those functions f in $BV(J)$ which are normalized by the requirement that f is continuous on the left on $(a, b]$.

Let $\mathscr{P}(J)$ be the subalgebra of $AC(J)$ consisting of the polynomials on J. $\mathscr{P}(J)$ is norm dense in $AC(J)$.

Let $T \in L(X)$. We define $p(T)$ in the natural way be setting

$$p(T) = \sum_{n=0}^{k} a_n T^n,$$

where $p(\lambda) = \sum_{n=0}^{k} a_n \lambda^n$. The map $p \to p(T)$ is an algebra homomorphism.

Definition 15.1. Let $T \in L(X)$. We say that T is *well-bounded* if there is a compact interval J and a real constant K such that

$$\| p(T) \| \leqslant K ||| p ||| \qquad (p \in \mathscr{P}(J)). \tag{1}$$

Observe that, if T is well-bounded then so is T^* (with the same J and K). Smart [**2**] introduced this definition and proved the following fundamental result.

LEMMA 15.2. *Let T be a well-bounded operator on X with natural algebra homomorphism $\phi : p \to p(T)$ from $\mathscr{P}(J)$ into $L(X)$. Let K and J be chosen so that* (1) *is satisfied. Then ϕ has a unique extension to an algebra homomorphism* (*also denoted by*) *$\phi : f \to f(T)$ from $AC(J)$ into $L(X)$ such that*

(i) $\| f(T) \| \leqslant K ||| f ||| \qquad (f \in AC(J))$,
(ii) *if* $S \in L(X)$ *and* $ST = TS$ *then*
$Sf(T) = f(T)S \qquad (f \in AC(J))$,
(iii) $f(T^*) = f(T)^* \qquad (f \in AC(J))$.

Proof. If f is absolutely continuous on J, then the derivative f' of f is in $L^1[a, b]$. There exists a sequence $\{q_n\}$ of polynomials such that

$$\int_J |q_n - f'| \to 0 \quad \text{as} \quad n \to \infty.$$

Define

$$p_n(t) = -\int_t^b q_n(u)du + f(b).$$

Obviously, p_n is a polynomial and

$$||| p_n - f ||| = \int_J |q_n - f'| \to 0 \quad \text{as} \quad n \to \infty.$$

Then

$$\| p_n(T) - p_m(T) \| \leqslant K ||| p_n - p_m ||| \to 0 \quad \text{as} \quad m, n \to \infty,$$

so that $\{p_n(T)\}$ converges in the norm topology of $L(X)$ to an operator (independent of the choice of the particular sequence $\{p_n\}$) which will be denoted by $f(T)$. Thus the extension of ϕ is well-defined. Moreover, it is clear that

$$\| f(T) \| \leqslant K ||| f ||| \qquad (f \in AC(J)).$$

Since ϕ is an algebra homomorphism from $\mathscr{P}(J)$ into $L(X)$ then by continuity it is also an algebra homomorphism from $AC(J)$ into $L(X)$. Since (ii) and (iii) are true for polynomials then, by the continuity of ϕ, these statements hold also for absolutely continuous functions on J. This completes the proof.

The notion of a decomposition of the identity was introduced by Ringrose [**3**].

Definition 15.3. A *decomposition of the identity* for X (on J) is a family $\{E(s): s \in \mathbf{R}\}$ of projections on X^* such that

(i) $E(s) = 0$ $(s < a)$, $E(s) = I$ $(s \geqslant b)$;

(ii) $E(s)E(t) = E(t)E(s) = E(s) \qquad (s \leqslant t)$;

(iii) there is a real constant K such that

$$\| E(s) \| \leqslant K \qquad (s \in \mathbf{R});$$

(iv) the function $s \to \langle x, E(s)y \rangle$ is Lebesgue measurable for x in X and y in X^*;

(v) if $x \in X$, $y \in X^*$, $a \leqslant s < b$, and if the function

$$t \to \int_a^t \langle x, E(u)y \rangle \, du$$

is right differentiable at s, then the right derivative at s is $\langle x, E(s)y \rangle$;

(vi) for each x in X, the map $y \to \langle x, E(\cdot)y \rangle$ from X^* into $L^\infty(a, b)$ is continuous when X^* and $L^\infty(a, b)$ are given their weak*-topologies (as duals of X and $L^1(a, b)$).

Definition 15.4. An operator T on X is said to be *decomposable* (on J) if there is a decomposition of the identity for X (on J) such that

$$\langle Tx, y \rangle = b\langle x, y \rangle - \int_a^b \langle x, E(t)y \rangle \, dt \tag{2}$$

for all x in X and y in X^*.

In this case, we say that the family $\{E(s): s \in \mathbf{R}\}$ is a decomposition of the identity for T.

The next main result is that T is well-bounded on J if and only if T is decomposable on J. Also if this is the case we can choose the family of projections $\{E(s): s \in \mathbf{R}\}$ so that

$$S^*E(s) = E(s)S^* \qquad (s \in \mathbf{R})$$

for all S in $L(X)$ satisfying $ST = TS$. Furthermore the algebra homomorphism of Lemma 15.2 is given by

$$\langle f(T)x, y\rangle = f(b)\langle x, y\rangle - \int_a^b \langle x, E(t)y\rangle f'(t)dt \tag{3}$$

for all x in X, y in X^* and f in $AC(J)$. We require several preliminary results.

LEMMA 15.5. *Let $g, f_1, f_2, \ldots, f_n$ be any $n + 1$ bounded linear functionals on X. Suppose that $x \in X$, $f_i(x) = 0$ $(i = 1, 2, \ldots, n)$ together imply that $g(x) = 0$. Then g is a linear combination of $f_1, f_2, \ldots, f_n$.*

Proof. Consider the linear mapping U from X into $\mathbf{C}^n$ defined by

$$U(x) = [f_1(x), \ldots, f_n(x)] \qquad (x \in X).$$

On the subspace UX of $\mathbf{C}^n$, define the mapping ϕ by

$$\phi(Ux) = \phi[f_1(x), \ldots, f_n(x)] = g(x).$$

The map ϕ is well-defined, since $U(x) = U(y)$ implies that $U(x - y) = 0$, so that by hypothesis $g(x) = g(y)$. It is obvious that ϕ is a linear functional on the subspace UX of $\mathbf{C}^n$. By the Hahn-Banach theorem it can be extended to a linear functional ϕ' on $\mathbf{C}^n$. Hence ϕ' has the form

$$\phi'[y_1, \ldots, y_n] = \sum_{r=1}^{n} \beta_r y_r \qquad (\beta_r \in \mathbf{C} \quad \text{for} \quad r = 1, \ldots, n).$$

Thus $g(x) = \sum_{r=1}^{n} \beta_r f_r(x)$, where $\beta_r \in \mathbf{C}$ for $r = 1, \ldots, n$ and the proof is complete.

THEOREM 15.6. *Let $\{E(t): t \in \mathbf{R}\}$ be a decomposition of the identity for X. Then there is a unique operator T on X which satisfies* (2).

Proof. The uniqueness of such an operator is trivial, and it is therefore sufficient to construct one. Let $L(x, y)$ be the bilinear form on $X \times X^*$ defined by

$$L(x, y) = b\langle x, y\rangle - \int_a^b \langle x, E(t)y\rangle \, dt \qquad (x \in X, y \in X^*). \tag{4}$$

We may deduce from Definition 15.3(iii) that

$$|L(x, y)| \leqslant \{|b| + K(b - a)\} \|y\| \|x\|. \tag{5}$$

We now choose and fix an element x of X and consider $L(x, y)$ as a linear functional on X^*. By virtue of condition (vi) of Definition 15.3

$$y \to \int_a^b \langle x, E(t)y\rangle \, dt$$

is a continuous function if we consider X^* with its weak*-topology. This is true of the mapping $y \to \langle x, y\rangle$. Hence $L(x, y)$ is a weak*-continuous linear functional on X^*. There exists a weak*-neighbourhood of 0 say

$$N(0, x_1, \ldots, x_n, \varepsilon) = \{y \in X^*: |\langle x_k, y\rangle| < \varepsilon \quad \text{for} \quad k = 1, \ldots, n\}$$

which is mapped by $L(x, \cdot)$ into the unit disc. For $z \in X^{**}$, define

$$H(z) = \{y \in X^*: \langle y, z\rangle = 0\},$$

and suppose that $y_0 \in \bigcap_{r=1}^{n} H(\tilde{x}_r)$ where for $r = 1, \ldots, n$, $\tilde{x}_r$ is the image of x_r under the natural embedding of X into X^{**}. Then $y_0 \in N(0, x_1, \ldots, x_n, \varepsilon)$ and hence $|L(x, y_0)| < 1$. Since $\bigcap_{r=1}^{n} H(\tilde{x}_r)$ is a linear space, it contains my_0 for every integer m. Hence

$$m|L(x, y_0)| = |L(x, my_0)| < 1 \qquad (m \in \mathbf{N})$$

from which we conclude that $L(x, y_0) = 0$. That is, $L(x, y_0) = 0$ whenever

$\langle x_r, y_0 \rangle = 0$ for $r = 1, \ldots, n$. It follows from Lemma 15.5 that $L(x, \cdot) = \sum_{r=1}^{n} \alpha_r \tilde{x}_r$ for some α_r in $\mathbf{C}$ $(r = 1, \ldots, n)$. Let $u = \sum_{r=1}^{n} \alpha_r x_r$. Then $L(u, y) = \langle y, \tilde{u} \rangle = \langle u, y \rangle$. Hence, for each x in X, there is an element $u = u(x)$ in X such that

$$L(x, z) = \langle u, z \rangle \qquad (z \in X^*). \tag{6}$$

It is obvious that u depends linearly on x and from (5) we deduce that

$$\| u \| \leqslant \{|b| + K(b - a)\} \| x \|.$$

Hence there is an operator T on X such that $Tx = u \, (x \in X)$. The required results now follow from (4) and (6).

THEOREM 15.7. *Let $\{E(t) : t \in \mathbf{R}\}$ be a decomposition of the identity for X, and let T be the associated decomposable operator defined by* (2). *Then*

(i) *T is well-bounded and satisfies* (1),

(ii) *if $f \to f(T)$ is the algebra homomorphism of Lemma* 15.2 *then*

$$\langle f(T)x, y \rangle = f(b)\langle x, y \rangle - \int_a^b \langle x, E(t)y \rangle f'(t) dt \qquad (x \in X, y \in X^*).$$

Proof. First, we shall show by induction that

$$\langle T^n x, y \rangle = b^n \langle x, y \rangle - \int_a^b \langle x, E(t)y \rangle \, nt^{n-1} dt, \tag{7}$$

for all x in X, y in X^* and n a positive integer. In the case $n = 1$, (7) follows immediately from (2). We now assume the validity of (7) for a particular positive integer n. Then

$$\begin{aligned}
\langle T^{n+1} x, y \rangle &= \langle T(T^n x), y \rangle \\
&= b\langle T^n x, y \rangle - \int_a^b \langle T^n x, E(t)y \rangle dt \\
&= b\{b^n \langle x, y \rangle - \int_a^b \langle x, E(t)y \rangle \, nt^{n-1} dt\} \\
&\quad - \int_a^b \{b^n \langle x, E(t)y \rangle - \int_a^b \langle x, E(s)E(t)y \rangle ns^{n-1} ds\} dt.
\end{aligned} \tag{8}$$

Now, by Fubini's theorem, we have

$$\int_a^b \int_a^b \langle x, E(s)E(t)y \rangle\, ns^{n-1} ds dt$$

$$= \int_a^b \left\{ \int_a^t \langle x, E(s)y \rangle\, ns^{n-1} ds + \int_t^b \langle x, E(t)y \rangle\, ns^{n-1} ds \right\} dt$$

$$= \int_a^b \langle x, E(s)y \rangle ns^{n-1}(b - s) ds + \int_a^b \langle x, E(t)y \rangle\, (b^n - t^n) dt.$$

By substituting this value for the last integral in (8), we obtain

$$\langle T^{n+1}x, y \rangle = b^{n+1}\langle x, y \rangle - \int_a^b \langle x, E(t)y \rangle\, (n + 1)t^n dt.$$

This completes the inductive proof of (7). It follows that, for any polynomial p,

$$\langle p(T)x, y \rangle = p(b)\langle x, y \rangle - \int_a^b \langle x, E(t)y \rangle p'(t) dt. \tag{9}$$

Thus

$$|\langle p(T)x, y \rangle| \leqslant \|x\|\,\|y\| \left\{ |p(b)| + K \int_a^b |p'(t)|\, dt \right\}$$
$$\leqslant K\|x\|\,\|y\|\,\{|p(b)| + \operatorname{var}(p, J)\}$$

and therefore $\|p(T)\| \leqslant K\{|p(b)| + \operatorname{var}(p, J)\}$. This proves the first part of the theorem.

Let x in X, y in X^* be fixed and define

$$L_1(f) = \langle f(T)x, y \rangle,$$

$$L_2(f) = f(b)\langle x, y \rangle - \int_a^b \langle x, E(t)y \rangle f'(t) dt \qquad (f \in AC(J)).$$

By virtue of (9) we have $L_1(p) = L_2(p)$, for every polynomial p. From the argument of the proof of Lemma 15.2, we know that $\mathscr{P}(J)$ forms a dense subset of $AC(J)$. Hence $L_1(f) = L_2(f)$ $(f \in AC(J))$ as required.

THEOREM 15.8. *Under the hypotheses of Theorem* 15.7 *define*

$$M_s = E(s)X^*, N_s = \bigcap_{t<s} (I - E(t))X^*$$

and let L_s *(respectively* R_s*) be the class of all functions in* $AC(J)$ *such that* $f(t) = 0$ *when* $t \leqslant s$ *(respectively* $t \geqslant s$*). Then, if* $f \to f(T)$ *is the algebra homomorphism of Lemma* 15.2

(i) $M_s = \{y \in X^*: f(T^*)y = 0$ *for all* f *in* $L_s\}$,
(ii) $N_s = \{y \in X^*: f(T^*)y = 0$ *for all* f *in* $R_s\}$,
(iii) $M_s \cap N_s = \{y \in X^*: T^*y = sy\}$,
(iv) $M_s = \bigcap_{t>s} M_t$,
(v) *the subspaces* M_s, N_s *are invariant under* T^* *and*

$$\sigma(T^*|M_s) \subseteq [a, s], \sigma(T^*|N_s) \subseteq [s, b] \text{ for } a \leqslant s \leqslant b.$$

Proof. (i) The result is obvious if $s \notin [a, b)$, since

$$L_s = \{0\} \ (s \geqslant b), L_s = AC(J) \ (s < a).$$

We may therefore suppose that $a \leqslant s < b$. Let $y \in X^*$. Then $y \in M_s$

$$\begin{aligned}
&\Leftrightarrow E(t)y = y \qquad (s \leqslant t \leqslant b)\\
&\Leftrightarrow \langle x, E(t)y\rangle = \langle x, y\rangle \ (s \leqslant t \leqslant b, x \in X)\\
&\Leftrightarrow \langle x, E(t)y\rangle = \langle x, y\rangle \text{ for almost all } t \text{ in } [s, b] \text{ and } x \text{ in } X\\
&\Leftrightarrow \langle x, y\rangle \int_a^b f'(t)dt - \int_a^b \langle x, E(t)y\rangle f'(t)dt = 0 \qquad (f \in L_s, x \in X)\\
&\Leftrightarrow f(b)\langle x, y\rangle - \int_a^b \langle x, E(t)y\rangle f'(t)dt = 0 \qquad (f \in L_s, x \in X)\\
&\Leftrightarrow \langle x, f(T^*)y\rangle = 0 \qquad (f \in L_s, x \in X)\\
&\Leftrightarrow f(T^*)y = 0 \qquad (f \in L_s).
\end{aligned}$$

This proves (i) The proof of (ii) is similar, and is omitted.

(iv) If $a \leqslant s < b$, the result follows from the fact that $y \in M_s$ if and only if $\langle x, E(t)y\rangle = \langle x, y\rangle$ for almost all t in $[s, b]$ and x in X. Since M_s is constant on each of the complementary intervals of $[a, b)$, the result is trivial when $s \notin [a, b)$.

(v) Since L_s and R_s are ideals in the Banach algebra $AC(J)$, it follows easily from (i) and (ii) that M_s and N_s are invariant under $f(T^*)$ $(f \in AC(J))$ and, in particular, under T^*.

Suppose that $r \notin [a, s]$ and that g in $AC(J)$ is such that

$$(t - r)g(t) = 1 \qquad (a \leqslant t \leqslant s).$$

Then the function $(t - r)g(t) - 1 = g(t)(t - r) - 1$ is in L_s and by (i)

$$(T^* - rI^*)g(T^*)y - y = g(T^*)(T^* - rI^*)y - y = 0 \qquad (y \in M_s).$$

Thus $T^* - rI^*$, as an operator acting on M_s, has the inverse $g(T^*)$. It follows that $r \notin \sigma(T^*|M_s)$. Hence $\sigma(T^*|M_s) \subseteq [a, s]$. The proof of the second inclusion is similar.

(iii) We first note that, by taking $s = b$ in (v), we obtain $\sigma(T^*) \subseteq [a, b]$. From this, it is obvious that

$$M_s \cap N_s = \{0\} = \{y \in X^* : T^*y = sy\}$$

when $s \notin [a, b]$. We may therefore suppose that $s \in [a, b]$. We claim that

$$M_s \cap N_s = \{y \in X^* : f(T^*)y = f(s)y \qquad (f \in AC(J))\}. \tag{10}$$

In fact, if $f(T^*)y = f(s)y$ $(f \in AC(J))$, then $f(T^*)y = 0$ when $f \in L_s \cup R_s$ and so $y \in M_s \cap N_s$. Conversely, if $y \in M_s \cap N_s$ and $f \in AC(J)$, then the function g defined by $g(t) = f(t) - f(s)$ can be expressed as the sum of some g_1 in L_s and g_2 in R_s. It follows that

$$f(T^*)y - f(s)y = g(T^*)y = g_1(T^*)y - g_2(T^*)y = 0.$$

This proves (10).

By taking $f(t) = t$, we deduce that $T^*y = sy$ whenever $y \in M_s \cap N_s$. On the other hand if $T^*y = sy$, then $f(T^*)y = f(s)y$ whenever f is a polynomial, and hence, by the continuity of the homomorphism $f \to f(T)$ of Lemma 15.2, the last equation holds for all f in $AC(J)$.

COROLLARY 15.9. *Under the hypotheses of Theorem* 15.7

$$\sigma(T) \subseteq [a, b].$$

Proof. We have already seen in the course of proving (iii) that $\sigma(T^*) \subseteq [a, b]$. The desired conclusion is immediate.

COROLLARY 15.10. *If T is a decomposable operator then the subspaces M_s, N_s depend only on T and not on the choice of the decomposition of the identity $\{E(t): t \in \mathbf{R}\}$.*

Proof. The result follows immediately from Theorem 15.8 (i) and (ii).

Now, let T be a well-bounded operator on X. Following Ringrose [3], we shall construct a decomposition of the identity $\{E(t): t \in \mathbf{R}\}$ for X such that (2) is satisfied. First we require some preliminary lemmas.

LEMMA 15.11. *Let L be a bounded linear functional on the Banach space $AC(J)$. Then there exist a constant m_L and a function w_L in $L^\infty(a, b)$ such that*

$$L(f) = m_L f(b) - \int_a^b w_L(t) df(t) \qquad (f \in AC(J)).$$

Moreover the norm of L is equal to $\max(|m_L|, \|w_L\|_\infty)$.

Proof. Observe that a complex function f is absolutely continuous on J if and only if its derivative f' (which exists a.e. on J) is in $L^1(J)$. In this case

$$f(b) - f(x) = \int_x^b f'(t) dt \qquad (a \leqslant x \leqslant b).$$

It follows that $AC(J)$ can be identified with the direct sum of $L^1(J)$ and a one-dimensional space. The first statement of the lemma follows immediately. Also it is clear that

$$\|L\| \leqslant \max(|m_L|, \|w_L\|_\infty).$$

To obtain an inequality in the opposite direction it is sufficient to consider separately the action of L on the subspace of constant functions and of functions in $AC(J)$ vanishing at b. This completes the proof.

LEMMA 15.12. *Given any x in X, ϕ in X^*, there exists a function $w_{x,\phi}$ in $L^\infty(a, b)$,* *uniquely determined to within a null function, such that*

$$\langle x, f(T^*)\phi\rangle = f(b)\langle x, \phi\rangle - \int_a^b w_{x,\phi}(\lambda)df(\lambda) \qquad (f \in AC(J)), \tag{11}$$

where $f \to f(T^*)$ is the homomorphism of $AC(J)$ into $L(X^*)$ associated as in Lemma 15.2 *with the well-bounded operator T^*. The function $w_{x,\phi}$ satisfies*

$$\| w_{x,\phi} \|_\infty \leqslant K \| \phi \| \, \| x \|,$$

and its equivalence class (modulo null functions) depends linearly on both x and ϕ.

Proof. For fixed x in X and ϕ in X^*, define

$$L_{x,\phi}(f) = \langle x, f(T^*)\phi\rangle \qquad (f \in AC(J)).$$

It is apparent from Lemma 15.2 (i) that $L_{x,\phi}$ is a continuous linear functional on $AC(J)$, and that

$$\| L_{x,\phi} \| \leqslant K \| \phi \| \, \| x \|.$$

Hence there exist a constant $m_{x,\phi}$ and a function $w_{x,\phi}$ in $L^\infty(a, b)$ such that

$$L_{x,\phi}(f) = m_{x,\phi} f(b) - \int_a^b w_{x,\phi}(\lambda)df(\lambda) \qquad (f \in AC(J)),$$

$$\max(|m_{x,\phi}|, \| w_{x,\phi} \|_\infty) \leqslant K \| \phi \| \, \| x \|.$$

By considering the function f identically equal to 1 we obtain $m_{x,\phi} = \langle x, \phi\rangle$. Hence

$$\langle x, f(T^*)\phi\rangle = f(b)\langle x, \phi\rangle - \int_a^b w_{x,\phi}(\lambda)df(\lambda) \qquad (f \in AC(J)).$$

Furthermore, the integral in this equation is uniquely determined and depends linearly on both x and ϕ, for any choice of f in $AC(J)$. Hence the equivalence class of $w_{x,\phi}$ is likewise uniquely determined and is a linear function of both x and ϕ.

Now, let $NBV_0(J)$ be the subalgebra of $NBV(J)$ consisting of those functions f in $NBV(J)$ whose continuous singular parts vanish identically; that is f can be written in the form $f = f_{ac} + f_b$ where $f_{ac} \in AC(J)$ and f_b is the limit of a uniformly convergent sequence of step-functions.

We shall attempt to use (11) as a method of defining an operator $f(T^*)$ for a general f in $NBV_0(J)$. If we choose s with $a \leqslant s \leqslant b$, then $\chi_{(-\infty, s]} \in NBV_0(J)$. Formal substitution in (11) yields

$$\langle x, \chi_{(-\infty, s]}(T^*)y \rangle = w_{x,y}(s).$$

Hence, in order that a linear operator $\chi_{(-\infty, s]}(T^*)$ can be defined in this way for each real s, it is necessary that the functions $w_{x,y}$ themselves (not merely their equivalence classes) shall depend linearly on both x in X and y in X^*. We are thus led to the problem of selecting representatives from the equivalence classes of L^∞ functions in such a way that a linear relation between equivalence classes implies the corresponding relation between the functions representing these classes.

This problem was solved by J. von Neumann with perfectly general algebraic relations. However, Ringrose [**3**] gave a less sophisticated construction of a set of representatives which has the properties required. In the following lemma, we use the symbol $=^\circ$ to denote equality a.e. on $[a, b)$, and $\equiv$ for equality everywhere on $[a, b)$.

LEMMA 15.13. *Let $\mathscr{F}$ be an ultrafilter on $(0, \infty)$ which converges to zero in the usual topology on the real line. Let u be any function which is essentially bounded and Lebesgue measurable on the interval $[a, b)$. Then for every λ in $[a, b)$,*

$$u_{\mathscr{F}}(\lambda) = \lim_{\substack{h \to 0 \\ \mathscr{F}}} \frac{1}{h} \int_0^h u(\lambda + t)dt$$

exists. Furthermore, if $u, v, w \in L^\infty(a, b)$, then

(i) *$u_{\mathscr{F}} =^0 u$, and if $u =^0 v$ then $u_{\mathscr{F}} \equiv v_{\mathscr{F}}$;*
(ii) *$u_{\mathscr{F}}$ is bounded and Lebesgue measurable on $[a, b)$ and*

$$\sup_{\lambda \in [a,b)} |u_{\mathscr{F}}(\lambda)| = \|u\|_\infty;$$

(iii) *if u is continuous on the right throughout $[a, b)$, then $u \equiv u_{\mathscr{F}}$;*

(iv) *if c, d are constants and* $cu + dv =^0 w$, *then* $cu_{\mathscr{F}} + dv_{\mathscr{F}} \equiv w_{\mathscr{F}}$;

(v) *if* $w =^{\circ} uv$ *and* $v_{\mathscr{F}}$ *is continuous on the right throughout* $[a, b)$, *then* $w_{\mathscr{F}} = u_{\mathscr{F}} v_{\mathscr{F}}$.

Proof. We may consider the interval $(0, \infty)$ as a subset of its Čech compactification Q. Then the ultrafilter $\mathscr{F}$ converges to some point α of Q.

We may define $u(\lambda) = 0$ when $\lambda \notin [a, b)$. For fixed u and λ, the function

$$r(h) = \frac{1}{h} \int_0^h u(\lambda + t)dt$$

is continuous on $(0, \infty)$ and is bounded since $|r(h)| \leqslant \| u \|_{\infty}$. Hence there is a unique continuous function r_0 on Q whose restriction to $(0, \infty)$ is r. Thus

$$\lim_{\substack{h \to 0 \\ \mathscr{F}}} r(h) = \lim_{h \to \alpha} r_0(h) = r_0(\alpha).$$

The proofs of (i), ..., (v) are straightforward, and are omitted.

We shall refer to $u_{\mathscr{F}}$ as the *$\mathscr{F}$-representative* of the equivalence class containing u.

LEMMA 15.14. (i) *Suppose that the functions* $w_{x,\phi}$ *of Lemma* 15.12 *are selected by taking in each case the $\mathscr{F}$-representative of the relevant equivalence class. Then, given any f in* $NBV_0(J)$, *there is a unique operator* $f_{\mathscr{F}}(T^*)$ *in* $L(X^*)$ *such that*

$$\langle x, f_{\mathscr{F}}(T^*)\phi \rangle = f(b)\langle x, \phi \rangle - \int_{[a,b)} w_{x,\phi}(\lambda) df(\lambda) \qquad (x \in X, \phi \in X^*). \quad (12)$$

Furthermore

$$\| f_{\mathscr{F}}(T^*) \| \leqslant K ||| f |||.$$

(ii) *The mapping* $f \to f_{\mathscr{F}}(T^*)$ *extends the homomorphism* $f \to f(T^*)$ *of Lemma* 15.2.

Proof. (i) Since $w_{x,\phi}$ is the $\mathscr{F}$-representative of an equivalence class, which depends linearly on both x and ϕ, we may deduce from Lemma 15.13 (iv) that the function $w_{x,\phi}$ itself has the same property. Furthermore

$$\sup_{\lambda \in [a,b)} |w_{x,\phi}(\lambda)| = \| w_{x,\phi} \|_\infty \leqslant K \| \phi \| \, \| x \|.$$

Since $w_{x,\phi}$ is Lebesgue measurable, it is f-measurable for every f in $NBV_0(J)$, and the equation

$$L(x, \phi) = f(b)\langle x, \phi \rangle - \int_{[a,b)} w_{x,\phi}(\lambda) df(\lambda)$$

defines a bilinear form L on $X \times X^*$. Also

$$\begin{aligned} |L(x, \phi)| &\leqslant \| \phi \| \, \| x \| \, \{| f(b) + K \operatorname*{var}_J f\} \\ &\leqslant K \| \phi \| \, \| x \| \, |||f|||. \end{aligned}$$

It easily follows from this that, given any ϕ in X^*, the equation

$$\langle x, f_{\mathscr{F}}(T^*)\phi \rangle = L(x, \phi)$$

defines uniquely a linear functional $f_{\mathscr{F}}(T^*)\phi$ on X, and that the operator $f_{\mathscr{F}}(T^*)$ has all the required properties.

(ii) When $f \in AC(J)$, we may deduce from (11) and (12) that $f_{\mathscr{F}}(T^*) = f(T^*)$.

LEMMA 15.15. *Suppose that $S \in L(X)$ and $ST = TS$. Then*

$$S^* f_{\mathscr{F}}(T^*) = f_{\mathscr{F}}(T^*)S^* \qquad (f \in NBV_0(J)).$$

Proof. For any x in X and ϕ in X^*,

$$\begin{aligned} \langle x, S^* f_{\mathscr{F}}(T^*)\phi \rangle &= \langle Sx, f_{\mathscr{F}}(T^*)\phi \rangle \\ &= f(b)\langle Sx, \phi \rangle - \int_{[a,b)} w_{Sx,\phi}(\lambda) df(\lambda) \end{aligned}$$

and

$$\langle x, f_{\mathscr{F}}(T^*)S^*\phi \rangle = f(b)\langle x, S^*\phi \rangle - \int_{[a,b)} w_{x,S^*\phi}(\lambda) df(\lambda).$$

It follows that $f_{\mathscr{F}}(T^*)S^* = S^*f_{\mathscr{F}}(T^*)$ if and only if

$$\int_{[a,b)} [w_{Sx,\phi}(\lambda) - w_{x,S^*\phi}(\lambda)]\, df(\lambda) = 0 \qquad (x \in X, \phi \in X^*). \tag{13}$$

By virtue of Lemma 15.2 (ii) and Lemma 15.14 (ii), (13) is satisfied whenever $f \in AC(J)$. It follows that

$$w_{Sx,\phi} \stackrel{0}{=} w_{x,S^*\phi}.$$

Thus these two functions are $\mathscr{F}$-representatives of the same equivalence class, and are therefore identically equal. Hence (13) is satisfied for every f in $NBV_0(J)$, as required.

LEMMA 15.16. *The mapping $f \to f_{\mathscr{F}}(T^*)$ from $NBV_0(J)$ into $L(X^*)$ is a homomorphism.*

Proof. It is immediate from (12) that this mapping is linear. It remains to establish the multiplicative property.

Let $f, g \in NBV_0(J)$, $x \in X$, $\phi \in X^*$. In the computations that follow, we shall omit the suffix $\mathscr{F}$; by virtue of Lemma 15.14 (ii) this causes no inconsistency. We have

$$\begin{aligned}\langle x, f(T^*)g(T^*)\phi\rangle &= f(b)\langle x, g(T^*)\phi\rangle - \int_{[a,b)} w_{x,g(T^*)\phi}(\lambda)df(\lambda)\\ &= f(b)g(b)\langle x, \phi\rangle - f(b)\int_{[a,b)} w_{x,\phi}(\lambda)dg(\lambda)\\ &\qquad - \int_{[a,b)} w_{x,g(T^*)\phi}(\lambda)df(\lambda)\end{aligned}$$

while

$$\langle x, (fg)(T^*)\phi\rangle = (fg)(b)\langle x, \phi\rangle - \int_{[a,b)} w_{x,\phi}(\lambda)d(fg)(\lambda)$$

$$= f(b)g(b)\langle x, \phi\rangle - \int_{[a,b)} w_{x,\phi}(\lambda)f(\lambda+)dg(\lambda)$$

$$- \int_{[a,b)} w_{x,\phi}(\lambda)g(\lambda+)df(\lambda).$$

It follows that

$$\langle x, f(T^*)g(T^*)\phi\rangle - \langle x, (fg)(T^*)\phi\rangle$$

$$= \int_{[a,b)} \{g(\lambda+)w_{x,\phi}(\lambda) - w_{x,g(T^*)\phi}(\lambda)\}df(\lambda) - f(b)\int_{[a,b)} w_{x,\phi}(\lambda)dg(\lambda)$$

$$+ \int_{[a,b)} f(\lambda)d\left(\int_{[a,\lambda]} w_{x,\phi}(\mu)dg(\mu)\right)$$

$$= \int_{[a,b)} \left\{g(\lambda+)w_{x,\phi}(\lambda) - w_{x,g(T^*)\phi}(\lambda) - \int_{[a,\lambda]} w_{x,\phi}(\mu)dg(\mu)\right\} df(\lambda). \tag{14}$$

By virtue of Lemma 15.2, the left-hand side of (14) is zero when $f, g \in AC(J)$. By fixing g and varying f we deduce that

$$g(\lambda+)w_{x,\phi}(\lambda) - w_{x,g(T^*)\phi}(\lambda) - \int_{[a,\lambda]} w_{x,\phi}(\mu)dg(\mu) =^0 0 \quad (x \in X, \phi \in X^*) \tag{15}$$

whenever $g \in AC(J)$. Since

$$g(\lambda+), \int_{[a,\lambda]} w_{x,\phi}(\mu)dg(\mu)$$

are both right continuous functions of λ, we may deduce from Lemma 15.13 (iii), (v) that each of the three terms in (15) is the $\mathscr{F}$-representative of its equivalence class. Hence, by Lemma 15.13 (iv), the left-hand side of (15) vanishes identically whenever $g \in AC(J)$. We may therefore deduce from (14) that

$$f(T^*)g(T^*) = (fg)(T^*) \cdot \quad (f \in NBV_0(J), g \in AC(J)). \tag{16}$$

However, when $g \in AC(J)$, $g(T)$ commutes with T and so, by Lemma 15.15, $g(T^*) = g(T)^*$ commutes with $f(T^*)$, for every f in $NBV_0(J)$. Since $(fg)(\lambda) \equiv (gf)(\lambda)$ we may rewrite (16) in the form

$$g(T^*)f(T^*) = (gf)(T^*) \qquad (f \in NBV_0(J), g \in AC(J))$$

which is equivalent to

$$f(T^*)g(T^*) = (fg)(T^*) \qquad (f \in AC(J), g \in NBV_0(J)).$$

We may now deduce from (14), by varying f in $AC(J)$, that (15) holds whenever $g \in NBV_0(J)$ The same argument as before shows that the left-hand side of (15) vanishes identically ($g \in NBV_0(J)$), and hence

$$f(T^*)g(T^*) = (fg)(T^*) \qquad (f, g \in NBV_0(J)).$$

LEMMA 15.17. *Let $E(\mu)$ be the image, under the homomorphism of Lemma* 15.16, *of the characteristic function e_μ of the interval $(-\infty, \mu]$. Then, in the lower weak topology of $L(X^*)$,*

$$E(\mu) = \lim_{\substack{h \to 0 \\ \mathscr{F}}} e(T^*; \mu, h), \tag{17}$$

where $e(\lambda; \mu, h)$ is the absolutely continuous function of λ which takes the values 1, 0 *on the intervals $(-\infty, \mu]$, $[\mu + h, \infty)$ respectively, and is linear on $[\mu, \mu + h]$.*

Proof. If $x \in X$ and $\phi \in X^*$, we have

$$\langle x, E(\mu)\phi \rangle = e_\mu(b)\langle x, \phi \rangle - \int_{[a,b)} w_{x,\phi}(\lambda) de_\mu(\lambda)$$

$$= \begin{cases} 0 & (\mu < a) \\ w_{x,\phi}(\mu) & (a \leqslant \mu < b) \\ \langle x, \phi \rangle & (\mu \geqslant b). \end{cases} \tag{18}$$

Since $w_{x,\phi}$ is the $\mathscr{F}$-representative of its equivalence class, it follows that, when $a \leqslant \mu < b$,

$$\langle x, E(\mu)\phi\rangle = \lim_{\substack{h \to 0 \\ \mathscr{F}}} \frac{1}{h} \int_0^h w_{x,\phi}(\mu + t)dt$$

$$= \lim_{\substack{h \to 0 \\ \mathscr{F}}} \left\{ e(b; \mu, h)\langle x, \phi\rangle - \int_{[a,b)} w_{x,\phi}(\lambda)de(\lambda; \mu, h) \right\}$$

since (for sufficiently small h) $e(b; \mu, h) = 0$, while

$$\frac{d}{d\lambda}[e(\lambda; \mu, h)] = \begin{cases} -1/h & (\mu < \lambda < \mu + h), \\ 0 & (\lambda \notin [\mu, \mu + h]). \end{cases}$$

Thus

$$\langle x, E(\mu)\phi\rangle = \lim_{\substack{h \to 0 \\ \mathscr{F}}} \langle x, e(T^*; \mu, h)\phi\rangle \qquad (x \in X, \phi \in X^*).$$

LEMMA 15.18. *The operators $\{E(\mu): \mu \in \mathbf{R}\}$ introduced in Lemma 15.17 form a decomposition of the identity for X. The associated decomposable operator is T.*

Proof. Since $e_\mu^2 = e_\mu$, it follows that $E(\mu)^2 = E(\mu)$, and so $E(\mu)$ is a projection on X^*. We now have to verify that conditions (i), ..., (vi) of Definition 15.3 are satisfied.

Now (i) and (iv) follow from (18), and (ii) is an immediate consequence of the corresponding relations for the functions e_λ, e_μ. Since

$$\| E(\lambda) \| \leqslant K |||e_\lambda||| \leqslant K,$$

(iii) is satisfied. Since $w_{x,\phi}$ is the $\mathscr{F}$-representative of its equivalence class, we may deduce from (18) that

$$\langle x, E(\lambda)\phi\rangle = \lim_{\substack{h \to 0 \\ \mathscr{F}}} \frac{1}{h} \int_0^h \langle x, E(\lambda + t)\phi\rangle\, dt \qquad (x \in X, \phi \in X^*),$$

from which property (v) follows at once.

It remains to prove (vi). From (11), (18), and Lemma 15.2 (iii) we deduce that

$$\langle f(T)x, \phi\rangle = f(b)\langle x, \phi\rangle - \int_a^b \langle x, E(\lambda)\phi\rangle f'(\lambda)d\lambda$$

$$(f \in AC(J), x \in X, \phi \in X^*). \qquad (19)$$

Given u in $L^1(a, b)$, set

$$f_u(\lambda) = \int_\lambda^b u(t)dt.$$

Then $f_u \in AC(J)$. For any fixed x in X, the mapping A from $L^1(a, b)$ into X defined by

$$Au = f_u(T)x$$

is clearly linear and continuous, and by use of (19) we obtain

$$\langle u, A^*\phi\rangle = \langle Au, \phi\rangle = \langle f_u(T)x, \phi\rangle = \int_a^b \langle x, E(\lambda)\phi\rangle u(\lambda)d\lambda.$$

Thus the mapping considered in condition (vi) of Definition 15.2 is A^*. Also, as the adjoint of a continuous linear mapping from $L^1(a, b)$ into X, it has the required continuity property.

This completes the proof that $\{E(\lambda): \lambda \in \mathbf{R}\}$ is a decomposition of the identity for X. By taking $f(\lambda) \equiv \lambda$ in (9) it follows that the associated decomposable operator is T.

The following theorem summarizes the main results of this section. Part (iii) is a consequence of Lemma 15.15.

THEOREM 15.19. *Let T be a well-bounded operator on X. Then*

(i) *T is decomposable;*

(ii) *if $\mathscr{F}$ is any ultrafilter on $(0, \infty)$ which converges to zero in the usual topology of the real line, then the operators $\{E(\mu): \mu \in \mathbf{R}\}$ defined by* (17) *are projections and form a decomposition of the identity for X, whose associated decomposable operator is T;*

(iii) *the decomposition of the identity $\{E(\lambda): \lambda \in \mathbf{R}\}$ constructed as* in (ii) *has the following property: if* $S \in L(X)$ *and* $ST = TS$, *then*

$$S^*E(\lambda) = E(\lambda)S^* \qquad (\lambda \in \mathbf{R}).$$

From Theorems 15.7 and 15.19 we deduce that an operator T is well-bounded if and only if it is decomposable. Observe that we have shown that the constant K in Definitions 15.1 and 15.3 can be chosen to be the same in the case that $\{E(\lambda): \lambda \in \mathbf{R}\}$ is a decomposition of the identity for X associated with the well-bounded operator T. Observe that Theorem 15.19 is the analogue for well-bounded operators of Theorem 5.21.

Given a decomposable operator T, the question arises as to whether or not the associated decomposition of the identity is unique. When this is the case, we shall say that T is *uniquely decomposable.*

Definition 15.20. Let $u \in L^1(a, b)$. We shall say that u is *C-limitable* on the right at a point s of $[a, b)$ if the indefinite integral of u is differentiable on the right at s.

This property is not affected if the values of u are altered on a set of measure zero, and depends only on the equivalence class of u. It is therefore meaningful to refer to this definition when u is known only to within a null function.

Definition 15.21. A function u in $L^1(a, b)$ is *C-continuous on the right at a point* s of $[a, b)$ if it is C-limitable on the right at s and the derivative of the indefinite integral of u at the point is equal to $u(s)$.

THEOREM 15.22. *Let* T *be a well-bounded operator on* X *satisfying* (1). *Then* T *is uniquely decomposable if and only if, for every* x *in* X *and* ϕ *in* X^*, *the function* $w_{x,\phi}$ *of Lemma* 15.12 *is C-limitable on the right throughout* $[a, b)$.

Proof. Amongst decompositions of the identity which give rise to the operator T, there is at least one for which

$$E(\lambda) = 0\,(\lambda < a), \qquad E(\lambda) = I(\lambda \geqslant b), \tag{20}$$

where $[a, b]$ is the interval J occurring in (1). It follows from the Corollary 15.10 that all such decompositions satisfy (20).

Suppose now that $\{E(\lambda): \lambda \in \mathbf{R}\}$ is one of these decompositions. From

Theorem 15.7 (ii) and Lemmas 15.12, 15.2 (iii) we may deduce that

$$\int_a^b \langle x, E(\lambda)\phi\rangle f'(\lambda)d\lambda = \int_a^b w_{x,\phi}(\lambda) f'(\lambda)d\lambda \qquad (f \in AC(J)),$$

and hence that

$$\langle x, E(\lambda)\phi\rangle = w_{x,\phi}(\lambda)$$

for almost all λ in $[a, b]$. If we assume that each of the functions $w_{x,\phi}$ is C-limitable on the right throughout $[a, b)$, then the same is true of the functions $\lambda \to \langle x, E(\lambda)\phi\rangle$. Property (v) of decompositions of the identity now implies that

$$\begin{aligned}\langle x, E(\mu)\phi\rangle &= \lim_{h\to 0+} \frac{1}{h}\int_0^h \langle x, E(\mu + t)\phi\rangle\, dt \\ &= \lim_{h\to 0+} \frac{1}{h}\int_0^h w_{x,\phi}(\mu + t)dt \qquad (x \in X, \phi \in X^*, a \leqslant \mu < b).\end{aligned}$$

We deduce that $E(\mu)$ is uniquely determined when $\mu \in [a, b)$; also, we have already proved this to be the case when $\mu \notin [a, b)$.

If, however, there exist x in X, ϕ in X^*, and μ in $[a, b)$ such that $w_{x,\phi}$ is not C-limitable on the right at μ, then we may choose ultrafilters $\mathscr{F}_1$, $\mathscr{F}_2$ on $(0, \infty)$, both convergent to zero, such that

$$\lim_{\substack{h\to 0\\ \mathscr{F}_1}} \frac{1}{h}\int_0^h w_{x,\phi}(\mu + t)dt \neq \lim_{\substack{h\to 0\\ \mathscr{F}_2}} \frac{1}{h}\int_0^h w_{x,\phi}(\mu + t)dt.$$

If $\{E(\lambda)^{(i)}\}$ is the decomposition of the identity obtained by taking $\mathscr{F} = \mathscr{F}_i$, then by (18) and Lemma 15.14

$$\langle x, E(\mu)^{(i)}\phi\rangle = \lim_{\substack{h\to 0\\ \mathscr{F}_i}} \frac{1}{h}\int_0^h w_{x,\phi}(\mu + t)dt \qquad (i = 1, 2).$$

Hence $E(\mu)^{(1)} \neq E(\mu)^{(2)}$, and there exist two distinct decompositions of the identity which give rise to the operator T.

COROLLARY 15.23. *Let $T \in L(X)$. Suppose that T is decomposable, but not* *uniquely. Then there exist two distinct associated decompositions of the identity, both having property* (iii) *of Theorem* 15.19.

Proof. The two decompositions constructed above have this property.

Note 15.24. It is of interest to ask whether the homomorphisms of Lemma 15.2 and Lemma 15.14 can be extended to a homomorphism from $NBV(J)$ into $L(X^*)$. This can be done by the method of Lemma 15.16 if the $\mathscr{F}$-representatives of the relevant equivalence classes are Borel measurable functions. This is certainly the case whenever T is uniquely decomposable since, by Theorem 15.22, in this case the $\mathscr{F}$-representative is the limit of the sequence $\{u_n\}$ of continuous functions on J where

$$u_n(\mu) = n \int_0^{1/n} w_{x,\phi}(\mu + t)dt.$$

We conclude this chapter with an example.

Example 15.25. Let $X = L^1(0, 1)$, and let T, in $L(X)$, be defined by the equation

$$(Tx)(t) = tx(t) + \int_0^t x(u)du \qquad (0 \leqslant t \leqslant 1).$$

A routine calculation shows that

$$(T^n x)(t) = t^n x(t) + nt^{n-1} \int_0^t x(u)du.$$

Hence, given any polynomial p,

$$(p(T)x)(t) = p(t)x(t) + p'(t) \int_0^t x(u)du.$$

It follows easily that

$$\begin{aligned} \|p(T)\| &\leqslant \sup_{t \in [0,1]} |p(t)| + \operatorname*{var}_{[0,1]} p \\ &\leqslant 2\{p(1) + \operatorname*{var}_{[0,1]} p\}. \end{aligned}$$

Thus T is a well-bounded operator. It is easily seen (by its uniqueness) that the homomorphism of Lemma 15.2 is determined by the equation

$$(f(T)x)(t) = f(t)x(t) + f'(t)\int_0^t x(u)du \qquad (x \in X, f \in AC(J)).$$

We shall now make the customary identification of X^* with $L^\infty(0, 1)$. When $\phi \in X^*$, $x \in X$, $f \in AC(J)$, we have

$$\begin{aligned}\langle f(T)x, \phi\rangle &= \int_0^1 \phi(t)f(t)x(t)dt + \int_0^1 \phi(t)f'(t)\left(\int_0^t x(u)du\right)dt \\ &= f(1)\langle x, \phi\rangle - \int_0^1 \left\{\int_0^t \phi(u)x(u)du - \phi(t)\int_0^t x(u)du\right\} f'(t)dt.\end{aligned}$$

Hence the functions $w_{x,\phi}$ of Lemma 15.12 are given by

$$w_{x,\phi}(t) = \int_0^t \phi(u)x(u)du - \phi(t)\int_0^t x(u)du.$$

The first term on the right-hand side of this equation is absolutely continuous, and therefore C-limitable on the right throughout $[0, 1)$. However, for suitably chosen x and ϕ, the second term will not have this property. For example, define

$$\begin{aligned} x(t) = 1\ (0 < t < \tfrac{1}{2}),\ x(t) &= 0\ (\tfrac{1}{2} \leqslant t < 1), \\ \phi(t) &= 2 + \sin \log |t - \tfrac{1}{2}|\ (0 < t < 1, t \neq \tfrac{1}{2}), \\ \phi(\tfrac{1}{2}) &= 0. \end{aligned}$$

It is easily verified that $\phi \in L^\infty(0, 1)$ and ϕ is not C-limitable on the right at $t = \frac{1}{2}$. It follows that $w_{x,\phi}$ is not C-limitable on the right throughout $[0, 1)$. We deduce from Theorem 15.22 that T is not uniquely decomposable. Note that the space X is weakly complete.

16. Some Special Classes of Well-bounded Operators

In this chapter we introduce three special classes of well-bounded operators. These are well-bounded operators decomposable in X and well-bounded operators of type (A) and type (B). The relationships between these classes and the class of uniquely decomposable well-bounded operators, introduced in the last chapter, are discussed. Examples are given to show that in general the four classes may be distinct.

Definition 16.1. Let $T \in L(X)$ and let T be decomposable. If the decomposition of the identity may be chosen so that each projection $E(\lambda)$ is the adjoint of some operator acting on X, we shall say that T is *decomposable* in X.

THEOREM 16.2. *Let T, in $L(X)$, be a well-bounded operator decomposable in X. Let $\{F(\lambda): \lambda \in \mathbf{R}\}$ be a family of projections on X whose adjoints $\{E(\lambda): \lambda \in \mathbf{R}\}$ form a decomposition of the identity for X and satisfy*

$$\langle Tx, \phi\rangle = b\langle x, \phi\rangle - \int_a^b \langle x, E(\lambda)\phi\rangle \, d\lambda \qquad (x \in X, \phi \in X^*).$$

Let $\mu \in [a, b]$. Then

(i) $$F(\mu)f(T) = f(T)F(\mu) \qquad (f \in AC(J)),$$

where $f \to f(T)$ is the homomorphism of Lemma 15.2;

(ii) $$\sigma(T|F(\mu)X) \subseteq [a, \mu], \sigma(T|(I - F(\mu))X) \subseteq [\mu, b].$$

Proof. (i) The equation established in Theorem 15.7 (ii) may be rewritten in the form

$$\langle f(T)x, \phi\rangle = \langle x, \phi\rangle - \int_a^b \langle F(\lambda)x, \phi\rangle f'(\lambda)d\lambda \qquad (x \in X, \phi \in X^*). \qquad (1)$$

By using this and the relation $F(\lambda)F(\mu) = F(\mu)F(\lambda)$, we obtain

$$\begin{aligned}\langle F(\mu)f(T)x, \phi\rangle &= \langle f(T)x, E(\mu)\phi\rangle \\ &= f(b)\langle x, E(\mu)\phi\rangle - \int_a^b \langle F(\lambda)x, E(\mu)\phi\rangle f'(\lambda)d\lambda \\ &= f(b)\langle F(\mu)x, \phi\rangle - \int_a^b \langle F(\lambda)F(\mu)x, \phi\rangle f'(\lambda)d\lambda \\ &= \langle f(T)F(\mu)x, \phi\rangle \qquad (x \in X, \phi \in X^*).\end{aligned}$$

Hence $F(\mu)f(T) = f(T)F(\mu)$.

(ii) Suppose that $\kappa \notin [\mu, b]$. Then there exists r in $AC(J)$ such that $(\lambda - \kappa)r(\lambda) = 1$ $(\mu \leqslant \lambda \leqslant b)$, and (as in the proof of Theorem 15.8 (v)) we have

$$(T^* - \kappa I^*)r(T^*)\phi = \phi = r(T^*)(T^* - \kappa I^*)\phi \qquad (\phi \in N_\mu).$$

Since $(I^* - E(\mu))X^* \subseteq N_\mu$, we deduce that

$$(T^* - \kappa I^*)r(T^*)(I^* - E(\mu)) = I^* - E(\mu) = r(T^*)(T^* - \kappa I^*)(I^* - E(\mu)).$$

From this, and by use of (i) and Lemma 15.2 (iii), we obtain

$$r(T)(T - \kappa I)(I - F(\mu)) = I - F(\mu) = (T - \kappa I)r(T)(I - F(\mu)).$$

Hence, if all the operators are restricted to the subspace $(I - F(\mu))X$, $r(T)$ is the inverse of $T - \kappa I$. Thus $\kappa \notin \sigma(T|(I - F(\mu))X)$, and

$$\sigma(T|(I - F(\mu))X) \subseteq [\mu, b].$$

The proof that $\sigma(T|F(\mu)X) \subseteq [a, b]$ is similar.

THEOREM 16.3. *Let $\{F(\lambda): \lambda \in \mathbf{R}\}$ be a family of projections on X whose adjoints form a decomposition of the identity for X. Then*

(i) *the associated decomposable operator T is uniquely decomposable*;
(ii) *if $S \in L(X)$ and $ST = TS$, then $F(\lambda)S = SF(\lambda)$ $(\lambda \in \mathbf{R})$*;
(iii) *given any x in X and ϕ in X^*, the function $\lambda \to \langle F(\lambda)x, \phi\rangle$ is everywhere C-continuous on the right.*

Proof. (i) Suppose that $\{E(\lambda): \lambda \in \mathbf{R}\}$ is any decomposition of the identity which gives rise to the operator T and has property (iii) of Theorem 15.19. Since, by Theorem 16.2 (i), $F(\lambda)$ commutes with T, it follows that $F(\lambda)^*$ commutes with $E(\lambda)$. Hence $E(\lambda)$, $F(\lambda)^*$ are commuting projections with the same range space, by Corollary 15.10, and are therefore equal. Thus $\{F(\lambda)^*: \lambda \in \mathbf{R}\}$ is the only decomposition of the identity of this type, and by Corollary 15.23 we deduce that T is uniquely decomposable.

(ii), (iii) Let $\mathscr{F}$ be an ultrafilter on $(0, \infty)$ which converges to zero in the usual topology of the real line and let $\{E(\lambda): \lambda \in \mathbf{R}\}$ be the corresponding decomposition of the identity. (See Lemmas 15.17, 15.18.) We may prove, by the method used above, that $E(\lambda) = F(\lambda)^*$ $(\lambda \in \mathbf{R})$. Part (ii) of the theorem now follows from Theorem 15.19 (iii). Furthermore, since T is uniquely decomposable, the functions $w_{x,\phi}$ are C-limitable on the right throughout $[a, b)$. Hence the function $\lambda \to \langle F(\lambda)x, \phi\rangle = \langle x, E(\lambda)\phi\rangle$, which is the $\mathscr{F}$-representative of the equivalence class containing $w_{x,\phi}$, is C-continuous on the right throughout $[a, b)$. It is constant, and hence C-continuous on the right, on each of the complementary intervals of $[a, b)$.

THEOREM 16.4. *Let X be weakly complete and let T be a decomposable operator on X. Then T is uniquely decomposable if and only if T is decomposable in X.*

Proof. The implication in one direction has already been established in Theorem 16.3 (i).

Suppose that T is uniquely decomposable. Then the functions $w_{x,\phi}$ are C-limitable on the right throughout $[a, b)$. In these circumstances, the argument used to prove Lemma 15.17 shows that if $\mu \in \mathbf{R}$, then

$$\begin{aligned}\langle x, E(\mu)\phi\rangle &= \lim_{h \to 0+} \langle x, e(T^*; \mu, h)\phi\rangle \\ &= \lim_{h \to 0+} \langle e(T; \mu, h)x, \phi\rangle \qquad (x \in X, \phi \in X^*).\end{aligned}$$

Since X is weakly complete, we deduce that, as $h \to 0+$ through any sequence, $e(T; \mu, h)$ converges in the weak operator topology of $L(X)$ to some operator $F(\mu)$ in $L(X)$. It is clear that $F(\mu)^* = E(\mu)$. Hence T is decomposable in X.

We now give an example of a Banach space X and a well-bounded operator on X which is uniquely decomposable but not decomposable in X.

Example 16.5. Let $X = C[0, 1]$. Define T, in $L(X)$, by

$$(Tx)(t) = tx(t) \qquad (x \in X, 0 \leqslant t \leqslant 1).$$

Obviously if $x \in X$ and $0 \leqslant t \leqslant 1$, then

$$(T^n x)(t) = t^n x(t) \qquad (n = 0, 1, 2, \ldots).$$

Hence, for any polynomial p,

$$(p(T)x) = p(t)x(t) \qquad (x \in X, 0 \leqslant t \leqslant 1).$$

It follows easily that

$$\| p(T) \| \leqslant \sup_{t \in [0, 1]} |p(t)| \leqslant |||p|||.$$

Thus T is a well-bounded operator. It is clear that the homomorphism of Lemma 15.2 is given by the equation

$$(f(T)x)(t) = f(t)x(t) \qquad (x \in X, f \in AC[0, 1]).$$

We shall now make the customary identification of X^* with the space of Radon measures on $[0, 1]$. When $y \in X^*$, $x \in X$, $f \in AC[0, 1]$, we have

$$\begin{aligned}\langle f(T)x, y\rangle &= \int_{[0, 1]} f(t)x(t)dy(t) \\ &= \int_{[0, 1]} f(t)d\left(\int_{[0, t]} x(u)dy(u)\right) \\ &= f(1) \int_{[0, 1]} x(t)dy(t) - \int_{[0, 1]} \int_{[0, t]} (x(u)dy(u))f'(t)dt.\end{aligned}$$

Hence the functions $w_{x,y}$ of Lemma 15.12 are given by

$$w_{x,y}(t) = \int_{[0,t]} x(u)dy(u) \qquad (0 \leqslant t \leqslant 1). \tag{2}$$

It is clear that $w_{x,y}$ is continuous and hence C-limitable on the right throughout $[0, 1)$. Hence T is uniquely decomposable by Theorem 15.22. From (2) we deduce that

$$\int_{[0,1]} x(u)d(E(t)y)(u) = \langle x, E(t)y\rangle = w_{x,y}(t) = \int_{[0,t]} x(u)dy(u).$$

Thus $E(t)y$ is the restriction of the measure y to the interval $[0, t]$.

Now, suppose that there is an operator $F(t)$ on X whose adjoint is $E(t)$. Then

$$\begin{aligned}\int_{[0,1]} (F(t)x)(u)dy(u) &= \langle F(t)x, y\rangle = \langle x, E(t)y\rangle \\ &= \int_{[0,t]} x(u)dy(u) \\ &= \int_{[0,t]} \chi_{[0,t]}(u)x(u)dy(u) \qquad (y \in X^*).\end{aligned}$$

It follows that

$$(F(t)x)(u) = \chi_{[0,t]}(u)x(u) \qquad (t, u \in [0, 1]).$$

However, if we take $t = \frac{1}{2}$ and $x(u) = 1$ $(0 \leqslant u \leqslant 1)$ then $\chi_{[0,t]}x \notin X$. This gives a contradiction. Hence T is uniquely decomposable but not decomposable in X.

Next, we give a sufficient condition for a decomposable operator on X to be decomposable in X.

THEOREM 16.6. *Let T be a decomposable operator on X. Suppose that T^* has no eigenvectors. Then T is decomposable in X.*

Proof. If $V \subseteq X$, define $V^{\perp} = \{\phi \in X^*: \langle y, \phi \rangle = 0$, for all y in $V\}$. With the notation of Theorem 15.13 let $s \in \mathbf{R}$ and

$$V_s = \text{clm}\{f(T)x: f \in R_s, x \in X\},$$
$$W_s = \text{clm}\{f(T)x: f \in L_s, x \in X\}.$$

Then by Theorem 15.13 (i), (ii) and Lemma 15.2 (iii), we have $V_s^{\perp} = N_s$ and $W_s^{\perp} = M_s$. Now,

$$E(s)X^* = M_s,$$
$$(I - E(s))X^* \subseteq N_s,$$
$$E(s)X^* + (I - E(s))X^* = X^*,$$
$$M_s \cap N_s = \{0\};$$

the last equation holds because $\sigma_p(T^*)$ is void (Theorem 15.8 (iii)). We deduce that

$$N_s = (I - E(s))X^*,$$
$$X^* = M_s + N_s.$$

Hence $E(s)$ is the projection from X^* onto M_s parallel to N_s.

It is clear that

$$(V_s + W_s)^{\perp} = V_s^{\perp} \cap W_s^{\perp} = M_s \cap N_s = \{0\}.$$

On the other hand, for every y in $V_s^{\perp} + W_s^{\perp}$, there exists ϕ in $V_s^{\perp}$ and ζ in $W_s^{\perp}$ such that $y = \phi + \zeta$. It follows that

$$\langle x, y \rangle = \langle x, \phi \rangle + \langle x, \zeta \rangle = 0 \qquad (x \in V_s \cap W_s)$$

and so $y \in (V_s \cap W_s)^{\perp}$. Therefore

$$(V_s \cap W_s)^{\perp} \supseteq V_s^{\perp} + W_s^{\perp} = M_s + N_s = X^*.$$

Thus $(V_s \cap W_s)^{\perp} = X^*$. It follows that $V_s \cap W_s = \{0\}$ and $V_s + W_s = X$. Let $F(s)$ be the projection from X onto V_s parallel to W_s. It is obvious that $F^*(s) = E(s)$. Since the above construction of $F(s)$ can be carried out for any

s in $[a, b)$ and since we can define

$$F(s) = 0\ (s < a),\ F(s) = I\ (s \geqslant b),$$

we deduce that T is decomposable in X.

We now introduce the classes of well-bounded operators of types (A) and (B). Let T be a well-bounded operator on X which is decomposable in X. Suppose that $\{F(\lambda): \lambda \in \mathbf{R}\}$ is the family of projections on X whose adjoints form a decomposition of the identity for X whose associated decomposable operator is T.

Definition 16.7. T is said to be of *type* (A) if the function $\lambda \to F(\lambda)x$ is continuous on the right for every x in X. That is $\lambda \to F(\lambda)$ is continuous on the right as a map from $\mathbf{R}$ into $L(X)$ endowed with the strong operator topology.

Definition 16.8. T is said to be of *type* (B) if T is of type (A) and in addition for each real μ, $\lim_{\lambda \to \mu -} F(\lambda)$ exists in the strong operator topology.

In order to study these classes of well-bounded operators we require a theory of vector-valued Riemann-Stieltjes integration.

We say that a finite set $u = \{u_k: k = 0, 1, \ldots, m\}$ is a *partition* of $J = [a, b]$ if $a = u_0 < u_1 < \ldots < u_m = b$. We write $u \geqslant u'$ and say that u is a *refinement* of u' if and only if each closed interval of the form $[u_r, u_{r+1}]\ (r = 0, \ldots, m-1)$ is contained in some interval of the form $[u'_{j-1}, u'_j]$. The family U_J of all partitions of J is directed by the relation $\geqslant$. We shall denote by $u + u'$ the totality of dividing points in both u and u' arranged in linear order.

Let f be a complex-valued function on J and let g be a function on J taking values in X. When $u \in U_J$, we define

$$\sum_u g\Delta f = \sum_{k=1}^{m} g(v_k)[f(u_k) - f(u_{k-1})],$$

where the v_k are chosen so that $u_{k-1} \leqslant v_k \leqslant u_k$.

Now, let I be any subinterval of J; that is $I = [c, d]$, where $a \leqslant c < d \leqslant b$. We define

$$w(g, I) = \sup\{\| g(t_1) - g(t_2) \| : t_1, t_2 \in I\}.$$

We call $w(g, I)$ the *oscillation of g on I*. Also we define

$$w(Sg\Delta f, I) = \sup\{\| \Sigma_u g\Delta f - \Sigma_{u'} g\Delta f \| : u, u' \in U_I\}.$$

We say that g is *Riemann-Stieltjes integrable with respect to* f, if $\lim_{U_J} \Sigma_u g\Delta f$ exists as a net limit in the norm of X and we define

$$\int_a^b gdf = \lim_{U_J} \Sigma_u g\Delta f.$$

LEMMA 16.9. *If g is a bounded function on J, f is of bounded variation on J and $I \subseteq J$ then*

$$w(Sg\Delta f, I) \leqslant w(g, I)\text{var}(f, I).$$

Proof. The proof is standard and is omitted.

THEOREM 16.10. *If f is a continuous function of bounded variation on J and g is a bounded function, then a sufficient condition that $\int_a^b gdf$ exist is that the set D of all discontinuities of g be countable.*

Proof. Since D is a countable set its points can be arranged as a sequence $\{t_n : n = 1, 2, 3, \ldots\}$. For each n and $\varepsilon > 0$ there exists an open interval $J_n \subseteq J$ with $t_n \in J_n$ such that $\text{var}(f, J_n) < 2^{-n}\varepsilon$ and $J_i \cap J_j = \varnothing$ whenever $i \neq j$. Hence there is a sequence $\{J_n\}$ of disjoint open intervals such that

$$\sum_n \text{var}(f, J_n) < \varepsilon.$$

Now $G = \bigcup_{n=1}^{\infty} J_n$ is an open set. Let $F = J \backslash G$. Then g is continuous on the compact set F and so is uniformly continuous there. Hence there is $\delta > 0$ such that

$$\| g(t) - g(s) \| < \varepsilon \qquad (t, s \in F)$$

whenever $|t - s| < \delta$. Furthermore $w(g, [s - \delta, s + \delta]) < 2\varepsilon$.

Consider now any partition u such that $|||u||| < \delta$, where

$$|||u||| = \max\{|u_k - u_{k-1}| : k = 1, \ldots, m\}.$$

Let $I_1', \ldots, I_r'$ be the closed intervals of the form $[u_{k-1}, u_k]$ that contain at least one point of F and $I_1'', \ldots, I_s''$ the remainder of the intervals of the form

$[u_{k-1}, u_k]$. Then

$$\sum_{i=1}^{s} \operatorname{var}(f, I_i'') \leqslant \sum_n \operatorname{var}(f, J_n) < \varepsilon.$$

Hence if $M = \sup_J \| g(t) \|$ then

$$\sum_{i=1}^{s} w(g, I_i'') \operatorname{var}(f, I_i'') < 2M\varepsilon.$$

On the other hand

$$\sum_{j=1}^{r} w(g, I'_j) \operatorname{var}(f, I_j') \leqslant 2\varepsilon \sum_{j=1}^{r} \operatorname{var}(f, I_j') \leqslant 2\varepsilon \operatorname{var}(f, J).$$

Consequently, for any partition u of J such that $|||u||| < \delta$,

$$\Sigma_u w(g, I) \operatorname{var}(f, I) < 2\varepsilon(M + \operatorname{var}(f, J)).$$

Moreover, if u' is any partition of J such that $u' \geqslant u$ then, by rearrangement of terms so as to bring together the terms in each subinterval of u, we find that

$$\begin{aligned} \Big\| \sum_{i=1}^{m} g(v_i)[f(u_i) - f(u_{i-1})] - \sum_{j=1}^{n} g(v_j')[f(u_j') - f(u_{j-1}')] \Big\| &\leqslant \sum_{i=1}^{m} w(Sg\Delta f, I_i) \\ &\leqslant \sum_{i=1}^{m} w(g, I_i) \operatorname{var}(f, I_i) \qquad \text{(by Lemma 16.9)} \\ &< 2\varepsilon(M + \operatorname{var}(f, J)). \end{aligned}$$

Finally, for any two partitions u', u'' such that $u' \geqslant u$ and $u'' \geqslant u$,

$$\begin{aligned} \| E_{u'} g\Delta f - \Sigma_{u''} g\Delta f \| &\leqslant \| \Sigma_{u'} g\Delta f - \Sigma_u g\Delta f \| + \| \Sigma_u g\Delta f - \Sigma_{u''} g\Delta f \| \\ &< 4\varepsilon(M + \operatorname{var}(f, J)). \end{aligned}$$

Hence the integral $\int_a^b g df$ exists.

THEOREM 16.11. *If* $\int_a^b gdf$ *exists then* $\int_a^b fdg$ *exists and*

$$\int_a^b fdg = g(b)f(b) - g(a)f(a) - \int_a^b gdf.$$

Proof. If $\int_a^b gdf$ exists then, given any $\varepsilon > 0$, there exists a partition u_ε of J such that

$$\| \int_a^b gdf - \Sigma_u g\Delta f \| < \varepsilon$$

for all partitions $u \geqslant u_J$. Choose one such partition u. Then

$$\begin{aligned} &\| \sum_{i=1}^{n} f(v_i)\,[g(u_i) - g(u_{i-1})] - g(b)f(b) + g(a)f(a) + \int_a^b gdf \| \\ &= \| \sum_{i=0}^{n} g(u_i)\,[f(v_{i+1}) - f(v_i)] - \int_a^b gdf \| \\ &= \| \sum_{i=0}^{n} \{g(u_i)\,[f(v_{i+1}) - f(v_i)] + g(u_i)\,[f(u_i) - f(v_i)]\} - \int_a^b gdf \| \\ &< \varepsilon. \end{aligned}$$

Since if v is the partition of J defined by

$$a = u_0 \leqslant v_1 \leqslant u_1 \leqslant v_2 \leqslant \ldots \leqslant v_{n+1} = u_n = b$$

then $v \geqslant u \geqslant u_\varepsilon$, it follows that $\int_a^b fdg$ exists and is equal to

$$f(b)g(b) - f(a)g(a) - \int_a^b gdf.$$

LEMMA 16.12. *Let h be a function continuous on the right throughout* **R** *and with values in a metric space (M, ρ). Then h has only a countable number of discontinuities.*

Proof. For each discontinuity t of h, define

$$d(t) = \overline{\lim_{u \to t}}\, \rho(h(u), h(t)).$$

Let $S_n = \{t \in \mathbf{R}: d(t) > 1/n\}$ and let s be any point of S_n. Since h is continuous on the right, there is $\varepsilon_n > 0$ such that if $s \in S_n$ then

$$d(t) \leqslant \frac{1}{n} \qquad (s < t < s + \varepsilon_n).$$

Choose a rational number in the interval $(s, s + \varepsilon_n)$. This maps S_n one-to-one onto a subset of the rationals. Hence each S_n is countable and therefore the set that concerns us $\bigcup_{n=1}^{\infty} S_n$ is also countable.

We note that the discontinuities of a function f of bounded variation are all of the first kind; that is $f(t+)$ and $f(t-)$ exist for each t. The set of discontinuities is countable.

Let $f \in NBV(J)$ and let $\{t_n : n = 1, 2, \ldots\}$ be the set of all discontinuities of f. Then for each positive integer m

$$\sum_{n=1}^{m} |f(t_n+) - f(t_n)| \leqslant \operatorname{var}(f, J).$$

It follows that the series

$$\sum_n [f(t_n+) - f(t_n)]\chi_{(t_n, b]}(t)$$

is absolutely and uniformly convergent on $[a, b]$. Define

$$f_b(t) = \sum_n [f(t_n+) - f(t_n)]\chi_{(t_n, b]}(t)$$

$$f_c(t) = f(t) - f_b(t) \qquad (a \leqslant t \leqslant b).$$

Then f_c is a continuous function of bounded variation on $[a, b]$. Hence f can be expressed in the form $f = f_c + f_b$, where f_c is a continuous function of bounded variation on $[a, b]$ and f_b is the limit of a uniformly convergent series of step functions.

We are now in a position to prove our main result on well-bounded operators of type (A).

THEOREM 16.13. *Let T be a well-bounded operator of type (A) on X and let $K, J = [a, b]$ be chosen so that*

$$\| p(T) \| \leqslant K |||p||| = K\{|p(b)| + \operatorname*{var}_J p\}.$$

Then the homomorphism of Lemma 15.2 *can be extended to a homomorphism ψ of the Banach algebra $NBV(J)$ into $L(X)$ such that*

$$\langle \psi(f)x, \phi \rangle = f(b)\langle x, \phi \rangle - \int_a^b \langle F(t)x, \phi \rangle \, df(t)$$

$$(x \in X, \phi \in X^*, f \in NBV(J))$$

where $\{F^*(t) : t \in \mathbf{R}\}$ is the decomposition of the identity for T. Furthermore

$$\| \psi(f) \| \leqslant K |||f||| \qquad (f \in NBV(J)).$$

If $S \in L(X)$ and $ST = TS$, then $S\psi(f) = \psi(f)S$ for all f in $NBV(J)$.

Proof. Since T is well-bounded of type (A), for each x in X the function $t \to F(t)x$ is continuous on the right and so by Lemma 16.12 its set of discontinuities $\{t_n : n = 1, 2, \ldots\}$ is countable; moreover

$$\| F(t)x \| \leqslant K \| x \| \qquad (t \in \mathbf{R}, x \in X).$$

It follows from Theorem 16.10 that

$$\int_a^b F(t)x \, df_c(t)$$

exists. For each n, the function $t \to f_n(t) = [f(t_n+) - f(t_n)]\chi_{(t_n, b]}(t)$ has a discontinuity at t_n. If u is a partition of J that contains the point t_n, then

$$\textstyle\sum_u F(\cdot)x\Delta f_n = [f(t_n+) - f(t_n)]F(t')x$$

$$(u_k = t_n \leqslant t' \leqslant u_{k+1}).$$

Hence

$$\int_a^b F(t)x\,df_n(t) = \lim_{U_J} \Sigma_u F(\cdot)x\Delta f_n$$
$$= \lim_{t' \to t_n+} [f(t_n+) - f(t_n)]F(t')x$$
$$= [f(t_n+) - f(t_n)]F(t_n)x.$$

It follows that

$$\int_a^b F(t)df_b(t) = \sum_n [f(t_n+) - f(t_n)]F(t_n)x$$

exists and therefore $\int_a^b F(t)x\,df(t)$ exists. Define

$$\psi(f)x = f(b)x - \int_a^b F(t)x\,df(t) \qquad (f \in NBV(J), x \in X).$$

By considering Riemann-Stieltjes sums we obtain

$$\langle \psi(f)x, \phi \rangle = f(b)\langle x, \phi \rangle - \int_a^b \langle F(t)x, \phi \rangle df(t) \qquad (x \in X, \phi \in X^*),$$

for every f in $NBV(J)$. By Note 15.24, ψ has the required properties and so the proof is complete.

Later in this chapter we shall give an example of a well-bounded operator on X which is decomposable in X but is not of type (A) and also an example of a well-bounded operator of type (A) but not of type (B). In order to discuss these examples it is convenient at this stage to introduce another class of well-bounded operators of which Example 16.5 is a special case.

Definition 16.14. Let T be a well-bounded operator on X. A decomposition of the identity $\{E(\lambda): \lambda \in \mathbf{R}\}$ for T is said to be of *bounded variation* if the function $\lambda \to \langle x, E(\lambda)\phi \rangle$ is of bounded variation on $\mathbf{R}$ for every x in X and ϕ in X^*.

THEOREM 16.15. *Suppose that $T \in L(X)$ and $\sigma(T) \subseteq \mathbf{R}$. Then the following three conditions are equivalent.*

(i) *T is a well-bounded operator with a decomposition of the identity of bounded variation.*

(ii) *There is a compact interval J and a constant M such that*

$$\|p(T)\| \leqslant 4M \sup_J |p(t)|$$

for every complex polynomial p.

(iii) T^* *is a scalar-type operator of class X.*

If (i) *holds, then T is a uniquely decomposable well-bounded operator.*

Proof. It is convenient to prove the second statement of the theorem first. Suppose that $\{E(\lambda): \lambda \in \mathbf{R}\}$ is a decomposition of the identity of bounded variation for T satisfying the conditions (i)–(vi) of Definition 15.3. Then if $x \in X$ and $\phi \in X^*$

$$\langle f(T)x, \phi\rangle = f(b)\langle x, \phi\rangle - \int_a^b \langle x, E(\lambda)\phi\rangle d\lambda \qquad (f \in AC(J)),$$

by Theorem 15.7. It follows that if $w_{x,\phi}$ is the function constructed in Lemma 15.12 then

$$w_{x,\phi}(\lambda) = \langle x, E(\lambda)\phi\rangle \qquad (\text{a.e. on } [a, b]).$$

Now since $\langle x, E(\cdot)\phi\rangle$ is in $BV[a, b]$, it is equal a.e. to a right-continuous function $\psi_{x,\phi}$ in $BV[a, b]$, and so

$$\psi_{x,\phi}(\lambda) = w_{x,\phi}(\lambda) \qquad (\text{a.e. on } [a, b]).$$

It follows that the indefinite integral of $w_{x,\phi}$ is differentiable on the right with right-hand derivative $\psi_{x,\phi}(\lambda)$ at every point λ of $[a, b)$. Hence, by Theorem 15.22, T is uniquely decomposable. Also by condition (v) of the definition of decomposition of the identity,

$$\psi_{x,\phi}(\lambda) = \langle x, E(\lambda)\phi\rangle \qquad (a \leqslant \lambda < b).$$

Since the function $\langle x, E(\cdot)\phi\rangle$ is constant on the intervals $(-\infty, a)$ and $[b, \infty)$, this function is continuous on the right at every point of $\mathbf{R}$.

We now prove the first statement of the theorem. Suppose that (i) holds.

Then we can find a decomposition of the identity $\{E(\lambda): \lambda \in \mathbf{R}\}$ for T, a compact interval $J = [a, b]$, and a constant K such that

$$E(\lambda) = 0\ (\lambda \leqslant a);\ E(\lambda) = I\ (\lambda \geqslant b);\ \| E(\lambda) \| \leqslant K\ (\lambda \in \mathbf{R}),$$

and for every x in X, ϕ in X^*, the function $\langle x, E(\cdot)\phi \rangle$ is a right-continuous function of bounded variation on $\mathbf{R}$. On the algebra Σ of subsets of $(a, b]$ expressible as a finite disjoint union of intervals of the form $(a_i, b_i]$ define μ as follows

$$\mu\left(\bigcup_{i=1}^{n} (a_i, b_i] \right) = \sum_{i=1}^{n} \{E(b_i) - E(a_i)\}.$$

It is easy to see that μ is well-defined and finitely additive on Σ. For x in X and ϕ in X^*

$$\begin{aligned} \left|\left\langle x, \mu\left(\bigcup_{i=1}^{n} (a_i, b_i] \right) \phi \right\rangle\right| &\leqslant \sum_{i=1}^{n} |\langle x, (E(b_i) - E(a_i))\phi \rangle| \\ &\leqslant \operatorname{var}\langle x, E(\cdot)\phi \rangle < \infty. \end{aligned}$$

Hence by the uniform boundedness principle there is a constant M such that

$$\| \mu(\sigma) \| \leqslant M \qquad (\sigma \in \Sigma).$$

Now

$$|\langle x, \mu(\sigma)\phi \rangle| \leqslant M \| x \| \, \| \phi \| \qquad (\sigma \in \Sigma)$$

and so it follows from Lemma III.1.5 of [**5**; p. 97] that

$$\operatorname{var}\langle x, E(\cdot)\phi \rangle \leqslant 4M \| x \| \, \| \phi \| \qquad (x \in X, \phi \in X^*).$$

Let p be a complex polynomial. Then

$$\begin{aligned} \langle p(T)x, \phi \rangle &= p(b)\langle x, \phi \rangle - \int_a^b \langle x, E(\lambda)\phi \rangle\, p'(\lambda) d\lambda \\ &= p(b)\langle x, \phi \rangle - \int_a^b \langle x, E(\lambda)\phi \rangle\, dp. \end{aligned}$$

On integrating by parts we obtain

$$\langle p(T)x, \phi\rangle = \int_a^b p(\lambda)d\langle x, E(\lambda)\phi\rangle \qquad (x \in X, \phi \in X^*),$$

$$|\langle p(T)x, \phi\rangle| \leqslant \sup_{\lambda \in [a,b]} |p(\lambda)| \operatorname{var}\langle x, E(\cdot)\phi\rangle$$

$$\leqslant 4M \|x\| \|\phi\| \sup_{\lambda \in [a,b]} |p(\lambda)|,$$

$$\|p(T)\| \leqslant 4M \sup_{\lambda \in [a,b]} |p(\lambda)|,$$

for every complex polynomial p. Hence (i) implies (ii). Now suppose that (ii) holds. By the Weierstrass polynomial theorem we may extend the map $p \to p(T)$ in the obvious way to get a continuous algebra homomorphism of $C(J)$ into $L(X)$ such that

$$\|f(T)\| \leqslant 4M \sup_J |f(t)| \qquad (f \in C(J))$$

and so, by Theorem 5.21, (ii) implies (iii).

Now suppose that (iii) holds. Let $\mathscr{G}(\cdot)$ be the resolution of the identity of class X for T^*. Let $J = [a, b]$ be a compact interval containing $\sigma(T)$, and let p be any complex polynomial. Then if $\|\mathscr{G}(\cdot)\| \leqslant M < \infty$, we have $\|p(T)\| = \|p(T^*)\|$, and

$$\|p(T)\| \leqslant 4M \sup_{t \in J} |p(t)| = 4M \sup_{t \in J} \left|p(b) - \int_t^b p'(\xi)d\xi\right|.$$

It follows that

$$\|p(T)\| \leqslant 4M |||p|||. \tag{3}$$

Hence T and T^* are well-bounded. Now let $f \in AC(J)$. Then, by the Weierstrass polynomial theorem, there is a sequence $\{p_n\}$ of polynomials converging uniformly to f on J. The inequality (3) shows that $\{p_n(T)\}$ converges in the norm of $L(X)$ to an operator $f(T)$, and, moreover, $f(T)$ is independent of the approximating sequence $\{p_n\}$ chosen. Also the map $f \to f(T)$ is multiplicative and

$$\|f(T)\| \leqslant 4M \sup_{t \in J} |f(t)| \leqslant 4M |||f||| \qquad (f \in AC(J)).$$

Hence, by the uniqueness clause in Lemma 15.2, the map $f \to f(T)$ is the homomorphism of $AC(J)$ into $L(X)$ described in the statement of that lemma. Define

$$E(\lambda) = \mathscr{G}((-\infty, \lambda]) \qquad (\lambda \in \mathbf{R}). \tag{4}$$

We shall show that $\{E(\lambda): \lambda \in \mathbf{R}\}$ is a decomposition of the identity of bounded variation for T. Let $x \in X$ and $\phi \in X^*$. Since $\mathscr{G}(\cdot)$ is of class X, it follows that $\langle x, E(\cdot)\phi\rangle$ is everywhere right continuous and of bounded variation on $\mathbf{R}$. The conditions (i), (ii), (iii) of Definition 15.3 follow immediately from (4). Conditions (iv) and (v) follow from the right-continuity of $\langle x, E(\cdot)\phi\rangle$. Observe that if $f \in AC(J)$

$$\begin{aligned}\langle f(T)x, \phi\rangle = \langle x, f(T^*)\phi\rangle &= \int_a^b f(\lambda) d\langle x, E(\lambda)\phi\rangle \\ &= f(b)\langle x, \phi\rangle - \int_a^b \langle x, E(\lambda)\phi\rangle f'(\lambda)\, d\lambda \end{aligned} \tag{5}$$

on integrating by parts, and so it remains to prove (vi). Fix x in X, and consider the map $A: L^1[a, b] \to X$ defined by

$$Au = f_u(T)x$$

where

$$f_u(\lambda) = \int_\lambda^b u(\xi) d\xi \qquad (u \in L^1[a, b]).$$

A is clearly bounded and linear. For u in $L^1[a, b]$ and ϕ in X^* we have

$$\begin{aligned}\langle u, A^*\phi\rangle = \langle Au, \phi\rangle &= \langle f_u(T)x, \phi\rangle = \langle x, f_u(T^*)\phi\rangle \\ &= \int_a^b f_u(\lambda) d\langle x, E(\lambda)\phi\rangle \\ &= \int_a^b \langle x, E(\lambda)\phi\rangle\, u(\lambda) d\lambda,\end{aligned}$$

using (5) and then integrating by parts. It follows that the map A^* from

X^* into $L^\infty[a, b]$ is given by

$$A^*\phi = \langle x, E(\cdot)\phi\rangle.$$

Since A is continuous, A^* is continuous when X^* and $L^\infty[a, b]$ are endowed with their weak*-topologies. This completes the proof that (iii) implies (i).

THEOREM 16.16. *Let X be weakly complete, and T in $L(X)$ satisfy $\sigma(T) \subseteq \mathbf{R}$. Then the following three conditions are equivalent.*

(i) *T is a well-bounded operator with a decomposition of the identity of bounded variation.*

(ii) *There is a compact interval J and a real constant M such that*

$$\| p(T) \| \leqslant 4M \sup_{t \in J} |p(t)|$$

for every complex polynomial p.

(iii) *T is a scalar-type spectral operator.*

If (i) *holds, then T is decomposable in X.*

Proof. The equivalence of (i) and (ii) was shown in the previous theorem. Also, by Theorem 6.13, (iii) implies (ii). That (ii) implies (iii) follows from the Weierstrass approximation theorem and Theorem 6.13. If (i) holds, then by Theorem 16.15, T is uniquely decomposable. Since X is weakly complete T is decomposable in X by Theorem 16.4.

THEOREM 16.17. *Let T be a scalar-type spectral operator on X with $\sigma(T) \subseteq \mathbf{R}$. Then T is a well-bounded operator of type* (B).

Proof. Let $\mathscr{G}(\cdot)$ be the resolution of the identity for T. Then T^* is a scalar-type operator with resolution of the identity $\mathscr{G}^*(\cdot)$ of class X by Theorem 6.9. Hence T is well-bounded with unique decomposition of the identity $\{E(\lambda): \lambda \in \mathbf{R}\}$ given by

$$E(\lambda) = \mathscr{G}^*((-\infty, \lambda]) \qquad (\lambda \in \mathbf{R}).$$

Observe that if $F(\lambda) = \mathscr{G}((-\infty, \lambda])$ then $F^*(\lambda) = E(\lambda)$ $(\lambda \in \mathbf{R})$. Hence T is decomposable in X. If $x \in X$, and λ (respectively μ) is a real number then for any strictly decreasing sequence $\{\lambda_n\}$ (respectively strictly increasing sequence $\{\mu_n\}$) of real numbers convergent to λ (respectively to μ), the following standard argument using the countable additivity of $\mathscr{G}(\cdot)$ may be used;

$$0 = \lim_{n\to\infty} \sum_{j=n}^{\infty} \mathscr{G}((\lambda_{j+1}, \lambda_n])x = \lim_{n\to\infty} \mathscr{G}((\lambda, \lambda_n])x.$$

Thus $F(\lambda_n)x \to F(\lambda)x$. Similarly,

$$\begin{aligned}\mathscr{G}((-\infty, \mu))x &= \lim_{n\to\infty}\{\mathscr{G}((-\infty, \mu_1]) + \sum_{j=1}^{n} \mathscr{G}((\mu_j, \mu_{j+1}])x\} \\ &= \lim_{n\to\infty} \mathscr{G}((-\infty, \mu_{n+1}])x.\end{aligned}$$

So $F(\mu_n)x \to \mathscr{G}((-\infty, \mu))x$. It follows that, in the strong operator topology, $\lim_{t\to\lambda+} F(t) = F(\lambda)$, $\lim_{t\to\mu-} F(t) = \mathscr{G}((-\infty, \mu))$, and so T is well-bounded of type (B).

Example 16.18. Let Σ denote the algebra of subsets of $(0, 1]$ generated by intervals of the form $(s, t]$ with $0 \leqslant s < t \leqslant 1$. Let X denote the Banach space of all limits of uniformly convergent sequences of finite complex combinations of characteristic functions of sets in Σ, under the norm

$$\| f \| = \sup_{t\in(0,1]} |f(t)| \qquad (f \in X).$$

Hewitt [**1**; Theorem 4.5] has characterized X as consisting of complex functions continuous on the left at every point of $(0, 1]$ and with a right-hand limit at each point of $[0, 1)$.

Let $ba\{(0, 1], \Sigma\}$ denote the Banach space of all finitely additive complex-valued set functions y defined on Σ for which $\text{var}_\Sigma(y, (0, 1])$ is finite and with the norm

$$\| y \| = \text{var}_\Sigma(y, (0, 1]).$$

It is well known that X^* and $ba\{(0, 1], \Sigma\}$ are isometrically isomorphic under the correspondence

$$x^*(f) = \int_{(0,1]} f(s)dy(s) \qquad (x^* \in X^*, y \in ba\{(0, 1], \Sigma\}).$$

(See, for example, Theorem IV.5.1 of [**5**; p. 258]). Also, by Theorem IV.9.9 of [**5**; p. 311], X^* is a weakly complete Banach space. Let m be Lebesgue measure on $\mathbf{R}$. Then X is isometrically isomorphic to a closed subspace Y_∞ of $L^\infty[0, 1]$ under the mapping which sends each function f in X into the equivalence class in $L^\infty[0, 1]$ that contains all extensions of f to $[0, 1]$. To see this, observe that this map is certainly isometric on the dense subspace of X consisting of finite linear combinations of characteristic functions of disjoint sets in Σ. Now $(L^\infty[0, 1])^*$ is isometrically isomorphic with the Banach space $ba\{[0, 1], \mathscr{L}, m\}$ consisting of all finitely additive complex-valued set functions ξ defined on $\mathscr{L}$, the σ-algebra of Lebesgue measurable sets, which vanish on sets of Lebesgue measure zero, and which have finite total variation on $[0, 1]$ with respect to $\mathscr{L}$, the norm of ξ being given by

$$\|\xi\| = \operatorname{var}_{\mathscr{L}}(\xi, [0, 1]).$$

(See, for example, Theorem IV.8.15 of [**5**; p. 296].) Also $(L^\infty([0, 1]))^*$ is weakly complete, since it is the dual space of a commutative C^*-algebra. It follows from the preceding discussion and the Hahn-Banach theorem that set functions in $ba\{(0, 1], \Sigma\}$ arise precisely from set functions in $ba\{[0, 1], \mathscr{L}, m\}$ by restriction to the subalgebra $\Sigma \subseteq \mathscr{L}$.

Let $X_0 = L^\infty[0, 1]$. Identify X with the closed subspace Y_∞ of X_0. Then $X_0^* = ba\{[0, 1], \mathscr{L}, m\}$. Also we may and shall identify $ba\{(0, 1], \Sigma\}$ and the quotient space $X_0^*/X^\perp$, where $X^\perp$ denotes the annihilator of X in X_0^*. Define S_0, in $L(X_0)$, by

$$(S_0 f)(t) = tf(t) \qquad (f \in L^\infty[0, 1], t \in [0, 1]).$$

Clearly X is invariant under S_0. Let $S = S_0 | X$. Then, if p is any complex polynomial,

$$\|p(S)\| \leqslant \|p(S_0)\| \leqslant \sup_{t \in [0, 1]} |p(t)|,$$

and hence

$$\|p(S^*)\| \leqslant \|p(S_0^*)\| \leqslant \sup_{t \in [0, 1]} |p(t)|.$$

It follows that S_0 and S (and hence S_0^* and S^*) are well-bounded operators. Also since $\sigma(S) = \sigma(S^*) = \sigma(S_0) = \sigma(S_0^*) = [0, 1]$, and since X^* and X_0^* are weakly complete, it follows from Theorem 16.16 that S_0^* and S^* are scalar-type spectral operators. The resolution of the identity $\mathscr{G}(\cdot)$ for S_0^* is given by

$$\mathscr{G}(\tau)\xi(\delta) = \xi(\delta \cap \tau) \qquad (\xi \in ba\{[0, 1], \mathscr{L}, m\}, \tau \in \Sigma_p, \delta \in \mathscr{L}).$$

Now S^* is a scalar-type spectral operator, and is the operator induced by S_0^* on the quotient space $X_0^*/X^\perp$. Hence, by Theorem 14.7, $\mathscr{G}(\tau)X^\perp \subseteq X^\perp$, for all τ in Σ_p, and the resolution of the identity of S^* is $(\mathscr{G}(\cdot))_{X^\perp}$, where $(\mathscr{G}(\tau))_{X^\perp}$ denotes the operator induced on the quotient space $X_0^*/X^\perp$ by $\mathscr{G}(\tau)$. It follows from this and the proof of Theorem 16.15 that if for each real λ we define

$$E(\lambda)y(\delta) = y(\delta \cap (-\infty, \lambda]) \qquad (\delta \in \Sigma, y \in ba\{(0, 1], \Sigma\}),$$

then $\{E(\lambda): \lambda \in \mathbf{R}\}$ is the unique decomposition of the identity for the well-bounded operator S.

For each real λ define a projection $F(\lambda)$ on X by

$$(F(\lambda)f)(t) = \chi_{(-\infty, \lambda]}(t)f(t) \qquad (f \in X: t \in (0, 1]),$$

and observe that $F^*(\lambda) = E(\lambda)$. Hence the well-bounded operator S is decomposable in X. However, consideration of the action of the family of operators $\{F(\lambda): \lambda \in \mathbf{R}\}$ on the function identically equal to 1 on $(0, 1]$ shows that the strong operator limits $\lim_{\lambda \to \mu +} F(\lambda)$ and $\lim_{\lambda \to \mu -} F(\lambda)$ fail to exist at any point μ of $(0, 1)$. Hence S is not a well-bounded operator of type (A).

Example 16.19. Let X be the Banach space of all convergent sequences $w = \{\beta_n\}$ of complex numbers under the norm

$$\|w\| = \sup_n |\beta_n|.$$

The pairing of X^* with l^1 given by

$$\langle w, f \rangle = \lambda_1 \lim_{n \to \infty} \beta_n + \sum_{n=1}^{\infty} \beta_n \lambda_{n+1},$$

where $f = \{\lambda_n\} \in l^1$, induces an isometric isomorphism of l^1 onto X^*.

Define T, in $L(X)$, by

$$T\{\beta_n\} = \left\{-\frac{1}{n}\beta_n\right\}.$$

Observe that if p is any complex polynomial then

$$\|p(T)\| \leqslant \sup\{|p(t)|: -1 \leqslant t \leqslant 0\}$$

and so T is well-bounded. Then if $f = \{\lambda_n\} \in l^1$ we have

$$\langle w, T^*f\rangle = \langle Tw, f\rangle = -\sum_{n=1}^{\infty} \frac{\beta_n}{n}\lambda_{n+1}$$

$$T^*\{\lambda_n\} = \{0, -\lambda_2, -\tfrac{1}{2}\lambda_3, -\tfrac{1}{3}\lambda_4, \ldots\}.$$

T^* is a scalar-type spectral operator. Also

$$\sigma(T) = \sigma(T^*) = \{0, -1, -\tfrac{1}{2}, -\tfrac{1}{3}, \ldots\}.$$

The resolution of the identity $\mathscr{G}(\cdot)$ for T^* may be calculated from the equations

$$\mathscr{G}(\{0\})\{\lambda_n\} = \{\lambda_1, 0, 0, \ldots\}$$

$$\mathscr{G}\left(\left\{-\frac{1}{r}\right\}\right)\{\lambda_n\} = \{0, \ldots, 0, \lambda_{r+1}, 0, \ldots\} \qquad (r = 1, 2, 3, \ldots).$$

Define for each λ in $\mathbf{R}$

$$E(\lambda) = ((-\infty, \lambda]).$$

Then the proof of Theorem 16.15 shows that $\{E(\lambda): \lambda \in \mathbf{R}\}$ is the unique decomposition of the identity for the well-bounded operator T. If $\lambda \geqslant 0$ then $E(\lambda) = I$. Observe that if $\lambda < 0$ then only a finite number of points, $\{-1, -\frac{1}{2}, \ldots, -1/m\}$ say, of $\sigma(T)$ lie in $(-\infty, \lambda]$, and if we define a projection $F(\lambda)$ on X by

$$F(\lambda)\{\beta_n\} = \{\beta_1, \ldots, \beta_m, 0, 0, \ldots\},$$

then $F^*(\lambda) = E(\lambda)$. It follows that T is decomposable in X. Moreover, if $\mu \in \mathbf{R}$, there is $\delta_\mu > 0$ such that the function $F(\cdot)$ is constant in $[\mu, \mu + \delta_\mu]$. Hence $F(\cdot)$ is continuous on the right in the strong operator topology. However, consideration of the action of the family of operators $\{F(\lambda): \lambda \in \mathbf{R}\}$ on the sequence all of whose terms are 1 shows that the strong operator limit $\lim_{\lambda \to 0-} F(\lambda)$ fails to exist. Hence the well-bounded operator T is of type (A), but not of type (B).

17. The Structure of Well-bounded Operators of Type (B)

The integrals described in this chapter are based on the modified Stieltjes integrals of Krabbe [3]. Following Spain [2] we shall apply this integration theory to establish various characterizations of well-bounded operators of type (B). We show that in the case of well-bounded operators of type (B) the algebra homomorphism from $AC(J)$ from $L(X)$ can be extended to an algebra homomorphism from $BV(J)$ into $L(X)$.

Let U_J be the family of all partitions of J. We recall that if $u = (u_k : k = 0, \ldots, m)$ and $v = (v_j : j = 0, \ldots, n)$, then $u \geqslant v$ if and only if each interval of the form $[u_{k-1}, u_k]$ is contained in some interval of the form $[v_{j-1}, v_j]$.

Let $M(u)$ be the family of finite sequences of the form $u^* = (u_k^* : k = 1, \ldots, m)$ such that

$$u_{k-1} \leqslant u_k^* \leqslant u_k \qquad (1 \leqslant k \leqslant m)$$

for each u in U_J.

A pair $\bar{u} = (u, u^*)$ with u in U_J and u^* in $M(u)$ is called a *marked partition* of J. We write π_J for the family of marked partitions of J and we define the preorder $\geqslant$ on π_J by setting $(u, u^*) \geqslant (v, v^*)$ if and only if $u \geqslant v$. Define

$$\pi_J^i = \{\bar{u} = (u, u^*) \in \pi_J : u_{k-1} < u_k^* < u_k \qquad (1 \leqslant k \leqslant m)\},$$
$$\pi_J^r = \{\bar{u} = (u, u^*) \in \pi_J : u_k^* = u_k \qquad (1 \leqslant k \leqslant m)\}.$$

The sets U_J, π_J, π_J^i and π_J^r are directed by $\geqslant$. Also π_J^i and π_J^r are cofinal in π_J.

Let Φ and Ψ be functions on J, one taking values in $\mathbf{C}$, the other taking values in $L(X)$. When $\bar{u} \in \pi_J$, we define

$$\sum \Phi(\Psi \Delta \bar{u}) = \sum_{k=1}^{m} \Phi(u_k^*)(\Psi(u_k) - \Psi(u_{k-1})).$$

The following integrals are defined as net limits in the strong operator topology of $L(X)$ when they exist.

$$\int_J \Phi d\Psi = \text{st} \lim_{\pi_J} \sum \Phi(\Psi \, \Delta \, \bar{u}). \tag{1}$$

This is the ordinary Stieltjes refinement integral which was discussed in Chapter 16.

$$\int_J^r \Phi \, d\Psi = \text{st} \lim_{\pi_J^r} \sum \Phi(\Psi \, \Delta \, \bar{u}). \tag{2}$$

This integral is called a *right Cauchy integral.*

$$\int_J^i \Phi \, d\Psi = \text{st} \lim_{\pi_J^i} \sum \Phi(\Psi \, \Delta \, \bar{u}). \tag{3}$$

This integral is called a *modified Stieltjes integral.*

Let $\mathscr{E}(J)$ be the family of functions $(E\cdot)$ from $\mathbf{R}$ into $L(X)$ satisfying

$$\begin{array}{lr} \text{(i) } E(s) = E(s+) = \text{st} \lim_{t \to s+} E(t) & (s \in \mathbf{R}), \\ \text{(ii) } E(s-) = \text{st} \lim_{t \to s-} E(t) \text{ exists} & (s \in \mathbf{R}), \\ \text{(iii) } E(s) = 0 & (s < a), \\ \text{(iv) } E(s) = E(b) & (s \geqslant b), \end{array}$$

where $J = [a, b]$.

LEMMA 17.1. *Let $E(\cdot) \in \mathscr{E}(J)$. Then* $\sup_{\mathbf{R}} \| E(s) \| = \sup_J \| E(s) \| < \infty$.

Proof. Let $x \in X$ and $s \in J$. Since $E(s+)$ and $E(s-)$ exist, $\| E(t)x \|$ is bounded for all t in some open neighbourhood of s. Since J is compact, a finite subfamily of these neighbourhoods covers J. It follows that $\sup_J \| E(s)x \|$ is finite. By the uniform boundedness principle, $\sup_J \| E(s) \|$ is finite. Clearly,

$$\sup_{\mathbf{R}} \| E(s) \| = \sup_J \| E(s) \| < \infty.$$

For T in $L(X)$ and $-\infty < c < d \leqslant \infty$, we have

$$T\chi_{[c,d)}(t) = \begin{cases} T & t \in [c, d) \\ 0 & t \notin [c, d). \end{cases}$$

We note that if $a \leqslant c < d \leqslant b$, then $T\chi_{[c,d)} \in \mathscr{E}(J)$ and $T\chi_{[b,\infty)} \in \mathscr{E}(J)$. Let $E \in \mathscr{E}(J)$ and $u = (u_k : k = 0, \ldots, m) \in U_J$. Define

$$E_u = \sum_{k=1}^{m} E(u_{k-1})\chi_{[u_{k-1}, u_k)} + E(b)\chi_{[b,\infty)}.$$

Then $E_u \in \mathscr{E}(J)$.

If g is any function in $BV(J)$ we define

$$\pi_J^g = \begin{cases} \pi_J & (g \in NBV(J)) \\ \pi_J^i & (g \in BV(J) \backslash NBV(J)). \end{cases}$$

The following integral is defined as a net limit in the strong operator topology when it exists.

$$\oint_J E\, dg = \operatorname{st} \lim_{\pi_J^g} \sum E(g \,\Delta\, \bar{u}) \qquad (g \in BV(J), E \in \mathscr{E}(J)).$$

It is easy to verify that if $\oint_J E_1\, dg$ and $\oint_J E_2\, dg$ exist, then $\oint_J (E_1 + E_2) dg$ also exists and

$$\oint_J (E_1 + E_2) dg = \oint_J E_1\, dg + \oint_J E_2\, dg.$$

We shall prove the existence of the integral in (4). First, we require some preliminary results.

LEMMA 17.2. $\lim_{U_J} \sup_J \| E(s)x - E_u(s)x \| = 0 \qquad (x \in X, E \in \mathscr{E}(J)).$

Proof. Let $E \in \mathscr{E}(J)$, $x \in X$ and let $\varepsilon > 0$ be given. For each s in $[a, b)$, there exists r_s with $s < r_s < b$ such that

$$\| E(t)x - E(t')x \| \leqslant \varepsilon \quad \text{whenever } t, t' \in [s, r_s),$$

since $E(s) = E(s+)$. For each s in $(a, b]$, there exists l_s with $a < l_s < s$ such that

$$\| E(t)x - E(t')x \| \leqslant \varepsilon \quad \text{whenever } t, t' \in [l_s, s),$$

since $E(s-)$ exists.

The sets $[a, r_a)$, $(l_b, b]$, (l_s, r_s) for $a \leqslant s \leqslant b$ form an open cover of J. By compactness there is a finite subcovering $[a, r_a)$, $(l_b, b]$, (l_{s_j}, r_{s_j}) $(j = 1, \ldots, n)$. Let v be the partition with points $a, b, r_a, l_b, l_{s_j}, r_{s_j}$, for $j = 1, \ldots, n$. If $u = (u_k : k = 0, \ldots, m)$ is any partition which refines v then each interval of the form $[u_{k-1}, u_k)$ is a subset of one of $[a, r_a)$, $[l_b, b)$, $[l_{s_j}, s_j)$ or $[s_j, r_{s_j})$, for some j, and so we deduce that

$$\sup_J \| E(s)x - E_u(s)x \| < \varepsilon.$$

Hence

$$\lim_{U_J} \sup_J \| E(s)x - E_u(s)x \| = 0 \qquad (x \in X, E \in \mathscr{E}(J)).$$

LEMMA 17.3. (i) $\oint_J T\chi_{[b, \infty)}\, dg = 0 \qquad (g \in BV(J), T \in L(X))$;

(ii) *if* $a \leqslant s < t \leqslant b$, *then* $\oint_J T\chi_{[s, t)} dg = (g(t) - g(s))T \quad (g \in BV(J), T \in L(X)))$;

(iii) $$\oint_J E_u dg = \sum_{k=1}^{m} E(u_{k-1})(g(u_k) - g(u_{k-1}))$$
$$= \operatorname{st}\lim_{\pi_J^g} \sum E_u(g \,\Delta\, \bar{v}) \qquad (g \in BV(J), E \in \mathscr{E}(J), u \in U_J).$$

Proof. (i) Let $\bar{u} \in \pi_J^g$. Then

$$\sum T\chi_{[b, \infty)}(g \,\Delta\, \bar{u}) = \sum_{k=1}^{m} T\chi_{[b, \infty)}(u_k^*)[g(u_k) - g(u_{k-1})]$$
$$= \begin{cases} [g(b) - g(u_{m-1})]T & (u_m^* = b) \\ 0 & (u_m^* < b). \end{cases}$$

Hence

$$\operatorname{st}\lim_{\pi_J^g} \sum T\chi_{[b, \infty)}(g \,\Delta\, \bar{u}) = 0.$$

(ii) Let $a < s \leqslant b$, $\bar{u} \in \pi_J^g$, $u \geqslant (a, s, b)$. Then $s = u_n$ for some n with

$1 \leqslant n \leqslant m$, and

$$\sum T\chi_{[a,s)}(g \,\Delta\, \bar{u}) = \sum_{k=1}^{n-1} [g(u_k) - g(u_{k-1})]T + T\chi_{[a,s)}(u_n^*)[g(s) - g(u_{n-1})]$$

$$= \begin{cases} [g(u_{n-1}) - g(a)]T & (u_n^* = u_n = s) \\ [g(s) - g(a)]T & (u_n^* < u_n = s). \end{cases}$$

It follows that

$$\text{st}\lim_{\pi_J^g} \sum T\chi_{[a,s)}(g \,\Delta\, \bar{u}) = [g(s) - g(a)]T.$$

Since $\chi_{[s,t)} = \chi_{[a,t)} - \chi_{[a,s)}$ $(a \leqslant s < t \leqslant b)$ we obtain

$$\oint_J T\chi_{[s,t)}dg = \oint_J T\chi_{[a,t)}dg - \oint_J T\chi_{[a,s)}dg$$
$$= (g(t) - g(a))T - (g(s) - g(a))T$$
$$= (g(t) - g(s))T.$$

(iii) From the definition of E_u, it is clear that the result follows directly from (i) and (ii).

THEOREM 17.4. *Let $g \in BV(J)$ and $E \in \mathscr{E}(J)$. Then $\oint_J Edg$ exists and*

$$\oint_J Edg = \text{st}\lim_{U_J} \oint_J E_u\, dg.$$

Also

$$\left\| \oint_J Edg \right\| \leqslant \text{var}(g, J) \sup_J \| E(s) \|, \tag{5}$$

$$\left\| \oint_J Edgx \right\| \leqslant \text{var}(g, J) \sup_J \| E(s)x \| \qquad (x \in X). \tag{6}$$

Proof. It is easy to verify that if $g \in BV(J)$ and $E, F \in \mathscr{E}(J)$, then

$$\left\| \sum E(g \,\Delta\, \bar{u}) - \sum F(g \,\Delta\, \bar{u}) \right\| \leqslant \text{var}(g, J) \sup_J \| E(s) - F(s) \| \tag{7}$$

and if $x \in X$, then

$$\left\| \sum E(g \, \Delta \, \bar{u})x - \sum F(g \, \Delta \, \bar{u})x \right\| \leqslant \operatorname{var}(g, J) \sup_J \| E(s)x - F(s)x \|. \tag{8}$$

Setting $F = 0$ in (7) and (8) we obtain

$$\left\| \sum E(g \, \Delta \, \bar{u}) \right\| \leqslant \operatorname{var}(g, J) \sup_J \| E(s) \|,$$

$$\left\| \sum E(g \, \Delta \, \bar{u})x \right\| \leqslant \operatorname{var}(g, J) \sup_J \| E(s)x \|.$$

Hence (5) and (6) follow immediately.

Now, let $u \in U_J$, let $\bar{v}, \bar{w} \in \pi_J^g$ and let $x \in X$. Then

$$\begin{aligned}
&\left\| \sum E(g \, \Delta \, \bar{v})x - \sum E(g \Delta \bar{w})x \right\| \\
&\quad \leqslant \left\| \sum E(g \, \Delta \, \bar{v})x - \sum E_u(g \, \Delta \, \bar{v})x \right\| + \left\| \sum E(g \Delta \bar{w})x - \sum E_u(g \, \Delta \bar{w})x \right\| \\
&\qquad\qquad + \left\| \sum E_u(g \, \Delta \, \bar{v})x - \sum E_u(g \, \Delta \, \bar{w})x \right\| \\
&\quad \leqslant 2 \operatorname{var}(g, J) \sup_J \| E(s)x - E_u(s)x \| + \left\| \sum E_u(g \, \Delta \, \bar{v})x - \sum E_u(g \, \Delta \, \bar{w})x \right\|.
\end{aligned}$$

Then Lemmas 17.2 and 17.3 show that the last term tends to zero as $\bar{v}$ and $\bar{w}$ increase in π_J^g. Therefore $\{\Sigma\, E(g \, \Delta \, \bar{v}) : \bar{v} \in \pi_J^g\}$ is a uniformly bounded strongly Cauchy net in $L(X)$ and so converges to its (unique) strong limit. Hence $\oint_J E dg$ exists. Moreover

$$\left\| \oint_J E_u \, dgx - \oint_J E \, dgx \right\| = \left\| \oint_J (E_u - E) dg \, x \right\|$$
$$\leqslant \operatorname{var}(g, J) \sup_J \| E_u(s)x - E(s)x \|.$$

It follows from Lemma 17.2 that

$$\oint_J E \, dg = \operatorname{st} \lim_{U_J} \oint_J E_u \, dg.$$

THEOREM 17.5. *Let $E \in \mathscr{E}(J)$, $g \in BV(J)$ and let $\{g_\alpha : \alpha \in \sigma\}$ be a net in $BV(J)$ with $\sup_\sigma \operatorname{var}(g_\alpha, J) < \infty$ and $g(s) = \lim g_\alpha(s)$ $(s \in J)$. Then*

$$\oint_J E \, dg = \operatorname{st} \lim_\sigma \oint_J E \, dg_\alpha.$$

Proof. Let $u \in U_J$. Then

$$\oint_J E\,dg - \int_J E\,dg_\alpha = \oint_J (E - E_u)dg - \oint_J (E - E_u)dg_\alpha + \oint_J E_u d(g - g_\alpha).$$

Let $x \in X$. It follows from Theorem 17.4 and Lemma 17.3(iii) that

$$\begin{aligned}
&\left\| \oint_J E\,dgx - \oint_J E\,dg_\alpha x \right\| \\
&\quad \leqslant \left\| \oint_J (E - E_u)dgx \right\| + \left\| \oint_J (E - E_u)dg_\alpha x \right\| + \left\| \oint_J E_u\,d(g - g_\alpha)x \right\| \\
&\quad \leqslant \operatorname{var}(g, J) \sup_J \| E(s)x - E_u(s)x \| + \sup_\sigma \operatorname{var}(g_\alpha, J) \| E(s)x - E_u(s)x \| \\
&\qquad + \sup_J \| E(s)x \| \sum_{k=1}^m |(g - g_\alpha)(u_k) - (g - g_\alpha)(u_{k-1})| \\
&\quad \leqslant (\operatorname{var}(g, J) + \sup_\sigma \operatorname{var}(g_\alpha, J)) \sup \| E(s)x - E_u(s)x \| \\
&\qquad + \sup_J \| E(s)x \| \sum_{k=1}^m |(g - g_\alpha)(u_k) - (g - g_\alpha)(u_{k-1})|.
\end{aligned}$$

This last expression can be made arbitrarily small by choosing u fine enough and then α suitably, by Lemma 17.2. It follows that

$$\oint_J E\,dg = \text{st}\lim_\sigma \oint_J E\,dg_\alpha$$

and the proof is complete.

For g in $BV(J)$ and E in $\mathscr{E}(J)$ we define

$$S(g, E) = g(b)E(b) - \oint_J E\,dg.$$

LEMMA 17.6. *Let $g \in BV(J)$, $E \in \mathscr{E}(J)$, $T \in L(X)$ and $s \in J$. Then*

(i) $S(g, \chi_{[s,\infty)}T) = g(s)T$,
(ii) $\| S(g, E) \| \leqslant |||g||| \sup_J \| E(s) \|$,
(iii) $\| S(g, E)x \| \leqslant |||g||| \sup_J \| E(s)x \| \qquad (x \in X)$,
(iv) $S(\chi_{[a,s]}, E) = E(s)$.

Proof. (i), (ii) and (iii) follow directly from Lemma 17.2 and Theorem 17.3.

(iv) If $s = b$, then

$$S(\chi_J, E) = E(b) - \oint_J E d\chi_J = E(b).$$

If $s < b$, then

$$S(\chi_{[a,s]}, E) = -\oint_J E d\chi_{[a,s]}$$

$$= -\operatorname{st}\lim_{U_J} \oint_J E_u d\chi_{[a,s]} \qquad \text{(Theorem 17.4)}$$

$$= -\operatorname{st}\lim_{U_J} \sum_{k=1}^{m} E(u_{k-1})(\chi_{[a,s]}(u_k) - \chi_{[a,s]}(u_{k-1}))$$

$$= -\operatorname{st}\lim_{U_J} (-E(u_{n-1})) \quad \text{where} \quad s \in [u_{n-1}, u_n)$$

$$= \operatorname{st}\lim_{U_J} E_u(s) = E(s)$$

by Lemma 17.2. The proof is complete.

LEMMA 17.7. *Let $g \in BV(J)$ and $\bar{u} \in \pi_J(g, r)$, where*

$$\pi_J(g, r) = \begin{cases} \pi_J & (g \in NBV(J)) \\ \pi_J^r & (g \in BV(J) \backslash NBV(J)). \end{cases}$$

Define

$$g_{\bar{u}} = g(a)\chi_{\{a\}} + \sum_{k=1}^{m} g(u_k^*)\chi_{(u_{k-1}, u_k]}.$$

Then $g_{\bar{u}} \in BV(J)$ and

$$g(s) = \lim_{\pi_J(g,r)} g_{\bar{u}}(s) \qquad (s \in J).$$

Also if g is real and monotonic increasing then

$$\operatorname{var}(g_{\bar{u}}, J) \leqslant 2 \sup_J |g(s)|.$$

Proof. It is obvious that $g_{\bar{u}} \in BV(J)$ and $g_{\bar{u}}(a) = g(a)$. If $a < s \leqslant b$ and $u \geqslant (a, s, b)$, then $s = u_n$ for some n with $1 \leqslant n \leqslant m$ and

$$g_{\bar{u}}(s) = g(u_n^*) = \begin{cases} g(u_n^*) & (g \in NBV(J)) \\ g(u_n) & (g \in BV(J) \backslash NBV(J)). \end{cases}$$

Therefore

$$\lim_{\pi_J(g, r)} g_{\bar{u}}(s) = g(s) \qquad (s \in J).$$

If g is real monotonic increasing, then $\operatorname{var}(g, J) \leqslant 2 \sup_J |g(s)|$ and $g_{\bar{u}}$ is also monotonic increasing. Hence

$$\operatorname{var}(g_{\bar{u}}, J) \leqslant 2 \sup_J |g_{\bar{u}}(s)| \leqslant 2 \sup_J |g(s)|.$$

THEOREM 17.8. *Let $E \in \mathscr{E}(J)$. Then*

$$S(g, E) = \begin{cases} g(a)E(a) + \oint_J g\, dE & (g \in NBV(J)) \\ g(a)E(a) + \oint_J^r g\, dE & (g \in BV(J) \backslash NBV(J)). \end{cases}$$

Proof. Let $\bar{u} \in \pi_J(g, r)$. Then

$$\begin{aligned} S(g_{\bar{u}}, E) &= g(a)S(\chi_{\{a\}}, E) + \sum_{k=1}^{m} g(u_k^*)[S(\chi_{[a, u_k]}, E) - S(\chi_{[a, u_{k-1}]}, E)] \\ &= g(a)E(a) + \sum g(E \,\Delta\, \bar{u}) \qquad \text{(Lemma 17.6).} \end{aligned}$$

Since ever g in $BV(J)$ can be expressed in the form

$$g = t_1 g_1 + t_2 g_2 + t_3 g_3 + t_4 g_4$$

where $t_i \in \mathbf{C}$ and g_i is real monotonic increasing ($i = 1, 2, 3, 4$), it is enough to prove the case in which g is real monotonic increasing. This result follows from Lemma 17.8 and Theorem 17.5.

We shall write $\int_J^{\oplus} g\, dE$ instead of $S(g, E)$ when $g \in BV(J)$ and $E \in \mathscr{E}(J)$.

Let Σ_J be the algebra of subsets of $[a, b)$ generated by sets of the form

$[s, t)$ $(a \leqslant s < t \leqslant b)$. Let $\mathscr{D}$ be the set of all finite linear combinations of characteristic functions of sets in Σ_J. Let Q_J be the closure of $\mathscr{D}$ in the supremum norm. Q_J is a Banach space consisting of all complex functions vanishing on $(-\infty, a)$ and on $[b, \infty)$, right continuous and left limitable on $\mathbf{R}$. This follows from Theorem 4.5 of Hewitt [**1**].

From the scalar version of Theorem 17.4 (that is obtained by taking $X = \mathbf{C}$), the integral $\int_J^i w dg$ exists for g in $BV(J)$ and w in Q_J. Moreover, if $w = \lim_n w_n$ in the supremum norm, where $\{w_n\} \subseteq \mathscr{D}$, then

$$\int_J^i w dg = \lim_n \int_J^i w_n dg \tag{9}$$

and from the scalar version of Lemma 17.3

$$\int_J^i \chi_{[s,t)} dg = g(t) - g(s). \tag{10}$$

Let $BV_0(J)$ denote the Banach subalgebra of $BV(J)$ consisting of functions which vanish at b. We use the notation above to prove the following well-known result.

THEOREM 17.9. *There is an isometric isomorphism between Q_J^* and $BV_0(J)$ determined by the identity*

$$\langle w, \phi \rangle = \int_J^i w\, dg \qquad (w \in Q_J, \phi \in Q_J^*, g \in BV_0(J)). \tag{11}$$

Proof. It follows from Theorem 17.4 that

$$\left| \int_J^i w\, dg \right| \leqslant \sup_J |w(s)|\, |||g||| \qquad (w \in Q_J, g \in BV_0(J)).$$

Hence, for each g in $BV_0(J)$, the map $w \to \int_J^i w\, dg$ defines a bounded linear functional ϕ on Q_J with $\| \phi \| \leqslant |||g|||$.

To show that every ϕ in Q_J^* is determined by some g in $BV_0(J)$ we define

$$g(s) = \langle -\chi_{[s,b)}, \phi \rangle \ (a \leqslant s < b),$$
$$g(b) = 0.$$

We shall show that g is of bounded variation. Consider any partition $(u_k: k = 0, \ldots, m)$ of J. Define

$$\lambda_k = \overline{\operatorname{sgn}}[g(u_k) - g(u_{k-1})] \qquad (k = 1, 2, \ldots, m),$$

where $\overline{\operatorname{sgn}}\, 0 = 0$ and $\overline{\operatorname{sgn}}\, c = \bar{c}|c|^{-1}$ if $c \neq 0$. Clearly

$$\lambda_k[g(u_k) - g(u_{k-1})] = |g(u_k) - g(u_{k-1})| \qquad (k = 1, 2, \ldots, m).$$

It follows that

$$\begin{aligned}\sum_{k=1}^{m} |g(u_k) - g(u_{k-1})| &= \sum_{k=1}^{m} \lambda_k[g(u_k) - g(u_{k-1})] \\ &= \sum_{k=1}^{m-1} \lambda_k[\langle -\chi_{[u_k, b)}, \phi\rangle \langle -\chi_{[u_{k-1}, b)}, \phi\rangle] \\ &\qquad\qquad - \lambda_m\langle -\chi_{[u_{m-1}, b)}, \phi\rangle \\ &= \left\langle \sum_{k=1}^{m} \lambda_k \chi_{[u_{k-1}, u_k)}, \phi \right\rangle \\ &\leqslant \|\phi\| \max\{|\lambda_1|, \ldots, |\lambda_m|\} \\ &\leqslant \|\phi\|.\end{aligned}$$

Hence g is of bounded variation and $\operatorname{var}(g, J) \leqslant \|\phi\|$. From (10),

$$\int_J^{i} \chi_{[s, b)} dg = -g(s) = \langle \chi_{[s, b)}, \phi \rangle.$$

Hence (11) holds for every function w in $\mathscr{D}$. Since $\mathscr{D}$ is dense in Q_J and since both sides of (11) are continuous in w, it follows that (11) holds for all w in Q_J. Moreover we have proved that $\|g\| = \|\phi\|$. Since the linearity of the correspondence between ϕ and g is clear, the theorem is proved.

LEMMA 17.10. *Let $\{g_\alpha : \alpha \in \sigma\}$ be a uniformly bounded net in $BV_0(J)$ and let $g \in BV_0(J)$. Then $g = \lim g_\alpha$ in the Q_J-topology of $BV_0(J)$ if and only if $g(s) = \lim_\alpha g_\alpha(s)$ $(a \leqslant s < b)$.*

Proof. From (10), we have

$$\langle -\chi_{[s,b)}, g\rangle = \int_J^i (-\chi_{[s,b)})\, dg = g(s) \qquad (a \leqslant s < b).$$

If $g = \lim_\sigma g_\alpha$ in the Q_J-topology of $BV_0(J)$, then

$$g(s) = \langle -\chi_{[s,b)}, g\rangle = \lim_\sigma \langle -\chi_{[s,b)}, g_\alpha\rangle = \lim_\sigma g_\alpha(s) \qquad (a \leqslant s < b).$$

Conversely, if $g(s) = \lim_\sigma g_\alpha(s)$ $(a \leqslant s < b)$, then it follows from Theorem 17.5 that

$$\langle w, g\rangle = \int_J^i w\, dg = \lim_\sigma \int_J^i w\, dg_\alpha = \lim_\sigma \langle w, g_\alpha\rangle \qquad (w \in Q_J).$$

Hence $g = \lim g_\alpha$ in the Q_J-topology of $BV_0(J)$.

THEOREM 17.11. *Let $g \in BV(J)$. There is a net $\{g_\alpha : \alpha \in \sigma\}$ of functions in $AC(J)$ such that $g = \lim_\sigma g_\alpha$ pointwise on J and* $\sup_\sigma |||g_\alpha||| \leqslant |||g|||$.

Proof. Since g can be written as $(g - g(b)\chi_J) + g(b)\chi_J$ we see that it suffices to show that if $g \in BV_0(J)$ then there is a net $\{g_\alpha : \alpha \in \sigma\}$ of functions in $AC(J)$ vanishing at b, such that $g = \lim_\sigma g_\alpha$ pointwise on $[a, b)$ and $\sup_\sigma \operatorname{var}(g_\alpha, J) \leqslant \operatorname{var}(g, J)$.

Now, for each w in Q_J, we define

$$P_w(f) = \int_J w\, df \qquad (f \in AC_0(J)),$$

where $AC_0(J)$ is the subalgebra of functions in $AC(J)$ vanishing at b. It is easy to verify that P_w is a bounded linear functional on $AC_0(J)$ and

$$\| P_w \| = \sup_J |w(s)|.$$

Therefore Q_J can be identified with a subspace of $AC_0^*(J)$.

By Theorem 17.9, each function g in $BV_0(J)$ defines a bounded linear functional L_g on Q_J by

$$L_g(w) = \int_J^i w\, dg \qquad (w \in Q_J)$$

and $\| L_g \| = \operatorname{var}(g, J)$. By the Hahn-Banach theorem, we can extend L_g to a linear functional (also denoted by) L_g on $AC_0^*(J)$ without increasing its norm. Hence $L_g \in AC_0^{**}(J)$. By Goldstine's theorem (see, for example, V.4.5 of [**5**; p. 424–5]) there is a net $\{g_\alpha : \alpha \in \sigma\}$ in $AC_0(J)$ converging to L_g in the $AC_0^*(J)$-topology of $AC_0^{**}(J)$ and satisfying

$$\operatorname{var}(g_\alpha, J) \leqslant \| L_g \| = \operatorname{var}(g, J).$$

Then $g = \lim_\sigma g_\alpha$ in the Q_J-topology of $BV_0(J)$ and so it follows from Lemma 17.10 that $g(s) = \lim g_\alpha(s)$ $(a \leqslant s < b)$.

We now apply the preceding integration theory to the study of well-bounded operators of type (B).

Definition 17.12. Let $M \subseteq X$. The *absolutely convex hull of* M (denoted by $aco(M)$) is the set of all linear combinations $\sum_{r=1}^{m} \alpha_r x_r$ of elements of x_r in M with $\alpha_r \in \mathbf{C}, |\alpha_r| \leqslant 1$ and $\sum_{r=1}^{m} |\alpha_r| = 1$. The *closed absolutely convex hull* of M is the closure of $aco(M)$ and is denoted by $\overline{aco}(M)$.

LEMMA 17.13. *Let $M \subseteq X$ be totally bounded. Then $\overline{aco}(M)$ is compact.*

Proof. The set $\overline{aco}(M)$, being a closed subspace of a complete metric space X is complete. Hence it suffices to show that $\overline{aco}(M)$ is totally bounded. Let $\varepsilon > 0$ be given. Since M is totally bounded, there is a finite subset $\{z_1, \ldots, z_n\}$ $\subseteq M$ such that $M \subseteq \bigcup_{t=1}^{n} B(z_r, \frac{1}{4}\varepsilon)$, where $B(z_r, \frac{1}{4}\varepsilon)$ denotes the open ball in X of centre z_r and radius $\frac{1}{4}\varepsilon$.

Let $N = aco(\{z_1, z_2, \ldots, z_n\})$. Now $\overline{aco}(M) \subseteq \cup \{B(x, \varepsilon/4) : x \in aco(M)\}$. However, if $y \in aco(M)$ then $y = \sum_{r=1}^{m} \alpha_r y_r$, where $y_r \in M$, $\alpha_r \in \mathbf{C}, |\alpha_r| \leqslant 1$ and $\sum_{r=1}^{m} |\alpha_r| = 1$. Let v be a function on M into $\{1, 2, \ldots, n\}$ with the property that if $x \in M$ then $\| x - z_{v(x)} \| < \frac{1}{4}\varepsilon$. We have

$$\left\| y - \sum_{r=1}^{m} \alpha_r z_{v(y_r)} \right\| = \left\| \sum_{r=1}^{m} \alpha_r (y_r - z_{v(y_r)}) \right\|$$
$$\leqslant \sum_{r=1}^{m} |\alpha_r| \, \| y_r - z_{v(y_r)} \| < \tfrac{1}{4}\varepsilon$$

and thus $\overline{aco}(M) \subseteq \cup \{B(x, \frac{1}{2}\varepsilon) : x \in N\}$. Now

$$N = \left\{ \sum_{r=1}^{n} \alpha_r z_r : \alpha_r \in \mathbf{C}, |\alpha_r| \leqslant 1 \quad \text{and} \quad \sum_{r=1}^{n} |\alpha_r| = 1 \right\}.$$

Let $Y = \{(\alpha_1, \alpha_2, \ldots, \alpha_n) \in \mathbf{C}^n : |\alpha_r| \leqslant 1, \sum_{r=1}^{n} |\alpha_r| = 1\}$. Then the mapping ϕ defined by

$$\phi(\alpha_1, \ldots, \alpha_n) = \sum_{r=1}^{n} \alpha_r z_r$$

is a continuous mapping of the compact set Y onto N. Thus N is compact and hence totally bounded. It follows that there is a finite subset $\{w_1, \ldots, w_n\}$ of N such that $N \subseteq \bigcup_{r=1}^{m} B(w_r, \frac{1}{2}\varepsilon)$. However $\overline{aco}(M) \subseteq \bigcup_{r=1}^{m} B(w_r, \varepsilon)$. Hence $\overline{aco}(M)$ is totally bounded and the proof is complete.

Now, we give three conditions on a well-bounded operator equivalent to the definition of type (B).

THEOREM 17.14. *Let T be a well-bounded operator on X and let $J = [a, b]$, K be chosen so that for every complex polynomial p*

$$\| p(T) \| \leqslant K\{|p(b)| + \operatorname{var}(p, J)\}.$$

Let $f \to f(T)$ be the algebra homomorphism from $AC(J)$ into $L(X)$ discussed in Lemma 15.2. Then the following conditions are equivalent.

(i) *T is of type (B).*

(ii) *For every x in X, $f \to f(T)x$ is a compact linear map of $AC(J)$ into X.*

(iii) *For every x in X, $f \to f(T)x$ is a weakly compact linear map of $AC(J)$ into X.*

(iv) *There is a family* $\{E(s): s \in \mathbf{R}\}$ *of projections on* X *such that*

$$\begin{aligned} &E \in \mathscr{E}(J),\ E(b) = I, \\ &E(t)E(s) = E(s)E(t) = E(s) \qquad (s \leqslant t), \\ &\| E(s) \| \leqslant K \qquad (s \in \mathbf{R}), \end{aligned}$$

and

$$T = \int_J^{\oplus} s\, dE(s).$$

Proof. We show that (i) $\Rightarrow$ (ii) $\Rightarrow$ (iii) $\Rightarrow$ (iv) $\Rightarrow$ (i).
(i) $\Rightarrow$ (ii). Let $\{E^*(s): s \in \mathbf{R}\}$ be the unique decomposition of the identity for T. By the definition of a type (B) operator, we have $E \in \mathscr{E}(J)$. We define a map ψ from $AC(J)$ into $L(X)$ by

$$\psi(f) = \int_J^{\oplus} f\, dE \qquad (f \in AC(J)).$$

The map ψ is linear and bounded; also if $f_0(s) = 1$ $(s \in J)$, then

$$\psi(f_0) = \int_J^{\oplus} f_0 dE = I.$$

Moreover, if $f_1(s) = s$ $(s \in J)$, then

$$\begin{aligned} \langle \psi(f_1)x, \phi \rangle &= \int_J^{\oplus} s d\langle E(s)x, \phi \rangle \\ &= b\langle x, \phi \rangle - \int_J \langle E(s)x, \phi \rangle\, ds \\ &= \langle Tx, \phi \rangle. \end{aligned}$$

It follows that $\psi(f_1) = T$. Since $E(t)E(s) = E(s)E(t) = E(s)$ $(s < t)$ we deduce that if $f, g \in AC(J)$ and if $\bar{u} \in \pi_J$, then

$$\begin{aligned} &\{f(a)E(a) + \textstyle\sum f(E\Delta\bar{u})\}\{g(a)E(a) + \textstyle\sum g(E\Delta\bar{u})\} \\ &\quad = (fg)(a)E(a) + \textstyle\sum (fg)(E\Delta\bar{u}). \end{aligned}$$

Hence $\psi(f)\psi(g) = \psi(fg)$ and so ψ is an algebra homomorphism. By Lemma 15.2

$$\psi(f) = f(T) \qquad (f \in AC(J)).$$

For each x in X, we define $\Omega_x = \{E(s)x : s \in \mathbf{R}\}$. Since $E(s) = 0$ $(s < a)$ and $E(s) = I$ $(s \geqslant b)$ it follows that for any $\delta > 0$

$$\Omega_x = \{E(s)x : s \in [a - \delta, b]\}.$$

Let $\varepsilon > 0$ be given. By the argument of the proof of Lemma 17.2 there exists $s_1 = a - \delta$, $s_2 = a$, $s_3, \ldots, s_n = b$ such that

$$\Omega_x \subseteq \bigcup_{m=1}^{n} B(E(s_r)x, \varepsilon).$$

Hence Ω_x is totally bounded. By Lemma 17.13, $\overline{aco}(\Omega_x)$ is compact. Further, let $f \in AC(J)$, $|||f||| \leqslant 1$ and $\bar{u} \in \pi_J$. Then

$$f(b)E(b)x - \sum E(f\Delta\bar{u})x$$
$$= f(b)x - \sum_{r=1}^{n} [f(u_r) - f(u_{r-1})]\, E(u_r^*)x \in \overline{aco}(\Omega_x).$$

Therefore $f(T)x \in \overline{aco}(\Omega_x)$. Hence, for each x in X, $f \to f(T)x$ is a compact linear map from $AC(J)$ into X.
(ii) $\Rightarrow$ (iii). This is trivial.
(iii) $\Rightarrow$ (iv). Let $\mathscr{F}$ be an ultrafilter on $(0, \infty)$ which converges to zero in the usual topology of the real line. For each x in X and ϕ in X^* we define

$$L_{x,\phi}(f) = \langle f(T)x, \phi \rangle \qquad (f \in AC(J)).$$

It is clear that $L_{x,\phi}$ is a bounded linear functional on $AC(J)$. By Lemmas 15.11 and 15.13 we have

$$L_{x,\phi}(f) = m_{x,\phi} f(b) - \int_a^b w_{x,\phi}(s) f'(s)\, ds \qquad (x \in X, \phi \in X^*, f \in AC(J)),$$

where $m_{x,\phi} \in \mathbf{C}$, $w_{x,\phi} \in L^\infty(J)$, and

$$w_{x,\phi}(s) = \lim_{\substack{h \to 0 \\ \mathscr{F}}} \frac{1}{h} \int_0^h w_{x,\phi}(s+t)dt \qquad (s \in J).$$

We have also

$$\langle x, \phi \rangle = L_{x,\phi}(f_0) = m_{x,\phi}.$$

We define the function $k_{s,h}$ on $\mathbf{R}$ for each real s and $h > 0$ by

$$k_{s,h}(t) = \begin{cases} 1 & (t \leqslant s) \\ 1 + h^{-1}(s-t) & (s \leqslant t \leqslant s+h) \\ 0 & (s+h \leqslant t). \end{cases}$$

Then $k_{s,h} \in AC(J)$ and $|||k_{s,h}||| \leqslant 1$. Also $\chi_{(-\infty, s]} = \lim_{h \to 0} k_{s,h}$ pointwise, for each real s.

The set $\mathscr{K}_x$, defined for each x in X by

$$\mathscr{K}_x = wk\{k_{s,h}(T)x\colon a \leqslant s < s+h < b\}$$

is compact in the weak topology of X. For a fixed s in $[a, b)$, we define a vector-valued function τ_s from $(0, \infty)$ into $\mathscr{K}_x$ by

$$\tau_s(h) = k_{s,h}(T)x \qquad (h > 0).$$

Since $\|\tau_s(h)\| = \|k_{s,h}(T)x\| \leqslant K\,|||k_{s,h}|||\,\|x\| \leqslant K\|x\|$, it follows that τ_s is bounded. Moreover, for every h, h' in $(0, \infty)$

$$\begin{aligned} \|\tau_s(h) - \tau_s(h')\| &= \|k_{s,h}(T)x - k_{s,h'}(T)x\| \\ &= \|(k_{s,h} - k_{s,h'})(T)x\| \\ &\leqslant K\,|||k_{s,h} - k_{s,h'}|||\,\|x\| \\ &\leqslant K\|x\| \frac{|h-h'|}{\max\{h, h'\}}. \end{aligned}$$

Hence, τ_s is continuous on $(0, \infty)$. It follows that there is a unique continuous function $\bar{\tau}_s$ on $\beta(0, \infty)$, the Stone-Čech compactification of $(0, \infty)$, such that the restriction of $\bar{\tau}_s$ to $(0, \infty)$ is τ_s. We have

$$\lim_{\substack{h \to 0 \\ \mathscr{F}}} \langle \tau_s(h), \phi \rangle = \langle \bar{\tau}_s(\alpha), \phi \rangle \qquad (\phi \in X^*),$$

for some α in $\beta(0, \infty)$. For each s in $[a, b)$ we define $E(s)x = \bar{\tau}_s(\alpha)$. Clearly, $E(s)x \in \mathscr{K}_x$ and since

$$\begin{aligned} \lim_{\substack{h \to 0 \\ \mathscr{F}}} \langle \tau_s(h), \phi \rangle &= \lim_{\substack{h \to 0 \\ \mathscr{F}}} \langle k_{s,h}(T)x, \phi \rangle \\ &= \lim_{\substack{h \to 0 \\ \mathscr{F}}} \frac{1}{h} \int_0^h w_{x,\phi}(s + t)dt \\ &= w_{x,\phi}(s) \qquad (\phi \in X^*), \end{aligned}$$

we deduce that

$$\langle E(s)x, \phi \rangle = w_{x,\phi}(s) \qquad (x \in X, \phi \in X^*). \tag{12}$$

Define $E(s) = 0$ $(s < a)$ and $E(s) = I$ $(s \geqslant b)$. Since

$$k_{s,h}k_{t,h'} = k_{t,h'}k_{s,h} = k_{s,h} \qquad (0 < h < t - s, 0 < h')$$

we have $E(s)E(t) = E(t)E(s) = E(s)$ when $s > t$. Thus $\{E(s): s \in \mathbf{R}\}$ is a naturally ordered net of operators. By Theorem 6.4 and the weak compactness of $\mathscr{K}_x$ it follows that the limits $E(s+)$ and $E(s-)$ exist in the strong operator topology for every real s. Also, since

$$\langle E(s)x, \phi \rangle = w_{x,\phi}(s) \qquad (a \leqslant s < b, x \in X, \phi \in X^*),$$

we have

$$\begin{aligned} \langle E(s)x, \phi \rangle &= \lim_{\substack{h \to 0 \\ \mathscr{F}}} \frac{1}{h} \int_0^h w_{x,\phi}(s + t)dt \\ &= w_{x,\phi}(s+) \\ &= \langle E(s+)x, \phi \rangle \qquad (a \leqslant s < b). \end{aligned}$$

Therefore $E(s) = E(s+)$ $(a \leqslant s < b)$; hence each $E(s)$ is a projection. Thus $E \in \mathscr{E}_J$. Moreover, since for each x in X and ϕ in X^*

$$\langle Tx, \phi\rangle = L_{x,\phi}(f_1) = b\langle x, \phi\rangle - \int_a^b w_{x,\phi}(s)ds$$

$$= b\langle x, \phi\rangle - \int_a^b \langle E(s)x, \phi\rangle\, ds$$

$$= \int_J^{\oplus} sd\langle E(s)x, \phi\rangle,$$

it follows that

$$T = \int_J^{\oplus} s\, dE(s).$$

Finally, since $\langle E(s)x, \phi\rangle = \lim_{\substack{h\to 0\\ \mathscr{F}}} \langle k_{s,h}(T)x, \phi\rangle$ and

$$|\langle k_{s,h}(T)x, \phi\rangle| \leqslant \| k_{s,h}(T)\|\, \| x\|\, \|\phi\| \leqslant K\| x\|\, \|\phi\|,$$

it follows that

$$|\langle E(s)x, \phi\rangle| \leqslant K\| x\|\, \|\phi\| \qquad (x \in X, \phi \in X^*)$$

and so $\| E(s)\| \leqslant K$ $(s \in \mathbf{R})$.

(iv) $\Rightarrow$ (i). It suffices to show that $\{E^*(s): s \in \mathbf{R}\}$ forms a decomposition of the identity for T. The conditions (i), (ii), (iii) of Definition 15.3 follow immediately. Since $E \in \mathscr{E}(J)$, it follows that for each x in X and ϕ in X^* the function $\lambda \to \langle x, E^*(\lambda)\phi\rangle = \langle E(\lambda)x, \phi\rangle$ is everywhere right-continuous on $\mathbf{R}$. Hence conditions (iv) and (v) are satisfied. Moreover, since $T = \int_J^{\oplus} sdE(s)$, we have

$$\langle Tx, \phi\rangle = \int_J^{\oplus} sd\langle E(s)x, \phi\rangle$$

$$= b\langle x, \phi\rangle - \int_a^b \langle x, E^*(s)\phi\rangle\, ds \qquad (x \in X, \phi \in X^*).$$

From the proof of Theorem 15.7, if $x \in X$, $\phi \in X^*$, then

$$\langle f(T)x, \phi\rangle = f(b)\langle x, \phi\rangle - \int_a^b \langle x, E^*(s)\phi\rangle f'(s)\, ds \qquad (f \in AC(J)).$$

It remains to verify condition (vi) of Definition 15.3. Fix x in X and consider the map A from $L^1(J)$ into X defined by $Au = f_u(T)x$, where

$$f_u(s) = \int_s^b u(t)dt \qquad (u \in L^1(J)).$$

A is clearly bounded and linear. For u in $L^1(J)$ and ϕ in X^*, we have

$$\begin{aligned}\langle u, A^*\phi\rangle &= \langle Au, \phi\rangle = \langle f_u(T)x, \phi\rangle \\ &= \int_a^b \langle x, E^*(s)\phi\rangle\, u(s)ds.\end{aligned}$$

It follows that the map A^* from X^* into $L^\infty(J)$ is given by

$$A^*\phi = \langle x, E^*(\cdot)\phi\rangle.$$

Since A is continuous, A^* is continuous when X^* and $L^\infty(J)$ are endowed with their weak*-topologies. This completes the proof that (iv) implies (i).

In our next theorem, we give some properties of well-bounded operators of type (B).

THEOREM 17.15. *Let T be a well-bounded operator of type (B) on X and let $J = [a, b]$, K be chosen so that for every complex polynomial p*

$$\| p(T) \| \leqslant K\{|p(b)| + \operatorname{var}(p, J)\}.$$

Let $\{F^*(t): t \in \mathbf{R}\}$ be the unique decomposition of the identity for T.

$$\text{(i) } f(T) = \int_{a-}^b f(t)dF(t) \qquad (f \in AC(J)),$$

where the integral exists as a strong limit of Riemann sums. (This is valid for well-bounded operators of type (A).)

(ii) *If for some s in $\mathbf{R}$ and x in X we have $(T - sI)^2 x = 0$, then also $(T - sI)x = 0$. (This is valid for an arbitrary well-bounded operator.)*

(iii) *For each real s, $F(s) - F(s-)$ is a projection on X whose range is given by*

$$\{F(s) - F(s-)\}X = \{x \in X : Tx = sx\}.$$

(iv) *The residual spectrum of T is empty.*

Proof. (i) For each x in X we have

$$f(T)x = f(b)x - \int_a^b F(t)x\, df(t) \qquad (f \in AC(J)),$$

where the vector-valued integral is a Riemann-Stieltjes integral of the type discussed in Chapter 16. This follows from the proof of Theorem 16.13. Let $\varepsilon > 0$ be given. We may assume that f is absolutely continuous on $[a - \varepsilon, b]$. It is easy to verify that

$$\int_{a-\varepsilon}^{a} F(t)x\, df(t) = 0.$$

Hence

$$f(T)x = f(b)x - \int_{a-\varepsilon}^{a} F(t)x\, df(t) + \int_a^b F(t)x\, df(t)$$
$$= f(b)x - \int_{a-\varepsilon}^{b} F(t)x\, df(t).$$

From Theorem 16.11 we deduce that $\int_{a-\varepsilon}^{b} f(t)dF(t)x$ exists and

$$f(T)x = f(b)x - f(b)x + f(a - \varepsilon)F(a - \varepsilon)x + \int_{a-\varepsilon}^{b} f(t)dF(t)x$$
$$= \int_{a-\varepsilon}^{b} f(t)dF(t)x.$$

Let $\varepsilon \to 0$. We obtain

$$f(T)x = \int_{a-}^{b} f(t)dF(t)x \qquad (x \in X, f \in AC(J)),$$

and so (i) is proved.

(ii) If $(T - sI)^2 x = 0$, then for any $M > 0$ we have

$$[I + M(T - sI)^2]x = x$$

so that

$$[I + M(T - sI)^2]^{-1}x = x,$$

the operators acting on the null-space of $(T - sI)^2$. Since, trivially, the restriction of a well-bounded operator to a closed invariant subspace is well-bounded, it follows that if $f(t) = (t - s)[1 + M(t - s)^2]^{-1}$ $(t \in J)$, then

$$\begin{aligned} \|(T - sI)x\| &= \|(T - sI)[I + M(T - sI)^2]^{-1}x\| \\ &\leqslant K\|x\| \, |||f||| \\ &\to 0 \quad \text{as} \quad M \to \infty. \end{aligned}$$

Hence $(T - sI)x = 0$.

(iii) Since $F(t)F(s) = F(s)F(t) = F(s)$ for $s < t$, we have

$$F(s)F(s-) = F(s-)F(s) = F(s-) \qquad (s \in \mathbf{R}).$$

Hence$F(s) - F(s-)$ is a projection. Now let $x \in [F(s) - F(s-)]X$ and let $\varepsilon > 0$ be given. Let $u = \{u_k : k = 0, 1, \ldots, m\}$ be a partition of $[a - \varepsilon, b]$. Then

$$\sum_{k=1}^{m} u_k^*[F(u_k)x - F(u_{k-1})x] = u_n^*[F(s)x - F(s-)x] = u_n^* x,$$

where $s \in (u_{n-1}, u_n]$. This is valid for any partition of $[a - \varepsilon, b]$. Hence it follows from (i) that $Tx = sx$.

Conversely, let $Tx = sx$ and $\theta > 0$. Then

$$TF(s - \theta)x = F(s - \theta)Tx = sF(s - \theta)x,$$

$$T[I - F(s + \theta)]x = [I - F(s + \theta)]Tx = s[I - F(s + \theta)]x.$$

By Theorem 16.2,

$$\sigma(T|F(s - \theta)X) \subseteq (-\infty, s - \theta],$$

$$\sigma(T|(I - F(s + \theta))X) \subseteq [s + \theta, \infty).$$

Hence $F(s - \theta)x = [I - F(s + \theta)]x = 0$ and so for any $\theta > 0$

$$x = [F(s + \theta) - F(s - \theta)]x.$$

Letting $\theta \to 0$ suffices to prove (iii).

(iv) It follows from Theorem 15.8 (iii) that

$$\begin{aligned}\{\phi \in X^*: T^*\phi = s\phi\} &= F^*(s)X^* \cap (I - F^*(s-))X^* \\ &= [F^*(s) - F^*(s-)]X^*.\end{aligned}$$

Hence, by (iii), s if an eigenvalue for T if and only if it is an eigenvalue for T^*, and so by Proposition 1.14 the residual spectrum of T is empty.

THEOREM 17.16. *Let T be a well-bounded operator of type (B) on X and let $J = [a, b]$, K be chosen so that for every complex polynomial p*

$$\| p(T) \| \leqslant K\{|p(b)| + \text{var}(p, J)\}.$$

The algebra homomorphism $f \to f(T)$ from $AC(J)$ into $L(X)$ can be extended to an algebra homomorphism ψ from $BV(J)$ into $L(X)$ such that

$$\|\psi(f)\| \leqslant K |||f||| \qquad (f \in BV(J)).$$

If $S \in L(X)$ and $ST = TS$, then $S\psi(f) = \psi(f)S$ $(f \in BV(J))$. Furthermore, let $\{g_\alpha: \alpha \in \sigma\}$ be a uniformly bounded net in $AC(J)$ converging pointwise to a function g in $BV(J)$. Then

$$\psi(g) = \text{st}\lim_{\sigma} \psi(g_\alpha),$$

$$\{\psi(g): g \in BV(J)\} \subseteq \{f(T): f \in AC(J)\}^s$$

where the superscript denotes closure in the strong operator topology.

Proof. Let $\{E^*(s): s \in \mathbf{R}\}$ be the unique decomposition of the identity for T. By the definition of a type (B) operator, $E \in \mathscr{E}(J)$. We define ψ from $BV(J)$ into $L(X)$ by

$$\psi(f) = \int_J^{\oplus} f\,dE \qquad (f \in BV(J)).$$

The argument in the proof of Theorem 17.14 ((i) ⇒ (ii)) shows that ψ is an algebra homomorphism from $BV(J)$ into $L(X)$ and, if $f \in AC(J)$ then $\psi(f) = f(T)$. Moreover, it follows from Lemma 17.6 that

$$\|\psi(f)\| = \left\| \int_J^{\oplus} f\,dE \right\| \leqslant \sup_{\mathbf{R}} \|E(s)\| \, |||f||| = K \, |||f|||$$

for all f in $BV(J)$. If $S \in L(X)$ and $ST = TS$, then by Theorem 16.3 (ii) we have $SE(s) = E(s)S$ $(s \in \mathbf{R})$. It follows from the definition of $\psi(f)$ that $\psi(f)S = S\psi(f)$ for all f in $BV(J)$.

Now let $\{g_\alpha : \alpha \in \sigma\}$ be a uniformly bounded net in $AC(J)$ converging pointwise to a function g in $BV(J)$. We deduce from Theorem 17.5 that

$$\operatorname{st}\lim_{\sigma} \psi(g_\alpha) = \operatorname{st}\lim_{\sigma} \int_J^{\oplus} g_\alpha dE = \int_J^{\oplus} g\,dE = \psi(g). \tag{13}$$

Moreover, given g in $BV(J)$, it follows from Lemma 17.10 that there is a uniformly bounded net $\{g_\alpha : \alpha \in \sigma\}$ in $AC(J)$ converging pointwise to g. By (13), $\psi(g) = \operatorname{st}\lim_{\sigma} g_\alpha(T)$, and this suffices to prove the last statement of the theorem. The proof is complete.

The next result follows immediately from Theorem 17.14 (iii).

THEOREM 17.17. *A well-bounded operator on a reflexive Banach space is of type (B).*

Note 17.18. In Example 16.19, we constructed a well-bounded operator which is of type (*A*) but not of type (*B*). It is easily verified that the point 0 lies in the residual spectrum of that operator. (See Example 11.29 for the full details of the argument.) It follows that Theorem 17.15 (iv) does not extend to well-bounded operators of type (*A*).

18. Well-bounded Operators on Hilbert Space

In this chapter we describe some work of Gillespie. A spectral operator has an unconditionally convergent spectral expansion while a well-bounded operator has a conditionally convergent spectral expansion. It is shown that a well-bounded operator on l^2 which is not a spectral operator can be constructed from a conditional basis on l^2. The sum and product of two commuting scalar-type spectral operators on a Hilbert space are themselves scalar-type spectral operators. (See Corollary 8.5.) It is shown that, in contrast, neither the sum nor the product of two commuting well-bounded operators on a Hilbert space is necessarily well-bounded. However, it is shown in a positive direction that, if one of the two well-bounded operators is in fact a scalar-type spectral operator, then their sum and product are well-bounded.

Definition 18.1. A sequence $\{b_n\}$ in X is said to be a *basis* of X if, for each element x of X, there is a unique sequence $\{\lambda_n\}$ of complex numbers such that $x = \sum_{n=1}^{\infty} \lambda_n b_n$.

Clearly, every Banach space that possesses a basis is separable. For many years it was not known whether every separable Banach space possessed a basis. However, in 1973 Enflo [**1**] showed that there exists a separable reflexive Banach space which does not possess a basis. Davie [**1**] showed that if $2 < p < \infty$, there is a closed subspace of l^p which does not possess a basis.

PROPOSITION 18.2. *Let $\{b_n\}$ be a sequence in X such that $X = \text{clm}\{b_n : n = 1, 2, 3, \ldots\}$. Suppose that there is a sequence $\{f_n\}$ in X^* such that $f_i(b_j) = \delta_{ij}$. Define*

$$P_n(x) = \sum_{i=1}^{n} f_i(x) b_i \qquad (x \in X).$$

If there exists a real number M such that $\| P_n \| \leqslant M$ *for all n, then* $\{b_n\}$ *is a basis of X.*

Proof. If $x = \sum_{j=1}^{\infty} \lambda_j b_j$, then the continuity of f_i gives

$$f_i(x) = \sum_{j=1}^{\infty} \lambda_i f_i(b_j) = \lambda_i.$$

Hence $f_i(x)$ is the only possible choice for λ_i. The result will follow as soon as we show that $P_n(x) \to x$ for each x. Fix x and let $\varepsilon > 0$ be given. By hypothesis there is an element $y = \alpha_1 b_1 + \ldots + \alpha_m b_m$ with $\| y - x \| < \varepsilon$. For $n \geqslant m$ we have $P_n(y) = y$ and so

$$\| P_n(x) - y \| = \| P_n(x - y) \| \leqslant M\varepsilon.$$

Hence $\| P_n(x) - x \| \leqslant (M + 1)\varepsilon$ and the proof is complete.

PROPOSITION 18.3. *Let* $\{b_n\}$ *be a sequence of non-zero elements of X such that* $\text{clm}\{b_n : n = 1, 2, 3, \ldots\} = X$. *Suppose that the following condition holds: there exists* $\delta > 0$ *such that for any sequence* $\{\lambda_n\}$ *of complex numbers we have*

$$\| \lambda_1 b_1 + \ldots + \lambda_p b_p \| \geqslant \delta \| \lambda_1 b_1 + \ldots + \lambda_n b_n \|$$

whenever $p \geqslant n$. *Then* $\{b_n\}$ *is a basis of X.*

Proof. If $\sum_{i=1}^{p} \lambda_i b_i = 0$ for some p and λ_i, then the condition tells us that $\sum_{i=1}^{n} \lambda_i b_i = 0$ for all $n < p$, and hence that $\lambda_i = 0$ for each i. Therefore the b_i are linearly independent, and each element x of Y, the linear span of the b_i, is uniquely expressible in the form $\Sigma \lambda_i b_i$ (where, of course, only finitely many of the λ_i are non-zero). For $x = \Sigma \lambda_i b_i$ in Y, we define $f_i(x) = \lambda_i$ and

$$P_n(x) = \sum_{i=1}^{n} \lambda_i b_i.$$

Then $f_i(b_j) = \delta_{ij}$, and the condition implies that the P_n are continuous, with norms not greater than δ^{-1}. Hence the f_n are also continuous on Y, and so can be extended to continuous linear functionals on X (which we also denote by f_n). Now define P_n on the whole of X by

$$P_n(x) = \sum_{i=1}^{n} f_i(x) b_i \qquad (n = 1, 2, 3, \ldots).$$

We obtain continuous linear mappings whose norms are not greater than δ^{-1} since Y is dense in X. By Proposition 18.2, $\{b_n\}$ is a basis for X.

Consider now the complex Hilbert space l^2. If e_n denotes the element of l^2 with 1 in its nth place and 0 elsewhere then $\{e_n\}$ constitutes an orthonormal basis for l^2. Define $\alpha_1 = 0$ and

$$\alpha_n = \frac{1}{n \log n} \qquad (n = 2, 3, 4, \ldots).$$

Observe that $\alpha_n \geqslant 0$ $(n = 1, 2, 3, \ldots)$, $\sum_{j=1}^{\infty} j\alpha_j^2 < \infty$ and $\sum_{j=1}^{\infty} \alpha_j = \infty$.

Proposition 18.4. *Sequences $\{x_n\}$ and $\{y_n\}$ of elements of l^2 are defined by*

$$x_{2n-1} = e_{2n-1} + \sum_{i=1}^{n} \alpha_{i-n+1} e_{2i},\ x_{2n} = e_{2n} \qquad (n = 1, 2, \ldots),$$

$$y_{2n-1} = e_{2n-1},\ y_{2n} = \sum_{i=1}^{n} (-\alpha_{n-i+1}) e_{2i-1} + e_{2n} \qquad (n = 1, 2, \ldots).$$

(i) *$\{x_n\}$ is a basis of l^2*
(ii) *$\langle x_n, y_m \rangle = \delta_{nm}$ for $n, m = 1, 2, \ldots$*
(iii) *$\langle e_{2n-1}, y_{2m-1} \rangle = \delta_{nm}$ for $n, m = 1, 2, \ldots$*
(iv) *$\left\| \sum_{j=1}^{n} n^{-1/2} x_{2j-1} \right\| \to \infty$ as $n \to \infty$.*

Proof. (i) For any finite sequence of scalars $\beta_1, \ldots, \beta_{2n}$ we have

$$\sum_{j=1}^{2n} \beta_j x_j = \sum_{j=1}^{n} \beta_{2j-1} e_{2j-1} + \sum_{j=1}^{n} \beta_{2j-1} \sum_{i=1}^{\infty} \alpha_{i-j+1} e_{2i} + \sum_{j=1}^{n} \beta_{2j} e_{2j}$$

$$= \sum_{j=1}^{n} \beta_{2j-1} e_{2j-1} + \sum_{j=1}^{n} \left(\sum_{k=1}^{j} \beta_{2k-1} \alpha_{j-k+1} + \beta_{2j} \right) e_{2j}$$

$$+ \sum_{j=n+1}^{\infty} \left(\sum_{k=1}^{n} \beta_{2k-1} \alpha_{j-k+1} \right) e_{2j},$$

whence

$$\left\| \sum_{j=1}^{2n} \beta_j x_j \right\|^2 = \sum_{j=1}^{n} |\beta_{2j-1}|^2 + \sum_{j=1}^{n} \left| \sum_{k=1}^{j} \beta_{2k-1}\alpha_{j-k+1} + \beta_{2j} \right|^2 + \sum_{j=n+1}^{\infty} \left| \sum_{k=1}^{n} \beta_{2k-1}\alpha_{j-k+1} \right|^2. \tag{1}$$

Since, by the Hölder inequality, we have

$$\begin{aligned} \sum_{j=n+1}^{\infty} \left| \sum_{k=1}^{n} \beta_{2k-1}\alpha_{j-k+1} \right|^2 &\leqslant \sum_{j=n+1}^{\infty} \left(\sum_{k=j-n+1}^{j} |\alpha_k|^2 \right) \sum_{i=1}^{n} |\beta_{2i-1}|^2 \\ &\leqslant \sum_{j=2}^{\infty} (j-1)|\alpha_j|^2 \sum_{i=1}^{n} |\beta_{2i-1}|^2 \\ &\leqslant \sum_{j=1}^{\infty} j\alpha_j^2 \sum_{i=1}^{n} |\beta_{2i-1}|^2, \end{aligned}$$

it follows that for any finite sequence of scalars $\beta_1, \ldots, \beta_m$ with $m > 2n$ we have

$$\begin{aligned} \left\| \sum_{j=1}^{2n} \beta_j x_j \right\|^2 &\leqslant \left(1 + \sum_{j=1}^{\infty} j\alpha_j^2\right) \sum_{j=1}^{n} |\beta_{2j-1}|^2 + \sum_{j=1}^{n} \left| \sum_{k=1}^{j} \beta_{2k-1}\alpha_{j-k+1} + \beta_{2j} \right|^2 \\ &\leqslant \left(1 + \sum_{j=1}^{\infty} j\alpha_j^2\right) \left\| \sum_{j=1}^{m} \beta_j x_j \right\|^2. \end{aligned}$$

Similarly, for any finite sequence of scalars $\beta_1, \ldots, \beta_m$ with $m > 2n - 1$ we obtain

$$\left\| \sum_{j=1}^{2n-1} \beta_j x_j \right\|^2 \leqslant \left(1 + \sum_{j=1}^{\infty} j\alpha_j^2 \right) \left\| \sum_{j=1}^{m} \beta_j x_j \right\|^2.$$

It follows from Proposition 18.3 that $\{x_n\}$ is a basis of l^2.

(ii) If m and n are positive integers we have

$$\langle x_{2n}, y_{2m-1} \rangle = \langle e_{2n}, e_{2m-1} \rangle = 0,$$

$$\langle x_{2n-1}, y_{2m-1} \rangle = \left\langle e_{2n-1} + \sum_{i=n}^{\infty} \alpha_{i-n+1} e_{2i}, e_{2m-1} \right\rangle = \delta_{mn},$$

$$\langle x_{2n}, y_{2m} \rangle = \left\langle e_{2n}, \sum_{i=1}^{m} (-\alpha_{m-i+1}) e_{2i-1} + e_{2m} \right\rangle = \delta_{mn},$$

$$\langle x_{2n-1}, y_{2m} \rangle = \left\langle e_{2n-1} + \sum_{i=n}^{\infty} \alpha_{i-n+1} e_{2i}, \sum_{i=1}^{m} (-\alpha_{m-i+1}) e_{2i-1} + e_{2m} \right\rangle$$

$$= \begin{cases} 0 & (m < n) \\ -\alpha_{m-n+1} \langle e_{2n-1}, e_{2n-1} \rangle + \alpha_{m-n+1} \langle e_{2m}, e_{2m} \rangle & (m \geqslant n). \end{cases}$$

These equations suffice to prove (ii). Statement (iii) is trivial.

(iv) From (1) we obtain

$$\left\| \sum_{j=1}^{n} \beta_{2j-1} x_{2j-1} \right\|^2 = \sum_{j=1}^{n} |\beta_{2j-1}|^2 + \sum_{j=1}^{n} \left| \sum_{k=1}^{j} \beta_{2k-1} \alpha_{j-k+1} \right|^2 + \sum_{j=n+1}^{\infty} \left| \sum_{k=1}^{n} \beta_{2k-1} \alpha_{j-k+1} \right|^2$$

$$\geqslant \sum_{j=1}^{n} \left| \sum_{k=1}^{j} \beta_{2k-1} \alpha_{j-k+1} \right|^2 + \sum_{j=1}^{n} |\beta_{2j-1}|^2.$$

We now take $\beta_{2j-1} = n^{-1/2}$ $(j = 1, \ldots, n)$. Observe that from our hypothesis $\left| \sum_{r=1}^{n} \alpha_r \right|^2 \to \infty$ as $n \to \infty$. We deduce that

$$\left\| \sum_{j=1}^{n} n^{-1/2} x_{2j-1} \right\|^2 \geqslant 1 + \frac{1}{n} \sum_{j=1}^{n} \left| \sum_{k=1}^{j} \alpha_{j-k+1} \right|^2 \to \infty \quad \text{as} \quad n \to \infty.$$

The proof is complete.

It follows from Proposition 18.4 (ii) that the coefficient functional associated with the element x_m of the basis $\{x_n\}$ is given by

$$x \to \langle x, y_m \rangle \qquad (x \in l^2).$$

If we define P_n in $L(l^2)$ by

$$P_n x = \langle x, y_n \rangle x_n \qquad (x \in l^2)$$

for $n = 1, 2, \ldots$ we see that each P_n is a projection, $P_n P_m = 0$ $(n \neq m)$, and

$$I = \text{st} - \sum_{n=1}^{\infty} P_n. \tag{2}$$

(This notation means that the series converges in the strong operator topology of $L(l^2)$.) Also, if $z_n = n^{-1/2}(e_1 + e_3 + \ldots + e_{2n-1})$, then z_n is a unit vector and so, using Proposition 18.4 (iii),

$$\left\| \sum_{j=1}^{n} P_{2j-1} \right\| \geqslant \left\| \sum_{j=1}^{n} P_{2j-1} z_n \right\| = \left\| \sum_{j=1}^{n} n^{-1/2} x_{2j-1} \right\|.$$

Hence, by Proposition 18.4 (iv),

$$\left\| \sum_{j=1}^{n} P_{2j-1} \right\| \to \infty \quad \text{as} \quad n \to \infty. \tag{3}$$

By (2) and the principle of uniform boundedness, the partial sums of the series $\sum_{n=1}^{\infty} P_n$ are bounded in norm, and so (3) implies that

$$\left\| \sum_{j=1}^{n} P_{2j} \right\| \to \infty \quad \text{as} \quad n \to \infty. \tag{4}$$

PROPOSITION 18.5. *Let $\{\lambda_n\}$ be a monotonic bounded sequence in* $\mathbf{R}$. *Then the series $\sum_{n=1}^{\infty} \lambda_n P_n$ converges strongly in $L(l^2)$ and its sum is a well-bounded operator.*

Proof. Setting $Q_n = \sum_{j=1}^{n} P_j$, Abel's summation lemma gives

$$\sum_{j=1}^{n} \lambda_j P_j = \lambda_{n+1} Q_n + \sum_{j=1}^{n} (\lambda_j - \lambda_{j+1}) Q_j \tag{5}$$

for $n = 1, 2, 3, \ldots$ Since $\{\lambda_n\}$ is convergent and $Q_n \to I$ strongly, the sequence $\{\lambda_{n+1} Q_n\}$ converges strongly in $L(l^2)$. As noted above, the sequence $\{Q_n\}$ is bounded in norm. This, together with the monotonicity and boundedness of $\{\lambda_n\}$, readily implies that $\sum_{j=1}^{\infty} (\lambda_j - \lambda_{j+1}) Q_j$ converges in the norm of $L(l^2)$. Thus the right-hand side of (5) converges strongly in $L(l^2)$ as $n \to \infty$, and so $\sum_{n=1}^{\infty} \lambda_n P_n$ converges strongly.

Let $T = \text{st} - \sum_{n=1}^{\infty} \lambda_n P_n$, let M be a positive constant such that $\| Q_n \| \leqslant M$

O

($n = 1, 2, \ldots$), let J be a compact interval containing $\{\lambda_n : n = 1, 2, 3, \ldots\}$ and let p be a polynomial. Since $P_n^2 = P_n$ and $P_n P_m = 0\,(m \neq n)$,

$$p\left(\sum_{j=1}^{n} \lambda_j P_j\right) = \sum_{j=1}^{n} p(\lambda_j) P_j;$$

and, by applying Abel's lemma again, this gives

$$p\left(\sum_{j=1}^{n} \lambda_j P_j\right) = p(\lambda_{n+1}) Q_n + \sum_{j=1}^{n} [p(\lambda_j) - p(\lambda_{j+1})] Q_j.$$

Therefore

$$\begin{aligned}\left\| p\left(\sum_{j=1}^{n} \lambda_j P_j\right)\right\| &\leqslant M\left\{|p(\lambda_{n+1})| + \sum_{j=1}^{n} |p(\lambda_j) - p(\lambda_{j+1})|\right\} \\ &\leqslant M\{\sup_{t \in J} |p(t)| + \operatorname{var}_J p\},\end{aligned}$$

for $n = 1, 2, \ldots$ Letting $n \to \infty$, it follows that

$$\|p(T)\| \leqslant M\{\sup_{t \in J} |p(t)| + \operatorname{var}_J p\},$$

showing that T is well-bounded and completing the proof.

We note that the definition of well-bounded operator used here is different from Definition 15.1 but is easily seen to be equivalent to it.

Define sequences $\{\lambda_n\}$ and $\{\mu_n\}$ by

$$\lambda_n = (n + 1)/n, \qquad \mu_{2n-1} = \mu_{2n} = (2n - 1)/2n$$

for $n = 1, 2, \ldots$, and let

$$S = \text{st} - \sum_{n=1}^{\infty} \lambda_n P_n, \qquad T = \text{st} - \sum_{n=1}^{\infty} \mu_n P_n.$$

Then S and T are well-bounded operators on l^2 by Proposition 18.5, and clearly $ST = TS$. We show that neither $S + T$ nor ST is well-bounded.

Suppose that $S + T$ is well-bounded. Then, by Theorem 17.15 (iii), there is a unique bounded projection Q mapping l^2 onto the eigenspace $\{x \in l^2 : (S + T)x = 2x\}$ such that Q belongs to the bicommutant of $S + T$. In particular, $QP_n = P_nQ$ for all n. Noting that

$$(S + T)P_n = (\lambda_n + \mu_n)P_n \qquad (n = 1, 2, \ldots)$$

and that

$$\lambda_n + \mu_n = 2\,(n \text{ even}),\ \lambda_n + \mu_n > 2\,(n \text{ odd}),$$

it is easily verified that

$$QP_n = P_n\,(n \text{ even}), \qquad QP_n = 0\,(n \text{ odd}).$$

Since $I = \text{st} - \sum_{n=1}^{\infty} P_n$, it follows that

$$Q = \text{st} - \sum_{n=1}^{\infty} QP_n = \text{st} - \sum_{n=1}^{\infty} P_{2n}.$$

Therefore the partial sums of the series $\sum_{n=1}^{\infty} P_{2n}$ are bounded in norm by the principle of uniform boundedness, contradicting (4). Hence $S + T$ is not well-bounded.

Since $\lambda_n\mu_n = 1$ (n odd) and $\lambda_n\mu_n < 1$ (n even), a similar argument shows that ST is not well-bounded, for otherwise (3) would be contradicted.

Definition 18.6. Let T be a well-bounded operator on H. We say that T *is implemented by* (M, J) if M is a positive constant and J a compact interval such that

$$\| p(T) \| \leqslant M\{\sup_{t \in J} |p(t)| + \operatorname*{var}_J p\}$$

for every complex polynomial p.

LEMMA 18.7. *Let T be a well-bounded operator on H implemented by (M, J), let $\lambda \in \mathbf{R}$, and let K be a compact interval such that $\lambda + t \in K$, $\lambda t \in K$ for all t in J. Then $\lambda I + T$ and λT are well-bounded operators and are implemented by (M, K).*

Proof. This follows readily from the definition of a well-bounded operator.

LEMMA 18.8. *Let T be a well-bounded operator on H implemented by (M, J), let S be a self-adjoint operator on H with finite spectrum such that $ST = TS$, and let K be a compact interval such that $\lambda + t \in K$, $\lambda t \in K$ whenever $t \in J$ and $\lambda \in \sigma(S)$. Then $S + T$ and ST are well-bounded operators, and are implemented by (M, K).*

Proof. Let $\sigma(S) = \{\lambda_1, \ldots, \lambda_n\}$, where $\lambda_1, \ldots, \lambda_n$ are distinct. Then there are self-adjoint projections $P_1, \ldots, P_n$ in $L(H)$ such that $I = \sum_{j=1}^{n} P_j$, $P_i P_j = 0$ $(i \neq j)$, and $S = \sum_{j=1}^{n} \lambda_j P_j$.

Let p be a polynomial. Since T commutes with S, $TP_j = P_j T$ for $j = 1, \ldots, n$, and hence

$$p(S + T) = \sum_{j=1}^{n} p(\lambda_j I + T) P_j.$$

Noting that $p(\lambda_j I + T) P_j H \subseteq P_j H$ for $j = 1, \ldots, n$ we see that

$$\| p(S + T)x \|^2 = \sum_{j=1}^{n} \| p(\lambda_j I + T) P_j x \|^2 \qquad (x \in H).$$

By Lemma 18.7,

$$\| p(\lambda_j I + T) P_j x \|^2 \leqslant M^2 \{\sup_{t \in K} |p(t)| + \operatorname*{var}_K p\}^2 \| P_j x \|^2$$

for x in H and $j = 1, \ldots, n$. Hence, for x in H,

$$\begin{aligned} \| p(S + T)x \|^2 &\leqslant M^2 \{\sup_{t \in K} |p(t)| + \operatorname*{var}_K p\}^2 \sum_{j=1}^{n} \| P_j x \|^2 \\ &= M^2 \{\sup_{t \in K} |p(t)| + \operatorname*{var}_K p\}^2 \| x \|^2 \end{aligned}$$

giving the required result for $S + T$. The result for ST is proved similarly.

THEOREM 18.9. *Let T be a well-bounded operator on H, and let S be a scalar-type spectral operator on H with real spectrum such that $ST = TS$. Then $S + T$ and ST are well-bounded.*

Proof. Since S has real spectrum it follows from Theorems 7.20 and 8.3 that

S is similar to a self-adjoint operator. Furthermore, it is easily seen that the property of well-boundedness is invariant under a similarity transformation. Consequently, we may assume without loss of generality that S is self-adjoint.

Let T be implemented by (M, J) and let K be a compact interval such that $\lambda + t \in K$, $\lambda t \in K$ whenever $\lambda \in \sigma(S)$ and $t \in J$. By the spectral theorem

$$S = \int_{\sigma(S)} \lambda F(d\lambda)$$

for some spectral measure $F(\cdot)$ on H with support $\sigma(S)$. Also, by Theorem 6.6, T commutes with $F(\cdot)$ since it commutes with S. By approximating the integrand in this integral by step functions, a standard argument gives the existence of a sequence $\{S_n\}$ of self-adjoint operators, each with finite spectrum contained in $\sigma(S)$ and each commuting with T, such that $\|S - S_n\| \to 0$.

Let p be a polynomial. By Lemma 18.8,

$$\|p(S_n + T)\| \leqslant M\{\sup_{t \in K}|p(t)| + \operatorname*{var}_K p\}$$

for all n. Letting $n \to \infty$, we obtain

$$\|p(S + T)\| \leqslant M\{\sup_{t \in K}|p(t)| + \operatorname*{var}_K p\},$$

which shows that $S + T$ is well-bounded. The well-boundedness of ST is obtained similarly.

19. Relationships Between Well-bounded and Prespectral Operators

The purpose of this chapter is to prove some results concerning relationships between the class of well-bounded operators and various classes of prespectral operators. We give an example of an operator which is both well-bounded and a prespectral operator but is not a scalar-type operator.

The concept of the single-valued extension property was introduced in Chapter 5. In connection with our first result recall Proposition 5.28: if an operator has nowhere dense spectrum then it has the single-valued extension property.

LEMMA 19.1. *Let T, in $L(X)$, have nowhere dense spectrum. Let Y be a closed subspace of X with $TY \subseteq Y$ such that $\sigma(T|Y)$ is nowhere dense. Then if $y \in Y$ and $\sigma_X(y)$, $\sigma_Y(y)$ denote the spectra of y with respect to T and $T|Y$, respectively, we have $\sigma_X(y) = \sigma_Y(y)$.*

Proof. We observe that by Theorem 1.29 the hypotheses of the lemma imply that $\sigma(T|Y) \subseteq \sigma(T)$. Let y_X and y_Y be the maximal X-valued and Y-valued analytic functions satisfying

$$(\zeta I - T)y_X(\zeta) = y \qquad (\zeta \in \rho_X(y))$$

$$(\zeta I - T)y_Y(\zeta) = y \qquad (\zeta \in \rho_Y(y))$$

where $\rho_X(y)$, $\rho_Y(y)$ are the domains of definition of these functions. Let $\mathbf{R}_X(y)$ and $\mathbf{R}_Y(y)$ be the closed subspaces defined by

$$\mathbf{R}_X(y) = \text{clm}\{(\zeta I - T)^{-1}y : \zeta \in \rho(T)\},$$

$$\mathbf{R}_Y(y) = \text{clm}\{(\zeta I - T)^{-1}y : \zeta \in \rho(T|Y)\}.$$

Since $\rho(T \mid Y) \supseteq \rho(T)$,

$$\mathbf{R}_Y(y) \supseteq \mathbf{R}_X(y). \tag{1}$$

Since a Y-valued function is also X-valued, the maximality of $\rho_X(y)$ implies that $\rho_Y(y) \subseteq \rho_X(y)$. Now $\rho(T)$ is dense in the plane, and so there is a sequence of points of $\rho(T)$ converging to each point in $\rho_X(y)$. Since y_X is analytic we have $y_X(\zeta) \in \mathbf{R}_X(y)$ for every ζ in $\rho_X(y)$. Hence

$$\mathbf{R}_X(y) = \operatorname{clm}\{y_X(\zeta): \zeta \in \rho_X(y)\},$$

$$\mathbf{R}_Y(y) = \operatorname{clm}\{y_Y(\zeta): \zeta \in \rho_Y(y)\}.$$

Since $\rho_Y(y) \subseteq \rho_X(y)$,

$$\mathbf{R}_Y(y) \subseteq \mathbf{R}_X(y). \tag{2}$$

From (1) and (2), $\mathbf{R}_Y(y) = \mathbf{R}_X(y)$, and so y_X is a Y-valued function. By the maximality of $\rho_Y(y)$ it now follows that $\rho_X(y) \subseteq \rho_Y(y)$. Hence $\rho_X(y) = \rho_Y(y)$, $\sigma_X(y) = \sigma_Y(y)$ and $y_Y(\zeta) = y_X(\zeta)$ $(\zeta \in \rho_X(y))$.

Our next two results are analogues for well-bounded operators of Theorem 5.33. In connection with the first of these we observe that if $T \in L(X)$ and $\sigma(T) \subseteq \mathbf{R}$, then $\sigma(T^*) \subseteq \mathbf{R}$, and so T^* has the single-valued extension property.

THEOREM 19.2. *Let T be a well-bounded operator on X and let $\{E(\lambda): \lambda \in \mathbf{R}\}$ be a decomposition of the identity for T. Then*

$$E(\lambda)X^* = \{y \in X^*: \sigma(y) \subseteq (-\infty, \lambda]\} \qquad (\lambda \in \mathbf{R}).$$

Proof. It was shown in Theorem 15.8 that for every real λ, $E(\lambda)X^*$ is a closed subspace of X^* invariant under T^*, and this subspace does not depend on the particular decomposition of the identity chosen. Since every well-bounded operator possesses some decomposition of the identity which commutes with T^* (Theorem 15.19) there is no loss of generality in assuming that $T^*E(\lambda) = E(\lambda)T^*$ $(\lambda \in \mathbf{R})$. It was shown in Theorem 15.8 (v) that $\sigma(T^* \mid E(\lambda)X^*) \subseteq (-\infty, \lambda]$ and in the course of proving Theorem 16.2 that $\sigma(T^* \mid (I^* - E(\lambda))X^*) \subseteq [\lambda, \infty)$. Let $y \in E(\lambda)X^*$ and $Y = E(\lambda)X^*$. Then by Lemma 19.1

$$\sigma(y) = \sigma_Y(y) \subseteq \sigma(T^* \mid E(\lambda)X^*) \subseteq (-\infty, \lambda]$$

where $\sigma_Y(y)$ denotes the spectrum of y with respect to the operator $T^*|Y$, so that

$$E(\lambda)X^* \subseteq \{y \in X^* : \sigma(y) \subseteq (-\infty, \lambda]\}.$$

Conversely, let $\sigma(y) \subseteq (-\infty, \lambda]$ and $\theta > 0$. Since

$$E(\lambda + \theta)T^* = T^*E(\lambda + \theta)$$

we have

$$\sigma((I^* - E(\lambda + \theta))y) \subseteq (-\infty, \lambda].$$

Let $Y = \{I^* - E(\lambda + \theta)\}X^*$. Then by Lemma 19.1

$$\begin{aligned}\sigma((I^* - E(\lambda + \theta))y) &= \sigma_Y((I^* - E(\lambda + \theta))y)\\ &\subseteq \sigma(T^*|(I^* - E(\lambda + \theta))X^*) \subseteq [\lambda + \theta, \infty).\end{aligned}$$

Hence $\sigma((I^* - E(\lambda + \theta))y)$ is void, and so $E(\lambda + \theta)y = y$ for all $\theta > 0$. Therefore by Theorem 15.8 (iv), $y \in E(\lambda)X^*$. Thus

$$E(\lambda)X^* \subseteq \{y \in X^* : \sigma(y) \subseteq (-\infty, \lambda]\}$$

and the proof is complete.

THEOREM 19.3. *Let T, in $L(X)$, be a well-bounded operator which is decomposable in X. If $\{F(\lambda) : \lambda \in \mathbf{R}\}$ is the family of projections on X whose adjoints form the unique decomposition of the identity for T we have*

$$F(\lambda)X = \{x \in X : \sigma(x) \subseteq (-\infty, \lambda]\} \qquad (\lambda \in \mathbf{R}).$$

Proof. By Theorem 16.2 (ii) we have $\sigma(T|F(\mu)X) \subseteq (-\infty, \mu]$ and $\sigma(T|(I - F(\mu))X) \subseteq [\mu, \infty)$ for all μ in $\mathbf{R}$. The argument of Theorem 19.2 then suffices to establish the present theorem, since with obvious modification the proof of Theorem 15.8 (iv) shows that

$$F(\mu)X = \bigcap_{\lambda > \mu} F(\lambda)X.$$

THEOREM 19.4. *Suppose that T, in $L(X)$, is both a spectral operator and a well-bounded operator. Then T is a scalar-type spectral operator.*

Proof. Let $\mathscr{G}(\cdot)$ be the resolution of the identity for T, and let

$$S = \int_{\sigma(T)} \lambda \mathscr{G}(d\lambda).$$

Since T is well-bounded, it possesses a decomposition of the identity $\{E(\lambda): \lambda \in \mathbf{R}\}$ with the following property: if $A \in L(X)$ and $AT = TA$, then

$$A^*E(\lambda) = E(\lambda)A^* \qquad (\lambda \in \mathbf{R}).$$

This follows from Theorem 15.19. By Theorem 19.2

$$E(\lambda)X^* = \{y \in X^*: \sigma(y) \subseteq (-\infty, \lambda]\} \qquad (\lambda \in \mathbf{R}).$$

Also, by Theorem 6.9, T^* is a prespectral operator on X^* with resolution of the identity $\mathscr{G}^*(\cdot)$ of class X. Hence, by Theorem 5.33,

$$\mathscr{G}^*((-\infty, \lambda])X^* = \{y \in X^*: \sigma(y) \subseteq (-\infty, \lambda]\}.$$

Therefore $E(\lambda)$ and $\mathscr{G}^*((-\infty, \lambda])$, are commuting projections with the same range and so are equal. It follows that for all x in X and y in X^* the function $\lambda \to \langle x, E(\lambda)y\rangle$ is everywhere continuous on the right and of bounded variation on $\mathbf{R}$. In the notation of Theorem 15.7

$$\langle Tx, y\rangle = b\langle x, y\rangle - \int_a^b \langle x, E(\lambda)y\rangle d\lambda \qquad (x \in X, y \in X^*).$$

On integrating by parts we obtain

$$\langle Tx, y\rangle = \int_a^b \lambda d\langle x, E(\lambda)y\rangle \qquad (x \in X, y \in X^*).$$

However, by Corollary 15.9, $\sigma(T) \subseteq [a, b]$ and so

$$\langle Sx, y\rangle = \int_a^b \lambda \mu(d\lambda) = \int_a^b \lambda d\langle x, E(\lambda)y\rangle \qquad (x \in X, y \in X^*),$$

where $\mu(\cdot) = \langle \mathscr{G}(\cdot)x, y\rangle$. It follows that $S = T$, and this completes the proof.

THEOREM 19.5. *Suppose that T, in $L(X)$, is both a prespectral operator of class Γ and a well-bounded operator decomposable in X. Then T is a scalar-type operator of class Γ.*

Proof. Let $\{F(\lambda): \lambda \in \mathbf{R}\}$ be the family of projections on X whose adjoints form the unique decomposition of the identity for T. Let

$$E(\lambda) = F^*(\lambda) \qquad (\lambda \in \mathbf{R}).$$

Let $\mathscr{G}(\cdot)$ be the resolution of the identity of class Γ for T, and let

$$S = \int_{\sigma(T)} \lambda \mathscr{G}(d\lambda).$$

By Theorem 19.3,

$$F(\lambda)X = \{x \in X : \sigma(x) \subseteq (-\infty, \lambda]\} \qquad (\lambda \in \mathbf{R}).$$

Also, by Theorem 5.33,

$$\mathscr{G}((-\infty, \lambda])X = \{x \in X : \sigma(x) \subseteq (-\infty, \lambda]\} \qquad (\lambda \in \mathbf{R}).$$

Now $\mathscr{G}((-\infty, \lambda])$ commutes with T and so by Theorem 15.19

$$\mathscr{G}^*((-\infty, \lambda])E(\lambda) = E(\lambda)\mathscr{G}^*((-\infty, \lambda]) \qquad (\lambda \in \mathbf{R}),$$
$$\mathscr{G}((-\infty, \lambda])F(\lambda) = F(\lambda)\mathscr{G}((-\infty, \lambda]) \qquad (\lambda \in \mathbf{R}).$$

Hence $\mathscr{G}((-\infty, \lambda])$ and $F(\lambda)$ are commuting projections with the same range and so are equal. It follows that for all x in X and y in Γ the function $\lambda \to \langle F(\lambda)x, y\rangle$ is everywhere continuous on the right and of bounded variation on $\mathbf{R}$. In the notation of Theorem 15.7

$$\begin{aligned}\langle Tx, y\rangle &= b\langle x, y\rangle - \int_a^b \langle x, E(\lambda)y\rangle \, d\lambda \\ &= b\langle x, y\rangle - \int_a^b \langle F(\lambda)x, y\rangle \, d\lambda \qquad (x \in X, y \in \Gamma).\end{aligned}$$

On integrating by parts we obtain

$$\langle Tx, y\rangle = \int_a^b \lambda d\langle F(\lambda)x, y\rangle \qquad (x \in X, y \in \Gamma).$$

However, by Corollary 15.9, $\sigma(T) \subseteq [a, b]$, and so

$$\langle Sx, y\rangle = \int_a^b \lambda\mu(d\lambda) = \int_a^b \lambda d\langle F(\lambda)x, y\rangle \qquad (x \in X, y \in \Gamma),$$

where $\mu(\cdot) = \langle \mathscr{G}(\cdot)x, y\rangle$. Since Γ is total, $S = T$, and the proof is complete.

Example 19.6. Let X be the complex Banach space $L^\infty[0, 1] \oplus L^1[0, 1]$ with the norm defined as follows. If $f \in L^\infty[0, 1]$ and $g \in L^1[0, 1]$

$$\|\{f, g\}\| = \operatorname*{ess-sup}_{[0,1]} |f| + \int_0^1 |g(t)|\, dt.$$

Define operators S, N and T on X by

$$S: \{f(t), g(t)\} \to \{tf(t), tg(t)\} \qquad (0 \leqslant t \leqslant 1),$$
$$N: \{f(t), g(t)\} \to \{0, f(t)\} \qquad (0 \leqslant t \leqslant 1),$$
$$T = S + N.$$

S is a scalar-type operator on X with resolution of the identity $E(\cdot)$ of class $L^1[0, 1] \oplus L^\infty[0, 1]$ given by

$$E(\tau): \{f(t), g(t)\} \to \{\chi_\tau(t)f(t), \chi_\tau(t)g(t)\} \qquad (\tau \in \Sigma_p).$$

Observe that $E(\tau)N = NE(\tau)$ $(\tau \in \Sigma_p)$ and so by Theorem 5.15 (ii), T is prespectral on X with resolution of the identity $E(\cdot)$ of class $L^1[0, 1] \oplus L^\infty[0, 1]$. $T = S + N$ is the Jordan decomposition of T, and so T is not of scalar type. We show that T is well-bounded. Observe that if p is any polynomial

$$p(T): \{f(t), g(t)\} \to \{p(t)f(t), p'(t)f(t) + p(t)f'(t)\} \qquad (0 \leqslant t \leqslant 1)$$

and

$$\begin{aligned}
\| \{pf, p'f + pg\} \| &= \operatorname*{ess-sup}_{[0,1]} |pf| + \int_0^1 |p'f + pg| \\
&\leqslant \left[\sup_{t \in [0,1]} |p(t)| \right] \left[\operatorname*{ess-sup}_{[0,1]} |f| \right] + \int_0^1 |p'f| + \int_0^1 |pg| \\
&\leqslant \left[\sup_{t \in [0,1]} |p(t)| \right] \left[\operatorname*{ess-sup}_{[0,1]} |f| \right] \\
&\qquad + \left[\operatorname*{ess-sup}_{[0,1]} |f| \right] \int_0^1 |p'| + \left[\sup_{t \in [0,1]} |p(t)| \right] \int_0^1 |g| \\
&\leqslant \left(\sup_{t \in [0,1]} |p(t)| + \operatorname*{var}_{[0,1]} p \right) \| \{f, g\} \| \\
&\leqslant \left(|p(1)| + 2 \operatorname*{var}_{[0,1]} p \right) \| \{f, g\} \| \\
&\leqslant 2\left(|p(1)| + \operatorname*{var}_{[0,1]} p \right) \| \{f, g\} \|.
\end{aligned}$$

Hence

$$\| p(T) \| \leqslant 2\left(|p(1)| + \operatorname*{var}_{[0,1]} p \right)$$

for every polynomial p, and so T is well-bounded. However, by Theorem 19.5, T is not decomposable in X.

We observe that this example shows that no analogue of Theorem 5.13 (iii) is valid for well-bounded operators.

20. Logarithms of L^p Translations

In this chapter we describe some work of Gillespie. Let G be a locally compact abelian group and let $1 \leqslant p \leqslant \infty$. Consider the complex Banach space $L^p(G)$, where the measure is Haar measure. The *translation operator* on $L^p(G)$ associated with an element x of G is the operator R_x on $L^p(G)$ defined by

$$(R_x f)(y) = f(y + x),$$

where $f \in L^p(G)$ and $y \in G$ a.e. (locally a.e. in the case when $p = \infty$). It is shown that the bilateral shift operator on $L^1(\mathbf{Z})$ has no logarithm. However, if $1 < p < \infty$, there is an operator W on $L^p(G)$ such that $\exp iW = R_x$. Moreover W is a well-bounded operator. It is shown that the operator W constructed lies in the bicommutant of R_x and from this we deduce that every operator V satisfying $\exp iV = R_x$ is well-bounded. We show that if W is a scalar-type spectral operator then so is any operator satisfying $\exp iV = R_x$. However this is the case if and only if either $p = 2$ or the element x has finite order in the group G. A corresponding result for R_x is proved. We give an example of an operator on a reflexive space which is well-bounded and hermitian but is not a spectral operator.

Example 20.1. We begin by showing that the bilateral shift operator on $L^1(\mathbf{Z})$ has no logarithm. For each integer n let e_n be the element of $L^1(\mathbf{Z})$ with 1 in the nth place and 0 elsewhere. Let U be the bilateral shift operator defined by

$$U\left(\sum_{n=-\infty}^{\infty} x(n)e_n\right) = \sum_{n=-\infty}^{\infty} x(n-1)e_n \qquad (x \in L^1(\mathbf{Z})).$$

Observe that $\{e_n : n \in \mathbf{Z}\}$ is a basis for $L^1(\mathbf{Z})$ and $\sum_{n=-\infty}^{\infty} |x(n)| < \infty$. Suppose that there exists an operator S on $L^1(\mathbf{Z})$ such that $\exp S = U$. Let $\hat{}$ denote the Fourier transform on $L^1(\mathbf{Z})$, so that

$$\hat{x}(e^{it}) = \sum_{n=-\infty}^{\infty} x(n)e^{int} \qquad (x \in L^1(\mathbf{Z}), t \in \mathbf{R}).$$

Since U commutes with S, we have

$$(Se_n)\hat{} = (SU^n e_0)\hat{} = (U^n Se_0)\hat{}$$

and so

$$(Se_n)\hat{}(e^{it}) = e^{int}.(Se_0)\hat{}(e^{it})$$

for n an integer and t real. It follows that

$$(Sx)\hat{}(e^{it}) = \hat{x}(e^{it}).(Se_0)\hat{}(e^{it}) \qquad (x \in L^1(\mathbf{Z})).$$

Therefore

$$(S^n x)\hat{}(e^{it}) = \hat{x}(e^{it})\,[(Se_0)\hat{}(e^{it})]^n$$

for $n = 0, 1, 2, \ldots$ and so

$$[(\exp S)x]\hat{}(e^{it}) = \hat{x}(e^{it}).\exp[(Se_0)\hat{}(e^{it})].$$

Taking $x = e_0$, and noting that

$$[(\exp S)x]\hat{}(e^{it}) = (Ux)\hat{}(e^{it}) = e^{it}\hat{x}(e^{it})$$

we obtain

$$e^{it} = \exp[(Se_0)\hat{}(e^{it})] \qquad (t \in \mathbf{R}),$$

showing that the function $(Se_0)\hat{}$ is a continuous logarithm on the unit circle. This gives a contradiction. Hence U has no logarithm.

Note 20.2. It has been shown by Gillespie and West **[2]** that, if x is an element of infinite order in the circle group $\mathbf{T}$, then the translation operator on $L^1(\mathbf{T})$ associated with x does not have a logarithm.

We now return to the general problem. Some preliminary results on p-multipliers are required.

Let Γ denote the dual group of the locally compact abelian group G. The value of γ in Γ at x in G is written as (x, γ). $\mathbf{T}$ denotes the circle group $\{z: |z| = 1\}$ with the usual topology and $\mathbf{T}_d$ denotes the circle group with the discrete topology. If ϕ is a function defined on a set E, then $\phi|F$ denotes the restriction of ϕ to a subset F of E.

Fix p in the range $1 \leqslant p \leqslant 2$ and let q be the index conjugate to p. The Fourier transform of a function f in $L^p(G)$ is denoted by $\hat{f}$, so that $f \to \hat{f}$ is a norm-decreasing mapping of $L^p(G)$ into $L^q(\Gamma)$ which agrees with the ordinary Fourier transform

$$\hat{f}(\gamma) = \int_G (-x, \gamma) f(x) dx \qquad (\gamma \in \Gamma)$$

on $L^p(G) \cap L^1(G)$. Note that $(R_x f)^\wedge = (x, \cdot)\hat{f}$ for x in G and f in $L^p(G)$.

A *p-multiplier* on Γ is a bounded measurable function ϕ from Γ into $\mathbf{C}$ with the property that, whenever $f \in L^p(G)$, there exists a (necessarily unique) g in $L^p(G)$ such that $\hat{g} = \phi\hat{f}$. In this case we write $g = T_\phi f$. It is well-known that T_ϕ is an operator on $L^p(G)$ which commutes with every translation operator R_x. Conversely, if an operator on $L^p(G)$ commutes with every R_x $(x \in G)$, then it is of the form T_ϕ for some p-multiplier ϕ on Γ. Let $M_p(\Gamma)$ denote the set of all p-multipliers on Γ. (For a general discussion on p-multipliers, see Chapters 0 and 4 of Larsen [**1**].)

Now let H be a closed subgroup of G and let Λ be the annihilator of H in Γ. The dual of H is identified with the quotient group Γ/Λ in the usual way. Let π_Λ be the quotient map of Γ onto Γ/Λ. The following result of Saeki ([**1**], Lemma 3.1) relates p-multipliers on Γ/Λ and Λ.

LEMMA 20.3. *If $\phi \in M_p(\Gamma/\Lambda)$, then $\phi \circ \pi_\Lambda \in M_p(\Gamma)$.*

When Γ is discrete, a similar result holds for the restrictions of p-multipliers. (See Lemma 3.1 of de Leeuw [**1**].)

LEMMA 20.4. *Let Γ be discrete and let $\phi \in M_p(\Gamma)$. Then $\phi|\Lambda \in M_p(\Lambda)$.*

Using these results, it is possible to construct p-multipliers on Γ from p-multipliers on the circle group.

LEMMA 20.5. *Let $x \in G$ and let $\phi \in M_p(\mathbf{T}) \cap M_p(\mathbf{T}_d)$. Then the function ψ from*

Γ into $\mathbf{C}$ defined by

$$\psi(\gamma) = \phi((x, \gamma)) \qquad (\gamma \in \Gamma)$$

is a p-multiplier on Γ.

Proof. Let H be the closed subgroup of G generated by x. Since H is monothetic (and closed), it is either discrete and isomorphic to $\mathbf{Z}$ or compact. Let Λ be the annihilator of H in Γ. Define a mapping v of Γ/Λ into $\mathbf{T}$ by

$$v(\gamma + \Lambda) = (x, \gamma).$$

Then v is a continuous algebraic isomorphism of Γ/Λ, the dual of H, onto a subgroup of $\mathbf{T}$. We show that $\phi \circ v \in M_p(\Gamma/\Lambda)$.

Case (a). Let H be discrete and isomorphic to $\mathbf{Z}$. Then Γ/Λ is compact and v is a topological and algebraic isomorphism of Γ/Λ onto $\mathbf{T}$. Since $\phi \in M_p(\mathbf{T})$, by hypothesis, it follows that $\phi \circ v \in M_p(\Gamma/\Lambda)$.

Case (b). Let H be compact. Then Γ/Λ is discrete and v is a homeomorphism of Γ/Λ onto a subgroup D of $\mathbf{T}_d$. Since $\phi \in M_p(\mathbf{T}_d)$, the result of Lemma 20.4 implies that $\phi | D \in M_p(D)$. Therefore $\phi \circ v \in M_p(\Gamma/\Lambda)$, since v is a topological and algebraic isomorphism of Γ/Λ onto D. It now follows from Lemma 20.3 that $\phi \circ v \circ \pi_\Lambda \in M_p(\Gamma)$. Since $\psi = \phi \circ v \circ \pi_\Lambda$, the proof is complete.

Some further results about p-multipliers on $\mathbf{T}$ and $\mathbf{T}_d$ will be required. Our next result follows from Hölder's inequality and Theorem A of Titchmarsh [**1**].

PROPOSITION 20.6. *Let B be the Hilbert form defined on $L^p(\mathbf{Z}) \times L^q(\mathbf{Z})$ for $1 < p < \infty$ by*

$$B(\xi, \lambda) = \sum_{m=-\infty}^{\infty} \sum_{n=-\infty}^{\infty}{}' \frac{\xi(m)\lambda(n)}{m - n},$$

where the dash implies the omission of the terms in which $m = n$. There is a constant B_p, depending on p such that

$$\left| \sum_{m=-\infty}^{\infty} \sum_{n=-\infty}^{\infty}{}' \frac{\xi(m)\lambda(n)}{m - n} \right| \leqslant B_p \| \xi \|_p \| \lambda \|_q. \tag{1}$$

From this we deduce the following result of Stečkin [**1**].

THEOREM 20.7. *Let f be a function of bounded variation on $[0, 2\pi]$ with $f(0) = f(2\pi)$.* Define

$$c_k = \frac{1}{2\pi}\int_0^{2\pi} f(t)e^{-ikt}\,dt \qquad (k \in \mathbf{Z}).$$

Then the map $F: \xi \to c * \xi$ is an operator on $L^p(\mathbf{Z})$ for $1 < p < \infty$, where $c = \{c_k\}$ and

$$(c * \xi)(n) = \sum_{m=-\infty}^{\infty} c_{n-m}\xi(m).$$

Proof. Let $\lambda \in L^q(\mathbf{Z})$. Consider

$$\begin{aligned} T_N &= \sum_{m=-N}^{N}\sum_{n=-N}^{N} c_{n-m}\xi(m)\lambda(n) \\ &= c_0 \sum_{m=-N}^{N} \xi(m)\lambda(n) + \sum_{m=-N}^{N}\sum_{n=-N}^{N}{}' c_{n-m}\xi(m)\lambda(n). \end{aligned}$$

If we denote the double sum by T'_N, then

$$\begin{aligned} |T'_N| &= \left|\sum_{m=-N}^{N}\sum_{n=-N}^{N}{}' c_{n-m}\xi(m)\lambda(n)\right| \\ &= \left|\sum_{m=-N}^{N}\sum_{n=-N}^{N}{}' \frac{1}{2\pi}\int_0^{2\pi} f(t)e^{-i(n-m)t}\,dt\,\xi(m)\lambda(n)\right| \\ &= \left|\frac{1}{2\pi}\int_0^{2\pi} f(t)\left[\sum_{m=-N}^{N}\sum_{n=-N}^{N}{}' \xi(m)\lambda(n)e^{-i(m-n)t}\right]dt\right|. \end{aligned}$$

On integrating by parts we obtain

$$\begin{aligned} |T'_N| &= \left|\frac{1}{2\pi i}\int_0^{2\pi}\left[\sum_{m=-N}^{N}\sum_{n=-N}^{N}{}' \frac{\xi(m)\lambda(n)e^{-i(n-m)t}}{n-m}\right]df(t)\right| \\ &\leqslant \frac{1}{2\pi}\sup_{t\in[0,2\pi]}\left|\sum_{n=-N}^{N}\sum_{n=-N}^{N}{}' \frac{\xi(m)e^{imt}\lambda(n)e^{-int}}{n-m}\right| \operatorname*{var}_{[0,2\pi]} f \\ &\leqslant \frac{B_p}{2\pi}\operatorname*{var}_{[0,2\pi]} f\,\|\xi\|_p\,\|\lambda\|_q, \end{aligned}$$

from (1). It follows that

$$|T_N| \leqslant |c_0| \, \|\xi\|_p \, \|\lambda\|_q + \frac{B_p}{2\pi}\Big(\operatorname*{var}_{[0,\,2\pi]} f\Big) \|\xi\|_p \, \|\lambda\|_q$$

$$= \Big(|c_0| + \frac{B_p}{2\pi} \operatorname*{var}_{[0,\,2\pi]} f\Big) \|\xi\|_p \, \|\lambda\|_q.$$

Since the last expression is independent of N, we let $N \to \infty$ and obtain

$$\left|\sum_{m=-\infty}^{\infty} \sum_{n=-\infty}^{\infty} c_{n-m}\xi(m)\lambda(n)\right| \leqslant \Big(|c_0| + \frac{B_p}{2\pi} \operatorname*{var}_{[0,\,2\pi]} f\Big) \|\xi\|_p \, \|\lambda\|_q.$$

Therefore, for all λ in $L^q(\mathbf{Z})$ we have

$$\left|\sum_{n=-\infty}^{\infty} (c * \xi)(n)\lambda(n)\right| \leqslant \Big(|c_0| + \frac{B_p}{2\pi} \operatorname*{var}_{[0,\,2\pi]} f\Big) \|\xi\|_p \, \|\lambda\|_q,$$

and this implies that $F(\xi) = c * \xi \in L^p(\mathbf{Z})$. Moreover

$$\|F(\xi)\|_p \leqslant \Big(|c_0| + \frac{B_p}{2\pi} \operatorname*{var}_{[0,\,2\pi]} f\Big) \|\xi\|_p.$$

Hence F is an operator on $L^p(\mathbf{Z})$ with

$$\|F\| \leqslant |c_0| + \frac{B_p}{2\pi} \operatorname*{var}_{[0,\,2\pi]} f$$

and the proof is complete.

Define a map u from $\mathbf{T}$ into $\mathbf{C}$ by

$$u(e^{it}) = t \qquad (0 \leqslant t < 2\pi)$$

and, for λ in $\mathbf{C}\backslash[0, 2\pi]$, define u_λ by

$$u_\lambda(e^{it}) = (\lambda - t)^{-1} \qquad (0 \leqslant t < 2\pi).$$

A simple reformulation of Stečkin's result asserts that if f is a function of bounded variation on $[0, 2\pi]$ and $g(e^{it}) = f(t)$ $(0 \leqslant t < 2\pi)$, then $g \in M_p(\mathbf{T})$ for $1 < p \leqslant 2$. The following lemma is a straightforward application of this.

LEMMA 20.8. *The functions u and u_λ belong to $M_p(\mathbf{T})$ whenever $1 < p \leqslant 2$ and $\lambda \in \mathbf{C}\backslash[0, 2\pi]$.*

A bounded measurable function ϕ from $\mathbf{T}$ into $\mathbf{C}$ is said to be *regulated* if, for every real x,

$$\frac{1}{2\varepsilon}\int_{x-\varepsilon}^{x+\varepsilon} \phi(e^{it})dt \to \phi(e^{ix}) \quad \text{as} \quad \varepsilon \to 0+.$$

It can be shown that, if ϕ is a regulated function in $M_p(\mathbf{T})$, then $\phi \in M_p(\mathbf{T}_d)$. This is valid for $1 \leqslant p \leqslant 2$ and is a special case of part of Corollary 4.5 of Saeki [**1**]. Using this, we obtain the following result for $\mathbf{T}_d$ corresponding to that of Lemma 20.8 for $\mathbf{T}$.

LEMMA 20.9. *The functions u and u_λ belong to $M_p(\mathbf{T}_d)$ whenever $1 < p \leqslant 2$ and $\lambda \in \mathbf{C}\backslash[0, 2\pi]$.*

Proof. Let $1 < p \leqslant 2$. Put $v = u + \pi\delta_1$, where δ_1 is the characteristic function of the singleton subset $\{1\}$ of $\mathbf{T}$. Observe that v is the regulated function obtained by redefining u to take the value π at the point 1. Since $u = v$ a.e. on $\mathbf{T}$ and $u \in M_p(\mathbf{T})$ by Lemma 20.8, it follows that $v \in M_p(\mathbf{T})$. Hence $v \in M_p(\mathbf{T}_d)$, by the preceding remarks, since v is regulated. The function δ_1 also belongs to $M_p(\mathbf{T}_d)$, and so $u = v - \pi\delta_1 \in M_p(\mathbf{T}_d)$. The result for u_λ is obtained similarly.

Recall that if $\phi \in M_p(\Gamma)$ the equation

$$(T_\phi f)\hat{} = \phi\hat{f} \qquad (f \in L^p(G))$$

defines an operator T_ϕ on $L^p(G)$, the norm of which is denoted by $\|\phi\|_{M_p(\Gamma)}$.

Let H be a closed subgroup of G and let Λ be the annihilator of H in Γ. Identify the dual of H with Γ/Λ and let π_Λ be the quotient map of Γ onto Γ/Λ.

LEMMA 20.10. *If $\phi \in M_p(\Gamma/\Lambda)$, then $\phi \circ \pi_\Lambda \in M_p(\Gamma)$ and*

$$\|\phi \circ \pi_\Lambda\|_{M_p(\Gamma)} = \|\phi\|_{M_p(\Gamma/\Lambda)}.$$

Proof. See Lemma 3.1 of Saeki [**1**].

LEMMA 20.11. *Let* Γ *be discrete and let* $\phi \in M_p(\Gamma)$. *Then* $\phi|\Lambda \in M_p(\Lambda)$ *and* $\|\phi|\Lambda\|_{M_p(\Lambda)} \leqslant \|\phi\|_{M_p(\Gamma)}$.

Proof. See Lemma 3.1 of deLeeuw [**1**] and its proof.

LEMMA 20.12. *Let* $x \in G$ *and let* $\phi \in M_p(\mathbf{T}) \cap M_p(\mathbf{T}_d)$. *Then the function* ψ *from* Γ *into* $\mathbf{C}$ *defined by*

$$\psi(\gamma) = \phi((x, \gamma)) \qquad (\gamma \in \Gamma)$$

belongs to $M_p(\Gamma)$. *Also*

$$\|\psi\|_{M_p(\Gamma)} \leqslant \max\{\|\phi\|_{M_p(\mathbf{T})}, \|\phi\|_{M_p(\mathbf{T}_d)}\}.$$

Proof. The fact that $\psi \in M_p(\Gamma)$ is the content of Lemma 20.5. Its proof, together with Lemmas 20.10, 20.11 easily gives the required norm inequality.

The next result is a simple consequence of Stečkin's result (Theorem 20.7). For λ in $[0, 2\pi]$, let χ_λ denote the characteristic function of the arc $\{e^{it}: 0 \leqslant t \leqslant \lambda\}$ of the unit circle.

LEMMA 20.13. *Let* $1 < p \leqslant 2$. *Then* $\chi_\lambda \in M_p(\mathbf{T})$ *for all* λ *in* $[0, 2\pi]$ *and there is a constant* α_p *(dependent on p but not on* λ*) such that*

$$\|\chi_\lambda\|_{M_p(\mathbf{T})} \leqslant \alpha_p \qquad (0 \leqslant \lambda \leqslant 2\pi).$$

A corresponding result holds with $\mathbf{T}$ replaced by $\mathbf{T}_d$.

LEMMA 20.14. *Let* $1 < p \leqslant 2$. *Then* $\chi_\lambda \in M_p(\mathbf{T}_d)$ *for all* λ *in* $[0, 2\pi]$ *and there is a constant* β_p *(dependent on p but not on* λ*) such that*

$$\|\chi_\lambda\|_{M_p(\mathbf{T}_d)} \leqslant \beta_p \qquad (0 \leqslant \lambda \leqslant 2\pi).$$

Proof. It is easily verified that χ_0 and $\chi_{2\pi}$ belong to $M_p(\mathbf{T}_d)$. Suppose that $0 < \lambda < 2\pi$. Put $\chi'_\lambda = \chi_\lambda - \frac{1}{2}(\delta_0 + \delta_\lambda)$, where δ_0 and δ_λ are the characteristic functions of the singleton subsets $\{1\}$ and $\{e^{i\lambda}\}$ respectively of $\mathbf{T}$. Then χ_λ equals χ'_λ a.e. on $\mathbf{T}$, so that $\chi'_\lambda \in M_p(\mathbf{T})$ and

$$\|\chi'_\lambda\|_{M_p(\mathbf{T})} \leqslant \alpha_p,$$

by Lemma 20.13. Now χ'_λ is regulated. Therefore by Corollary 4.5 of Saeki [**1**], $\chi'_\lambda \in M_p(\mathbf{T}_d)$ and

$$\|\chi'_\lambda\|_{M_p(\mathbf{T}_d)} = \|\chi'_\lambda\|_{M_p(\mathbf{T})} \leqslant \alpha_p.$$

Finally, $\chi_\lambda - \chi'_\lambda = \frac{1}{2}(\delta_0 + \delta_\lambda)$ belongs to $M_p(\mathbf{T}_d)$ and

$$\|\chi_\lambda - \chi'_\lambda\|_{M_p(\mathbf{T}_d)} \leqslant 1.$$

Hence $\chi_\lambda \in M_p(\mathbf{T}_d)$ and $\|\chi_\lambda\|_{M_p(\mathbf{T}_d)} \leqslant \alpha_p + 1$. This gives the required result.

LEMMA 20.15. *Let $1 < p \leqslant 2$ and let $x \in G$. For λ in $[0, 2\pi]$, define a map ψ_λ from Γ into $\mathbf{C}$ by*

$$\psi_\lambda(\gamma) = \chi_\lambda((x, \gamma)) \qquad (\gamma \in \Gamma).$$

Then $\psi_\lambda \in M_p(\Gamma)$ and there is a constant K_p (dependent on p, but independent of λ, x and G) such that

$$\|\psi_\lambda\|_{M_p(\Gamma)} \leqslant K_p \qquad (0 \leqslant \lambda \leqslant 2\pi).$$

Proof. Apply Lemmas 20.12, 20.13 and 20.14.

In the following result the Haar measure on Γ is chosen so that the inversion formula for the Fourier transform is valid.

LEMMA 20.16. *Let $\phi \in M_p(\Gamma)$ and let k be a complex function defined and integrable over Γ. Then $k * \phi \in M_p(\Gamma)$ and*

$$\|k * \phi\|_{M_p(\Gamma)} \leqslant \|k\|_1,$$

where $\|k\|_1$ denotes the L^1 norm of k.

Proof. See Lemma 4.7.1 of Larsen [**1**; p. 125].

The proof of the next result is straightforward and so is omitted.

LEMMA 20.17. *Let Γ be discrete, let D be a subgroup of Γ, and let $\psi \in M_p(D)$. If ϕ is a complex function on Γ defined by*

$$\phi|D = \psi, \qquad \phi|(\Gamma \backslash D) = 0,$$

then $\phi \in M_p(\Gamma)$.

We now consider the problem of the existence of logarithms of the operator R_x, defined at the beginning of this chapter. If the element x has finite order n in the group G then $R_x^n = I$ and so, by the spectral mapping theorem, the spectrum of R_x is a finite subset of the unit circle. The situation in the case that x has infinite order in G is quite different as the following result shows.

THEOREM 20.18. *Let $1 \leqslant p \leqslant \infty$ and let x in G have infinite order. Then the spectrum of the operator R_x on $L^p(G)$ equals* $\mathbf{T}$.

Proof. Suppose first that $1 \leqslant p \leqslant 2$. The inclusion $\sigma(R_x) \subseteq \mathbf{T}$ is clear since R_x is an invertible isometry. To obtain the reverse inclusion, let $\lambda \in \mathbf{C} \backslash \sigma(R_x)$. Then there is an operator S on $L^p(G)$ such that

$$(\lambda I - R_x)S = S(\lambda I - R_x) = I. \tag{2}$$

Since R_x commutes with each R_y $(y \in G)$, so also does S. Hence $S = T_\phi$ for some ϕ in $M_p(\Gamma)$. Using the uniqueness parts of Theorem 0.1.1 (when $p = 1$) and Corollary 4.1.2 (when $1 < p \leqslant 2$) of Larsen [**1**], equation (2) gives

$$(\lambda - (x, \gamma))\phi(\gamma) = 1 \qquad (\gamma \in \Gamma \text{ locally a.e}). \tag{3}$$

Let $\alpha = \text{ess sup}\,\{|\phi(\gamma)| : \gamma \in \Gamma\}$. By (3), $\alpha > 0$ and

$$|\lambda - (x, \gamma)| \geqslant \alpha^{-1} \qquad (\gamma \in \Gamma \text{ locally a.e.}).$$

Hence the set

$$\{\gamma \in \Gamma : |\lambda - (x, \gamma)| < \alpha^{-1}\}$$

is locally null. However this is an open subset of Γ and is therefore empty. We have thus proved that

$$|\lambda - (x, \gamma)| \geqslant \alpha^{-1} > 0 \qquad (\gamma \in \Gamma). \tag{4}$$

Since x has infinite order, given ω in $\mathbf{T}$ there exists a (not necessarily continuous) character χ on G with $\chi(x) = \omega$. Since χ can be approximated pointwise on G by continuous characters (see Rudin [**2**; 1.8.3]), it follows that $\{(x, \gamma) : \gamma \in \Gamma\}$ is a dense subset of $\mathbf{T}$. Therefore (4) implies that $\lambda \notin \mathbf{T}$. This completes the proof when $1 \leqslant p \leqslant 2$.

Finally, suppose that $2 < p \leqslant \infty$ and let q be the index conjugate to p. It is easily verified that the adjoint of R_{-x} on $L^q(G)$ is R_x on $L^p(G)$. Since $1 \leqslant q < 2$, the first part of the proof gives $\sigma(R_{-x}) = \mathbf{T}$. Hence $\sigma(R_x) = \mathbf{T}$, completing the proof.

THEOREM 20.19. *Let* $1 < p < \infty$ *and let* $x \in G$. *Then there exists an operator* A_x *on* $L^p(G)$ *such that* $\exp(iA_x) = R_x$ *and* $\sigma(A_x) \subseteq [0, 2\pi]$.

Proof. Suppose first that $1 < p \leqslant 2$ and let ψ be the complex function defined by

$$\psi(\gamma) = u((x, \gamma)) \qquad (\gamma \in \Gamma).$$

By Lemmas 20.5, 20.8 and 20.9, ψ is a p-multiplier on Γ. We show that the corresponding operator $A_x = T_\psi$ has the required properties.

Fix f in $L^p(G)$ and for $n = 0, 1, 2, \ldots$ define

$$f_n = \sum_{r=0}^{n} \frac{(iT_\psi)^r}{r!} f.$$

Then $f_n \to (\exp iT_\psi) f$ in $L^p(G)$ and so $\hat{f}_n \to \{(\exp iT_\psi) f\}\hat{}$ in $L^q(\Gamma)$, where q is the index conjugate to p. Hence there is a subsequence $\{f_{n_k}\}$ such that

$$\hat{f}_{n_n}(\gamma) \to \{(\exp iT_\psi) f\}\hat{}(\gamma) \qquad (\gamma \in \Gamma \text{ a.e.}).$$

However, for almost all γ in Γ,

$$\hat{f}_n(\gamma) = \sum_{r=0}^{n} \frac{(i\psi(\gamma))^r}{r!} \hat{f}(\gamma) \to (\exp i\,\psi(\gamma)) \hat{f}(\gamma)$$

and, from the definitions of ψ and u,

$$(\exp i\psi(\gamma)) \hat{f}(\gamma) = (x, \gamma) \hat{f}(\gamma) = (R_x f)\hat{}(\gamma).$$

Therefore

$$\{\exp iT_\psi) f\}\hat{}(\gamma) = (R_x f)\hat{}(\gamma) \qquad (\gamma \in \Gamma \text{ a.e.})$$

and so $(\exp iT_\psi) f = R_x f$. Hence $\exp iT_\psi = R_x$.

Let $\lambda \in \mathbf{C} \backslash [0, 2\pi]$ and let ψ_λ be the complex function defined by

$$\psi_\lambda(\gamma) = u_\lambda((x, \gamma)) \qquad (\gamma \in \Gamma).$$

By Lemmas 20.5, 20.8 and 20.9, $\psi_\lambda \in M_p(\Gamma)$. Since

$$\psi_\lambda(\gamma)[\lambda - \psi(\gamma)] = [\lambda - \psi(\gamma)]\psi_\lambda(\gamma) = 1 \qquad (\gamma \in \Gamma),$$

it follows that

$$T_{\psi_\lambda}(\lambda I - T_\psi) = (\lambda I - T_\psi)T_{\psi_\lambda} = I.$$

Hence $\lambda \notin \sigma(T_\psi)$ and so $\sigma(T_\psi) \subseteq [0, 2\pi]$.

Suppose now that $2 < p < \infty$. Since the conjugate index q satisfies $1 < q < 2$, we can apply the result above to obtain an operator B on $L^q(G)$ satisfying $\exp iB = R_{-x}$ and $\sigma(B) \subseteq [0, 2\pi]$. Putting $A_x = B^*$, we obtain an operator on $L^p(G)$ with the required properties.

Next, we show that the operator A_x, constructed in the last theorem, lies in the bicommutant of R_x. The first lemma, which is basic to the discussion, follows easily from the proofs of Lemmas 20.12 and 20.5.

Lemma 20.20. *Let $1 \leqslant p < \infty$ and let $x \in G$. Let $\phi \in M_p(\mathbf{T}_x)$, where $\mathbf{T}_x$ denotes $\mathbf{T}_d$ if the closed subgroup of G generated by x is compact and $\mathbf{T}_x$ denotes $\mathbf{T}$ if the closed subgroup of G generated by x is not compact. Then if $\tilde{x}$ denotes the function $(x, \cdot)$ on Γ, $\phi \circ \tilde{x} \in M_p(\Gamma)$ and*

$$\| \phi \circ \tilde{x} \|_{M_p(\Gamma)} \leqslant \| \phi \|_{M_p(\mathbf{T}_x)}.$$

This result gives sense to the statement of the following lemma.

Lemma 20.21. *Let $x \in G$ and let $\mathbf{T}_x$ be as above. Then*

$$\{R_x\}'' \subseteq \{T_{\phi \circ \tilde{x}} : \phi \in M_p(\mathbf{T}_x)\}.$$

Proof. Let $S \in \{R_x\}''$. Since R_x commutes with R_y for each y in G, S does so also. Consequently, there exists ψ in $M_p(\Gamma)$ such that $S = T_\psi$.

Let H be the closed subgroup of G generated by x and let Λ be the annihilator of H in Γ. Given λ in Λ, define an operator S_λ on $L^p(G)$ by setting

$$S_\lambda f = \lambda . f \qquad (f \in L^p(G)).$$

Each S_λ commutes with R_x and hence with $S = T_\psi$. It follows easily that for each λ in Λ we have

$$\psi(\gamma - \lambda) = \psi(\gamma) \qquad (\gamma \in \Gamma \text{ locally a.e.}).$$

Hence, by §3.6.5 of Reiter [**1**; p. 82], there exists a bounded measurable function ψ' on Γ/Λ such that

$$\psi(\gamma) = \psi'(\gamma + \lambda) \qquad (\gamma \in \Gamma \text{ locally a.e.}).$$

Then $\psi' \in M_p(\Gamma/\Lambda)$ by Corollary 3.5 of Saeki [**1**].

Identify Γ/Λ with the dual group of H in the usual way and define a map v from Γ/Λ into $\mathbf{T}$ by

$$v(\gamma + \Lambda) = (x, \gamma).$$

If H is a compact group, v is an isomorphism of Γ/Λ onto a subgroup D of $\mathbf{T}_d$. Then $\psi' \circ v^{-1}$ is a p-multiplier on D which, by Lemma 20.17, is the restriction to D of a p-multiplier ϕ on $\mathbf{T}_d$. Hence

$$\psi(\gamma) = \psi'(\gamma + \Lambda) = (\psi' \circ v^{-1})((x, \gamma)) = \phi((x, \gamma))$$

for γ in Γ locally a.e showing that $S = T_\psi = T_{\phi \circ \hat{x}}$ and giving the required result in the case that H is compact.

If H is not compact, v is an isomorphism of Γ/Λ onto $\mathbf{T}$ and the required result follows similarly (although no appeal to Lemma 20.17 is necessary in this case).

THEOREM 20.22. *Let $1 \leqslant p \leqslant 2$ and let x be an element of G such that the closed subgroup of G generated by x is not compact. Then*

$$\{R_x\}'' = \{T_{\phi \circ \hat{x}} : \phi \in M_p(\mathbf{T})\} = \mathscr{A}(R_x, R_{-x}),$$

where $\mathscr{A}(R_x, R_{-x})$ denotes the weakly closed subalgebra of operators on $L^p(G)$ generated by R_x and R_{-x}.

Proof. From Lemma 20.21 and the inclusion $\mathscr{A}(R_x, R_{-x}) \subseteq \{R_x\}''$ it suffices to prove that

$$T_{\phi\circ\tilde{x}} \in \mathscr{A}(R_x, R_{-x}) \tag{5}$$

for all ϕ in $M_p(\mathbf{T})$. Fix ϕ in $M_p(\mathbf{T})$. When $p = 1$, ϕ has an absolutely convergent Fourier series, say $\sum_{n\in Z} a_n e^{nit}$. Then

$$T_{\phi\circ\tilde{x}} = \sum_{n\in \mathbf{Z}} a_n (R_x)^n,$$

where this series converges unconditionally in the operator norm, giving (5).

Suppose now that $1 < p \leqslant 2$ and let K_n denote the nth Fejér kernel defined on $\mathbf{T}$. For each positive integer n, $K_n * \phi$ is a trigonometric polynomial and

$$\| K_n \quad \phi \|_{M_p}(\mathbf{T}) \leqslant \| \phi \|_{M_p(\Gamma)} \tag{6}$$

by Lemma 20.16, since each K_n has unit L^1 norm. Let $\psi_n = (K_n * \phi) \circ \tilde{x}$, a polynomial in $\tilde{x}$ and $\tilde{x}^{-1}$. Then each ψ_n is a p-multiplier on Γ, with T_{ψ_n} a polynomial in R_x and R_{-x}. Also, (6) and the norm inequality of Lemma 20.20 show that the set $\{T_{\psi_n} : n = 1, 2, \ldots\}$ of operators is uniformly bounded in norm. We show that $T_{\psi_n} \to T_{\phi\circ\tilde{x}}$ in the weak operator topology. This will give (5).

Let Λ and ν be as in the proof of Lemma 20.21, with the Haar measures on Γ/Λ and Λ adjusted so that Weil's formula is valid. (See §3.3.3 (i), p. 59 and §3.4.5 (iii), p. 70 of Reiter [**1**].) Let $A_c(G)$ be the space of continuous complex-valued functions on G with compact supports and Fourier transforms in $L^1(\Gamma)$. Let $f, g \in A_c(G)$. Considering f as an element of $L^p(\mathrm{G})$ and g as an element of $L^q(\mathrm{G})$, where q is the index conjugate to p,

$$\begin{aligned}\langle T_{\psi_n} f, g\rangle &= \int_\Gamma (T_{\psi_n} f)\hat{}(\gamma) g(-\gamma) d\gamma \\ &= \int_{\Gamma/\Lambda} h(\gamma + \Lambda) d(\gamma + \Lambda)\end{aligned} \tag{7}$$

where, for almost all $\gamma + \Lambda$,

$$h(\gamma + \Lambda) = \int_\Lambda \psi_n(\gamma + \lambda) \hat{f}(\gamma + \lambda) \hat{g}(-\gamma - \lambda) d\lambda$$

$$h(\gamma+\Lambda)=(K_n*\phi)(v(\gamma+\Lambda))\int_\Lambda \hat{f}(\gamma+\lambda)\hat{g}(-\gamma-\lambda)d\lambda. \tag{8}$$

Since v is a topological and algebraic isomorphism,

$$(K_n*\phi)\circ v\rightarrow\phi\circ v$$

in the weak*-topology on $L^\infty(\Gamma/\Lambda)$. Since the function defined a.e by

$$\gamma+\Lambda\rightarrow\int_\Lambda f(\gamma+\lambda)\hat{g}(-\gamma-\lambda)d\lambda$$

belongs to $L^1(\Gamma/\Lambda)$, it follows from (7) and (8) that $\langle T_{\psi_n}f,g\rangle$ tends to

$$\int_{\Gamma/\Lambda}\phi(v(\gamma+\Lambda))\left\{\int_\Lambda \hat{f}(\gamma+\lambda)\hat{g}(-\gamma-\lambda)d\lambda\right\}d(\gamma+\Lambda)$$
$$=\int_\Gamma(\phi\circ\tilde{x})(\gamma)\hat{f}(\gamma)\hat{g}(-\gamma)d\gamma=\langle T_{\phi\circ\tilde{x}}f,g\rangle.$$

Thus

$$\langle T_{\psi_n}f,g\rangle\rightarrow\langle T_{\phi\circ\tilde{x}}f,g\rangle\qquad(f,g\in A_c(G)).$$

Since $A_c(G)$ is norm dense in each of $L^p(G)$, $L^q(G)$ and $\{T_{\psi_n}: n=1,2,\ldots\}$ is a uniformly bounded set of operators, it follows that $T_{\psi_n}\rightarrow T_{\phi\circ\tilde{x}}$ in the weak operator topology, completing the proof.

In order to obtain the corresponding description of $\{R_x\}''$ in the case in which the element x generates a compact subgroup of G, we need to use the following classical result of Lorch [**3**; p. 54–6].

THEOREM 20.23. *Let X be reflexive and let V, in $L(X)$, have the property*

$$\|V^n\|\leqslant k<\infty\qquad(n=0,1,2,\ldots).$$

Let T_n be the operator defined by

$$T_n=n^{-1}(I+V+V^2+\ldots+V^{n-1}).$$

Then the sequence $\{T_n\}$ *converges strongly to a projection* P *for which* $\| P \| \leqslant k$. *Let* Y *be the null space of* $I - V$ *and let* Z *be the closure of the range of* $I - V$. *Then* $X = Y \oplus Z$, $PX = Y$ *and* $(I - P)X = Z$.

Proof. We observe that $Y = \{f: Vf = f\}$ and Z is the closure of the set $\{f: f = g - Vg, g \in X\}$. Also

$$\|(V^*)^n\| = \|(V^n)^*\| = \|V^n\| \leqslant k \qquad (n = 0, 1, 2, \ldots),$$

$$T_n^* = n^{-1}(I^* + V^* + \ldots + (V^*)^{n-1}),$$

$Z^\perp$ is the null space of the operator $(I - V)^* = I^* - V^*$, and $Y^\perp$ is the closure of the range of $I^* - V^*$.

Suppose now that $f \in Y$; thus $Vf = f$. Then for each positive integer n we have $V^n f = f$ and $T_n f = f$. Thus on Y, $T_n f \to f$. Now let $f \in Z$. This means that for $\varepsilon > 0$, there exist vectors g and h such that $f = g - Vg + h$ and $\| h \| < \varepsilon$. It is easily shown that

$$T_n f = n^{-1}(g - V^n g + h + Vh + \ldots + V^{n-1} h).$$

Bearing in mind that $\| V^r \| \leqslant k$ and $k \geqslant 1$ we obtain

$$\| T_n f \| \leqslant 2n^{-1} k \| g \| + k\varepsilon.$$

It is now clear that if $f \in Z$, then $T_n f \to 0$.

The argument above shows that $Y \cap Z = \{0\}$. Suppose $f_1 \in Y$ and $f_2 \in Z$. Then $T_n(f_1 + f_2) \to f_1$. Since $\| T_n \| \leqslant k$ we have $\| f_1 \| \leqslant k \| f_1 + f_2 \|$. We have therefore shown that $\{T_n\}$, restricted to vectors in $Y \oplus Z$, converges to a projection P on Y for which $\| P \| \leqslant k$. It remains to prove that $X = Y \oplus Z$.

By virtue of our preliminary remarks, we see that T_n^*, restricted to the subspace $Y^\perp \oplus Z^\perp$, converges to a projection on $Y^\perp$. In particular $Y^\perp \cap Z^\perp = \{0\}$. Now suppose there is an $f \neq 0$ such that $f \notin Y \oplus Z$. Then there exists an element ϕ in X^* such that $\phi(f) = 1$ and $\phi(g) = 0$ for all g in $Y \oplus Z$; hence $\phi \neq 0$. However $\phi \in Y^\perp$ and $\phi \in Z^\perp$. Hence $\phi = 0$. This contradiction shows that $X = Y \oplus Z$ and so the theorem is proved.

In order to prove a complete analogue of Theorem 20.22 for the case in which the closed subgroup of G generated by x is compact, a lot of extra machinery would be required. The following weaker result, pertaining to the operator A_x constructed in Theorem 20.19, is adequate for our purpose.

THEOREM 20.24. *Let $1 < p \leqslant 2$ and let x be an element of G such that the closed subgroup of G generated by x is compact. Then $A_x \in \{R_x\}''$.*

Proof. Let H be the closed subgroup of G generated by x and let Λ be the annihilator of H in Γ. Identify Γ/Λ with the dual group of H in the usual way and define a map v from Γ/Λ into $\mathbf{T}$ by $v(\gamma + \Lambda) = (x, \gamma)$. Since H is compact, v is an isomorphism of Γ/Λ onto a subgroup D of $\mathbf{T}_d$. Let $K_n : \mathbf{T} \to \mathbf{C}$ be the nth Fejér kernel and let $u_n = K_n * u$, where

$$u(e^{it}) = t \qquad (0 < t < 2\pi),\ u(1) = \pi.$$

Hence $u_n \circ v \to u \circ v$ in the weak*-topology of $L^\infty(\Gamma/\Lambda)$. Observe that $u_n \circ \tilde{x}$ is a polynomial in $\tilde{x}$ and $\tilde{x}^{-1}$ and so is a p-multiplier on Γ, with the corresponding operator on $L^p(G)$ being the same polynomial in R_x and R_x^{-1}. Let T_n be this corresponding operator. Thus in the weak operator topology of $L(L^p(G))$, $T_n \to T$, where T is the operator on $L^p(G)$ corresponding to the p-multiplier $u \circ \tilde{x}$. This follows by the argument of the proof of Theorem 20.22.

Let $u = v + \pi\chi_1$, where χ_1 is the characteristic function of the subset $\{1\}$ of $\mathbf{T}$. Since A_x corresponds to $v \circ \tilde{x}$ we thus have $T = A_x + \pi P$, where P corresponds to the p-multiplier $\chi_1 \circ \tilde{x}$ (that is the characteristic function of Λ). Since T is the weak limit of polynomials in R_x, R_x^{-1}, to get $A_x \in \{R_x\}''$ it suffices to show that $P \in \{R_x\}''$. By Theorem 20.23,

$$n^{-1} \sum_{k=1}^{n} R_x^k \to Q \quad \text{as} \quad n \to \infty$$

in the strong operator topology of $L(L^p(G))$. Observe that if $f \in L^p(G)$, then

$$n^{-1} \sum_{n=1}^{n} R_x^k f \to Qf \quad \text{as} \quad n \to \infty$$

in the norm of $L^p(G)$, and hence

$$n^{-1} \sum_{n=1}^{n} \tilde{x}^k \hat{f} \to (Qf)\hat{} \quad \text{as} \quad n \to \infty$$

in the $L^q(\Gamma)$ norm. However, for γ in Γ we have

$$n^{-1}\sum_{k=1}^{n}[\tilde{x}(\gamma)]^k \to \begin{cases} 1 & \text{if } \tilde{x}(\gamma) = 1 \\ 0 & \text{if } \tilde{x}(\gamma) \neq 1 \end{cases}$$

and so

$$n^{-1}\sum_{k=1}^{n}\tilde{x}^k \to \chi_1 \circ \tilde{x}$$

pointwise on Γ. Thus

$$(Pf)\hat{} = (\chi_1 \circ \tilde{x})\hat{f} = (Qf)\hat{} \qquad (f \in L^p(G)).$$

Hence $P = Q$. Clearly $Q \in \{R_x\}''$ and so the proof is complete.

THEOREM 20.25. *Let $1 < p < \infty$ and let $x \in G$. Then there exists an operator A_x on $L^p(G)$ such that $\exp(i A_x) = R_x$, $\sigma(A_x) \subseteq [0, 2\pi]$ and A_x lies in the bicommutant of R_x.*

Proof. For $1 < p \leqslant 2$ this result follows from Theorems 20.19, 20.22 and 20.24. Suppose now that $2 < p < \infty$. Since the conjugate index q satisfies $1 < q < 2$, we can apply the result above to obtain an operator B on $L^q(G)$ satisfying $\exp iB = R_{-x}$, $\sigma(B) \subseteq [0, 2\pi]$ and B lies in the second commutant of R_{-x}. Putting $A_x = B^*$ we obtain an operator on $L^p(G)$ with the required properties.

Next, we show that the operator A_x constructed is well-bounded.

THEOREM 20.26. *Let $1 < p < \infty$ and let $x \in G$. Then there is a family $\{E(\lambda): \lambda \in \mathbf{R}\}$ of projections in $L(L^p(G))$ satisfying*

(i) $\|E(\lambda)\| \leqslant K < \infty \quad (\lambda \in \mathbf{R})$,

(ii) $E(\lambda)E(\mu) = E(\mu)E(\lambda) = E(\lambda) \quad (\lambda, \mu \in \mathbf{R}, \lambda \leqslant \mu)$,

(iii) *$E(\lambda) \to E(\mu)$ strongly as $\lambda \to \mu+ \quad (\mu \in \mathbf{R})$,*

(iv) *$E(\lambda)$ converges in the strong operator topology of $L(L^p(G))$ as $\lambda \to \mu-$ for each real μ,*

(v) *$E(\lambda) = 0$ ($\lambda < 0$) and $E(\lambda) = I$ ($\lambda \geqslant 2\pi$),*

(vi) $A_x = \int_0^{2\pi} \lambda \, dE(\lambda)$ *and* $R_x = \int_{0-}^{2\pi} e^{i\lambda} dE(\lambda)$.

In particular, A_x is a well-bounded operator.

Proof. Suppose first that $1 < p \leqslant 2$. For each λ in $[0, 2\pi]$, the function

$$\psi_\lambda(\gamma) = \chi_\lambda((x, \gamma)) \qquad (\gamma \in \Gamma)$$

is a p-multiplier on Γ by Lemma 20.15. Since $\psi_\lambda^2 = \psi_\lambda$, the corresponding operators T_{ψ_λ} are projections. For each real λ define $E(\lambda)$ by

$$E(\lambda) = \begin{cases} I & \text{if} \quad \lambda > 2\pi \\ T_{\psi_\lambda} & \text{if} \quad 0 \leqslant \lambda \leqslant 2\pi \\ 0 & \text{if} \quad \lambda < 0. \end{cases}$$

Then $\{E(\lambda): \lambda \in \mathbf{R}\}$ is a family of projections in $L(L^p(G))$ which, by the norm inequality of Lemma 20.15, satisfy (i) with $K = K_p$. It is easily verified that $E(\cdot)$ also satisfies (ii) and (v).

We now show that $E(\cdot)$ satisfies (iii). This is clear when $\mu < 0$ or $\mu \geqslant 2\pi$, and so we assume that $0 \leqslant \mu < 2\pi$. Fix f in $L^p(G)$. Since $L^p(G)$ is reflexive its closed unit ball is weakly compact and so, by Theorem 6.4, properties (i) and (ii) imply that

$$E(\lambda)f \to g \quad \text{as} \quad \lambda \to \mu+$$

for some g in $L^p(G)$, the convergence being in the L^p norm. Taking Fourier transforms we obtain

$$\psi_\lambda \hat{f} \to \hat{g} \quad \text{as} \quad \lambda \to \mu+$$

in $L^q(\Gamma)$ (where q is the index conjugate to p), and hence

$$\psi_{\lambda_n}(\gamma)\hat{f}(\gamma) \to \hat{g}(\gamma) \quad \text{as} \quad n \to \infty \qquad (\gamma \in \Gamma \text{ a.e}) \tag{9}$$

for some sequence $\{\lambda_n\}$ in $(\mu, 2\pi)$ converging to μ. Now $\chi_{\lambda_n} \to \chi_\mu$ pointwise on $\mathbf{T}$, so that $\psi_{\lambda_n} \to \psi_\mu$ pointwise on Γ. Thus (9) shows that

$$\psi_\mu(\gamma)\hat{f}(\gamma) = \hat{g}(\gamma) \qquad (\gamma \in \Gamma \text{ a.e}).$$

Hence $g = E(\mu)f$, and therefore $E(\lambda)f \to E(\mu)f$ in $L^p(G)$ as $\lambda \to \mu+$. Thus $E(\cdot)$ satisfies (iii). Property (iv) also follows from Theorem 6.4.

Let A be the well-bounded operator $\int_{-\varepsilon}^{2\pi} \lambda \, dE(\lambda)$, where $\varepsilon > 0$. We show that $A = A_x$. To this end, fix f in $L^p(G)$. By Theorem 16.11,

$$Af = 2\pi f - \int_0^{2\pi} E(\lambda) f \, d\lambda,$$

where the integral on the right exists as an $L^p(G)$-valued Riemann-Stieltjes integral. Approximating this integral by Riemann sums we see that

$$f_n = 2\pi n^{-1} \sum_{r=1}^{n} E\left(\frac{2\pi r}{n}\right) f \to 2\pi f - Af$$

in the L^p norm as $n \to \infty$. Hence there is a subsequence $\{n_k\}$ such that, for almost all γ in Γ,

$$\begin{aligned} \hat{f}_{n_k}(\gamma) &= 2\pi n_k^{-1} \sum_{r=1}^{n_k} \psi_{2\pi r/n_k}(\gamma) \hat{f}(\gamma) \\ &\to 2\pi \hat{f}(\gamma) - (Af)\hat{}(\gamma) \end{aligned} \tag{10}$$

as $k \to \infty$. For each γ in Γ,

$$2\pi n_k^{-1} \sum_{r=1}^{n_k} \psi_{2\pi r/n_k}(\gamma) = 2\pi n_k^{-1} \sum_{r=1}^{n_k} \chi_{2\pi r/n_k}((x, \gamma)). \tag{11}$$

Observe that the right-hand side of (11) is an approximating Riemann sum for the integral

$$\int_0^{2\pi} \chi_\lambda((x, \gamma)) d\lambda,$$

which has the value $2\pi - u((x, \gamma))$, where

$$u(e^{it}) = t \qquad (0 \leqslant t < 2\pi).$$

Hence

$$2\pi n_k^{-1} \sum_{r=1}^{n_k} \psi_{2\pi r/n_k}(\gamma) \to 2\pi - u((x, \gamma))$$

as $k \to \infty$ for each γ in Γ. Combining this with (10), we see that

$$(Af)\hat{}(\gamma) = u((x, \gamma))\hat{f}(\gamma) \qquad (\gamma \in \Gamma \text{ a.e}).$$

Therefore $A = T_\psi$, where ψ is the p-multiplier on Γ defined by $\psi(\gamma) = u((x, \gamma))$. However $A_x = T_\psi$ from the proof of Theorem 20.19 and so $A = A_x$.

Suppose now that $2 < p < \infty$. The adjoint of R_x on $L^p(G)$ is R_{-x} on $L^q(G)$ and, from the proof of Theorem 20.19, the adjoint of A_x on $L^p(G)$ is A_{-x} on $L^q(G)$. Since $1 < q < 2$, we can apply the first part of the present proof to obtain a family $\{E_0(\lambda): \lambda \in \mathbf{R}\}$ in $L(L^q(G))$ satisfying (i)–(v) with E_0 in place of E and such that

$$A_{-x} = \int_{-\varepsilon}^{2\pi} \lambda \, dE_0(\lambda).$$

Putting $E(\lambda) = E_0(\lambda)^*$ for each real λ, it is easily verified that the family $\{E(\lambda): \lambda \in \mathbf{R}\}$ has the required properties. (To see (iii) and (iv) use Theorem 6.4.)

By Theorem 16.13,

$$\exp(iA_x) = \int_{-\varepsilon}^{2\pi} e^{i\lambda} dE(\lambda),$$

where the left-hand side of the equation denotes the usual Banach algebra infinite sum. Now, for every $\varepsilon > 0$

$$A_x = \int_{-\varepsilon}^{2\pi} \lambda \, dE(\lambda)$$

and we can therefore write

$$A_x = \int_{0-}^{2\pi} \lambda \, dE(\lambda). \tag{12}$$

In fact,

$$A_x = \int_{0}^{2\pi} \lambda \, dE(\lambda),$$

since the integrand in (12) vanishes at 0. Similarly,

$$R_x = \int_{0-}^{2\pi} e^{i\lambda} dE(\lambda)$$

but $0-$ cannot in general be replaced by 0 here as can be seen by considering the case when $R_x = I$. The proof is complete.

Note 20.27. The bilateral shift operator on $L^1(\mathbf{Z})$ cannot have an integral representation of the form

$$\int_{a-}^{b} e^{i\lambda} dE(\lambda)$$

with $E(\cdot)$ satisfying conditions (i)–(iv) of Theorem 20.26. For if such a representation did exist, then the operator $i \int_{a-}^{b} \lambda dE(\lambda)$ would be a logarithm for this operator, contradicting Example 20.2.

THEOREM 20.28. *Let* $1 < p < \infty$. *If* T *is an operator on* $L^p(G)$ *such that* $\exp iT = R_x$, *then* T *is well-bounded.*

Proof. Observe that, by Theorem 20.25, $\exp iT = \exp iA_x = R_x$ and, moreover, since T commutes with R_x we have $TA_x = A_xT$. It follows from Theorem 1.23 that

$$\exp i(T - A_x) = I.$$

By Proposition 10.6 and the spectral mapping theorem, $T - A_x$ is a scalar-type spectral operator of the form $\sum_{r=1}^{n} \lambda_r F(\lambda_r)$, where each λ_r is real and $F(\lambda_r)$ is the spectral projection corresponding to the open-and-closed subset $\{\lambda_r\}$ of $\sigma(T - A_x)$. Also T, A_x and $T - A_x$ commute. Hence these operators commute with each projection $F(\lambda_r)$. Since A_x is well-bounded so also is the restriction of A_x to the range of $F(\lambda_r)$. Since T differs from A_x by a real multiple of the identity on this space, the restriction of T to the range of $F(\lambda_r)$ is well-bounded. Clearly, by Theorems 17.15 and 17.17, the direct sum of a finite number of well-bounded operators on reflexive spaces is well-bounded. The proof is complete.

Next, we consider conditions under which the logarithms of the translation operator R_x on $L^p(G)$ $(1 < p < \infty)$ are spectral operators. A preliminary result is required.

PROPOSITION 20.29. *Let $1 < p < \infty$. If T is an operator on $L^p(G)$ such that $\exp iT = R_x$, then T is a scalar-type spectral operator if and only if A_x is a scalar-type spectral operator.*

Proof. Observe that, by Theorem 20.25, $\exp iT = \exp iA_x = R_x$ and, since T commutes with R_x, we have $TA_x = A_xT$. It follows from Theorem 1.23 that $\exp i(T - A_x) = I$. By Proposition 10.6, $T - A_x$ is a scalar-type spectral operator of the form $\sum_{r=1}^{n} \lambda_r F(\lambda_r)$, where $F(\lambda_r)$ is the spectral projection corresponding to the open-and-closed subset $\{\lambda_r\}$ of $\sigma(T - A_x)$. Since T, A_x and $T - A_x$ commute, the required result follows from Theorem 10.5.

In other words, if one logarithm of R_x is a scalar-type spectral operator, then all logarithms of R_x are scalar-type spectral operators. In the next two theorems we give necessary and sufficient conditions for R_x and its logarithms to be spectral operators.

THEOREM 20.30. *Let $1 \leqslant p \leqslant \infty$, let $x \in G$ and let R_x act on $L^p(G)$. Then*

(i) *R_x is a scalar-type spectral operator if either $p = 2$ or x has finite order in G;*

(ii) *R_x is not a spectral operator if $p \neq 2$ and x has infinite order in G.*

Proof. (i) If $p = 2$, then R_x is a unitary operator on the Hilbert space $L^2(G)$ and so is a scalar-type spectral operator by the spectral theorem. If x has finite order n in G, then $(R_x)^n = I$ and so R_x is a scalar-type spectral operator by Theorem 10.10 (ii).

(ii) Suppose that $p \neq 2$ and that x has infinite order in G. Since an invertible isometry on a Banach space is a spectral operator only if it is of scalar type by Theorem 10.17, it suffices to show that R_x is not of scalar type. This will be done by showing that R_x does not satisfy the following necessary condition for R_x to be a scalar-type spectral operator; for every polynomial f, there is a real constant M such that

$$\| f(R_x) \| \leqslant M \sup\{| f(z)| : z \in \sigma(R_x)\}. \tag{13}$$

Suppose first that $1 \leqslant p < 2$. For n a positive integer, let f_n be the nth

Rudin-Shapiro polynomial. Then there is a sequence $\{\varepsilon_n\}$ of numbers, each equal to ± 1, such that

$$f_n(z) = \sum_{r=1}^{n} \varepsilon_r z^r.$$

Also,

$$\sup\{|f_n(z)|: z \in \mathbf{T}\} \leqslant 5n^{1/2} \qquad (n \in \mathbf{N}). \tag{14}$$

The existence of such a sequence of polynomials was proved in Rudin [**1**].

For n a positive integer, let U_n be a compact neighbourhood of 0 in G for which the sets

$$-x + U_n, -2x + U_n, \ldots, -nx + U_n \tag{15}$$

are disjoint, such U_n's existing because x has infinite order in G. Let ξ_n be the characteristic function of U_n. Then for $r = 1, 2, \ldots, n$, $(R_x)^n \xi_n$ is the characteristic function of $-rx + U_n$. Using the disjointness of the sets (15), it follows that

$$\int_G |f_n(R_x)\xi_n|^p = \int_{V_n} 1 = nc_n,$$

where

$$V_n = \bigcup_{r=1}^{n} (-rx + U_n)$$

and c_n is the measure of U_n. Therefore

$$\| f_n(R_x) \| \geqslant \| f_n(R_x)(c_n^{-1/p}\xi_n) \|_p = n^{1/p}. \tag{16}$$

Since $\sigma(R_x) = \mathbf{T}$ by Theorem 20.18, (14) shows that

$$\sup\{|f_n(z)|: z \in \sigma(R_x)\} \leqslant 5n^{1/2}. \tag{17}$$

Since $1 \leqslant p < 2$, the inequalities (16) and (17) show that there does not exist a real constant M for which (13) holds with $f = f_n$ $(n = 1, 2, 3, \ldots)$. Hence R_x is not a scalar-type spectral operator when $1 \leqslant p < 2$.

A simple duality argument, based on the inequalities (16) and (17), shows that R_x does not satisfy the necessary condition (13) for it to be a scalar-type spectral operator when $2 < p \leqslant \infty$. This completes the proof of (ii).

The corresponding result for A_x follows easily.

THEOREM 20.31. *Let $1 < p < \infty$, let $x \in G$ and let A_x act on $L^p(G)$. Then*

(i) *A_x is a scalar-type spectral operator if either $p = 2$ or x has finite order in G;*

(ii) *A_x is not a spectral operator if $p \neq 2$ and x has infinite order in G.*

Proof. This is immediate from Theorem 20.30 and the fact that an operator A is spectral (resp. scalar-type spectral) if and only if $\exp iA$ is spectral (resp. scalar-type spectral). See Theorems 10.9, 5.16 and Proposition 10.8.

COROLLARY 20.32. *Let $1 < p < \infty$, let $x \in G$ and let R_x act on $L^p(G)$. Let T be an operator on $L^p(G)$ such that $\exp iT = R_x$. Then*

(i) *T is a scalar-type spectral operator if either $p = 2$ or x has finite order in G;*

(ii) *T is not a spectral operator if $p \neq 2$ and x has infinite order in G.*

Proof. This follows from Theorem 20.31 and Proposition 20.29.

It is well-known that scalar-type spectral operators with real spectra are hermitian-equivalent and well-bounded. It is also known that hermitian-equivalent operators need not be spectral. Indeed in Chapter 9 we gave an example of a hermitian operator on a reflexive Banach space which is not spectral. However, the question arises whether every hermitian-equivalent well-bounded operator is spectral. Theorem 20.31 shows that this is not the case, even when the underlying Banach space is reflexive.

To see this, note that, since

$$\|\exp(in\, A_x)\| = \|(R_x)^n\| = 1 \qquad (n \in \mathbf{Z}),$$

it follows that

$$\sup\{\|\exp(it\, A_x)\| : t \in \mathbf{R}\} = \sup\{\|\exp(it\, A_x)\| : t \in [0, 1]\}.$$

Hence

$$\sup\{\|\exp(it\, A_x)\| : t \in \mathbf{R}\} < \infty.$$

Therefore A_x is hermitian equivalent and also well-bounded by Theorem 20.26, but is not a spectral operator if $p \neq 2$ and x has infinite order in G.

Example 20.33. Take $G = \mathbf{Z}$ and $x = -1$. Theorems 20.30 and 20.31 show that, on $L^p(\mathbf{Z})$, neither the bilateral shift R_{-1} (for $1 \leqslant p \leqslant \infty$) nor $A_{-1} = \pi I + iH$ (for $1 < p < \infty$) is a spectral operator if $p \neq 2$. Also $\pi I + iH$ is a hermitian-equivalent well-bounded operator on $L^p(\mathbf{Z})$ which is not spectral if $1 < p < 2$ or $2 < p < \infty$.

Notes and Comments on Part Five

By virtue of the countable additivity of the resolution of the identity, spectral operators have important expansion theorems associated with them. For example, if T is a spectral operator and its spectrum is countable, then every x in X has an unconditionally convergent expansion of the type $x = \sum_n x_n = \sum_{\lambda \in \sigma(T)} E(\lambda)x$, where the spectrum of x_n consists of just one point λ_n, so that x_n is a kind of "generalized eigenvector" associated with λ_n. If T is a scalar-type spectral operator, then the generalized eigenvectors are simply eigenvectors in the ordinary sense. If T is a spectral operator of type m, then the generalized eigenvectors x_n satisfy the equations $(\lambda_n I - T)^{m+1} x_n = 0$ $(n = 1, 2, \ldots)$. Thus, as long as T has a countably additive resolution of the identity, we are not far from the simple situation characteristic of compact normal operators on a separable Hilbert space. If the countable additivity of the spectral resolution fails, so do many other convenient eigenvalue expansion properties.

It is by no means the case that all the familiar eigenvalue expansions of classical analysis are unconditionally convergent. Indeed, there are many examples, such as Fourier series expansions in $L^p[0, 2\pi]$ with $1 < p < \infty$, $p \neq 2$, where the expansion converges, but only conditionally. In the spaces L^p with $1 < p < \infty$, $p \neq 2$, the most important operators, that is those integral and differential operators which on L^2 would be self-adjoint, have eigenfunction expansions which converge (Zygmund [**1**; §7.3, 12.42], Hille and Tamarkin [**2**], Rutovitz [**1**], Smart [**1**]) but only conditionally. See Zygmund [**1**; §9.5].

Let X be a reflexive (or even weakly complete) Banach space and let T be an operator on X such that $\sigma(T) \subseteq [a, b]$. We have seen in Chapter 6 that T is a scalar-type spectral operator if and only if there is a positive real constant K such that

$$\| p(T) \| \leqslant K \sup_{t \in [a,b]} |p(t)|$$

for every polynomial p. Analogously, in 1960 Smart [2] defined an operator T to be *well-bounded* if there exists a positive real constant K such that

$$\| p(T) \| \leqslant K \, |||p|||$$

for every polynomial p, where $||| \quad |||$ denotes the norm on the space $BV[a, b]$.

Smart [2] proved that if X is reflexive, then for any real number t there exists a unique projection $E(t)$ in $L(X)$ such that:

(i) $E(t)$ commutes with any operator commuting with T;
(ii) $\| E(t) \| \leqslant 2K \qquad (t \in \mathbf{R})$;
(iii) $E(t) = 0$ for $t < a$ and $E(t) = I$ for $t \geqslant b$;
(iv) $E(s) = E(s)E(t) = E(t)E(s)$ for $s \leqslant t$;
(v) $\lim_{t \to s+} E(t)x = E(s)x \qquad (x \in X, s \in \mathbf{R})$;
(vi) $\sigma(T | E(t)X) \subseteq (-\infty, t] \cap \sigma(T)$ and $\sigma(T | (I - E(t))X) \subseteq [t, \infty) \cap \sigma(T)$, for each real t.

Smart also proved that the operator

$$T - \int_{a+}^{b} t\,dE(t)$$

was quasinilpotent. Ringrose [1] showed that the operator was 0.

The approach used by Smart and Ringrose was based in part on the fact that a well-bounded operator admits an operational calculus for absolutely continuous functions, but was basically "constructive" in character. Subsequently, in 1966, Sills [1] presented a different method of attack, which we now describe. Let $AC_0(J)$ be the Banach algebra of absolutely continuous functions on $J = [a, b]$ vanishing at a. The method consists of introducing Arens multiplication into $[AC_0(J)]^{**}$ and investigating the larger Banach algebra (which is neither commutative nor semi-simple) for a suitable family of idempotents to serve as candidates for spectral projections associated with T. Idempotents in $[AC_0(J)]^{**}$ are mapped into these projections by means of a homomorphism extension technique which extends the original operational calculus of $AC_0(J)$ into $L(X)$ to a bounded homomorphism of $[AC_0(J)]^{**}$ into $L(X)$. The extended homomorphism is defined on a quotient algebra of $[AC_0(J)]^{**}$. This quotient algebra turns out to be a copy of all functions of bounded variation on J which vanish at a. The operational calculus for T extends to functions of bounded variation on J in the obvious way.

Earlier, in 1963, Ringrose [3] discussed well-bounded operators on a general Banach space X. It turns out that the well-boundedness of an operator T on X is equivalent to the existence of a family of projections $\{F(t): t \in \mathbf{R}\}$ in $L(X^*)$ (called a "decomposition of the identity for T") satisfying certain natural properties and such that the equation

$$T^* = I^* - \int_a^b F(t)dt$$

holds in the (lower) weak operator topology. (Note that this equation is obtained from the usual spectral theorem by formally integrating by parts.) In this case, the family $\{F(t): t \in \mathbf{R}\}$ is not necessarily unique; however, if each $F(t)$ is the adjoint of an operator on X, then uniqueness does hold.

In 1970, Berkson and Dowson [2] considered well-bounded operators possessing a family $\{E(t): t \in \mathbf{R}\}$ of projections in $L(X)$ such that $F(t) = E(t)^*$, for every t in $[a, b]$. Using the work of Ringrose and Sills, they considered well-bounded operators of type (B). These have the additional property that (i) E is strongly continuous on the right and (ii) for each real μ, the strong limit $\lim_{\lambda \to \mu -} E(\lambda)$ exists (denoted by $E(\mu -)$). (These conditions are automatically fulfilled if X is reflexive.) In this case, if $f \in AC(J)$ we have

$$f(T) = \int_{a-}^b f(\lambda)dE(\lambda)$$

where the Riemann integral exists in the strong operator topology; moreover, $E(\mu) - E(\mu -)$ is a projection of X onto $\{x: Tx = \mu x\}$. Also, the residual spectrum of T is empty. They further show that an operator T on X with real spectrum has adjoint a scalar-type operator on X^* of class X if and only if T is well-bounded and the function $\lambda \to \langle x, F(\lambda)y \rangle$ is of bounded variation on J for every x in X and y in X^*. Also a well-bounded spectral operator is of scalar type.

The results of Chapter 15 are due to Ringrose [3]. Theorems 16.2, 16.3, 16.4, 16.6 and Example 16.5 are also due to Ringrose [3]. Other results in this chapter are due to Berkson and Dowson [2].

Spain [2], using the modified Stieltjes integrals of Krabbe [2], [3] introduced a new approach to the theory of well-bounded operators of type (B). He gave much simpler proofs of many of the theorems of Berkson and Dowson [2] and obtained some new results as well.

The results of Chapter 19 are due to Berkson and Dowson [**2**]. Boon-Hee Lim [**1**] organised the material in Chapters 15, 16, 17, 19 in the order presented here, simplifying many of the original proofs. The results of Chapter 18 are due to Gillespie [**3**]. His construction is based on Example 14.5 of Singer [**1**; p. 429–31], introduced for another purpose.

The results in the final chapter are based on three papers of Gillespie [**1**], [**2**], [**3**] and two of his unpublished manuscripts. Earlier, Dowson and Spain [**1**] had shown that if H is the Hilbert transform on $L^p(\mathbf{Z})$, then $T = \pi I + iH$ is a well-bounded operator for $1 < p < \infty$, but is not a scalar-type spectral operator except when $p = 2$. Their proof of the second statement is based on the classical result that if $p \neq 2$ there exists a continuous function whose sequence of Fourier coefficients does not belong to l^p. (See Zygmund [**1**; p. 190].) For related earlier work, see Krabbe [**1**], [**2**] and Fixman [**1**].

Bibliography and Author Index

We have divided the bibliography into two sections. The works in "General background reading" are referred to in the text by a number alone. Other references are cited by giving the author's name followed by a number. The page references at the end of the entries indicate the places in the text where the entry is quoted; other references to an author are listed after his name.

General Background Reading

1. F. F. Bonsall and J. Duncan, *Numerical ranges of operators on normed spaces and of elements of normed algebras,* London Math. Soc. Lecture Note Series No. 2 (Cambridge University Press, 1971).
2. F. F. Bonsall and J. Duncan, *Numerical ranges II*, London Math. Soc. Lecture Note Series No. 10 (Cambridge University Press, 1973).
3. F. F. Bonsall and J. Duncan, *Complete normed algebras* (Springer-Verlag, 1973).
4. A. L. Brown and A. Page, *Elements of functional analysis* (Van Nostrand Reinhold, 1970).
5. N. Dunford and J. T. Schwartz, *Linear operators, Part I* (1958), *Part II* (1963), *Part III* (1971) (Wiley-Interscience).
6. P. R. Halmos, *Measure theory* (Van Nostrand, 1950).
7. P. R. Halmos, *Introduction to Hilbert space and the theory of spectral multiplicity* (Chelsea, New York, 1951).
8. P. R. Halmos, *Finite-dimensional vector spaces* (Van Nostrand-Reinhold, 1958).
9. P. R. Halmos, *A Hilbert space problem book* (Van Nostrand, 1967).
10. K. Hoffman, *Banach spaces of analytic functions* (Prentice-Hall, 1962).
11. M. H. A. Newman, *Topology of plane sets,* 2nd Edition (Cambridge University Press, 1961).
12. C. E. Rickart, *General theory of Banach algebras* (Van Nostrand, 1960).
13. J. R. Ringrose, *Lecture notes on von Neumann algebras* (University of Newcastle upon Tyne, 1966–67).
14. W. Rudin, *Real and complex analysis* (McGraw-Hill, 1966).
15. A. E. Taylor, *Introduction to functional analysis* (Wiley, 1958).

Other References

N. I. Ahiezer and I. M. Glazman

1. *Theory of linear operators on Hilbert space* (Ungar, New York, 1961). [277]

FREDA E. ALEXANDER
1. Compact and finite rank operators on subspaces of l^p, *Bull. London Math. Soc.* **6** (1974) 341-342. [49]

J. C. ALEXANDER
1. Compact Banach algebras, *Proc. London Math. Soc.* (3) **18** (1968) 1-18. [97]
2. On Riesz operators, *Proc. Edinburgh Math. Soc.* (2) **16** (1969) 227-232. [97]

M. Š. ALTMAN
1. The Fredholm theory of linear equations in locally convex topological spaces, *Bull. Acad. Polon. Sci.* Cl. III **2** (1954) 267–269. [96]

N. ARONSZAJN
1. Approximation methods for eigenvalues of completely continouus symmetric operators, *Proceedings of the Symposium on Spectral Theory and Differential Problems,* Oklahoma Agricultural and Mechanical College, Stillwater, Oklahoma (1951) 179-202. [96]
2. The Rayleigh-Ritz and A. Weinstein methods for approximation of eigenvalues I,II *Proc. Nat. Acad. Sci. U.S.A.* **34** (1948) 474-480, 594-601. [96]

N. ARONSZAJN AND K. T. SMITH [56, 57, 60, 96]
1. Invariant subspaces of completely continuous operators, *Ann. of Math.* (2) **60** (1954) 345-350. [56, 96]

W. G. BADE
1. Weak and strong limits of spectral operators, *Pacific J. Math.* **4** (1954) 393-413. [276, 279]
2. On Boolean algebras of projections and algebras of operators, *Trans. Amer. Math. Soc.* **80** (1955) 345-360. [279]
3. A multiplicity theory for Boolean algebras of projections in Banach spaces, *Trans. Amer. Math. Soc.* **92** (1959) 508-530. [279]

W. G. BADE AND P. C. CURTIS JR. [139]
1. The Wedderburn decomposition of commutative Banach algebras, *Amer. J. Math.* **82** (1960) 851-866. [140, 276]

S. BANACH
1. *Théorie des opérations linéaires,* Monografje Matematyczne (Warsaw, 1932). [95]

J. Y. BARRY
1. On the convergence of ordered sets of projections, *Proc. Amer. Math. Soc.* **5** (1954) 313–314. [159, 276]

F. L. BAUER
1. On the field of values subordinate to a norm. *Numer. Math.* **4** (1962) 103–111. [113]

E. BERKSON [276]

1. A characterization of scalar type operators on reflexive Banach spaces, *Pacific J. Math.* **13** (1963) 365-373. [275, 276]
2. Some characterizations of C*-algebras, *Illinois J. Math.* **10** (1966) 1-8. [114]
3. Semi-groups of scalar type operators and a theorem of Stone, *Illinois J. Math.* **10** (1966) 345-352. [276, 277]

E. BERKSON AND H. R. DOWSON

1. Prespectral operators, *Illinois J. Math.* **13** (1969) 291-315. [144, 275, 276]
2. On uniquely decomposable well-bounded operators, *Proc. London Math. Soc.* (3) **22** (1971) 339-358. [401, 402]
3. On reflexive scalar-type spectral operators, *J. London Math. Soc.* (2) **8** (1974) 652-656. [279]

E. BERKSON, H. R. DOWSON AND G. A. ELLIOTT

1. On Fuglede's theorem and scalar-type operators, *Bull. London Math. Soc.* **4** (1972) 13-16. [114, 275]

A. BEURLING

1. Sur les intégrales de Fourier absolument convergentes et leur application à une transformation fonctionnelle, *Proc. IX Congrès de Math. Scandinaves,* Helsingfors (1938) 345-366. [40]

B. BOLLABÁS

1. A property of hermitian elements, *J. London Math. Soc.* (2) **4** (1971) 379-380. [214]
2. The spectral decomposition of compact hermitian operators on Banach spaces, *Bull. London Math. Soc.* **5** (1973) 29-36. [114]

F. F. BONSALL

1. Compact linear operators from an algebraic standpoint, *Glasgow Math. J.* **8** (1967) 41-49. [95, 97]
2. The numerical range of an element of a normed algebra, *Glasgow Math. J.* **10** (1969) 68-72. [113]

F. F. BONSALL AND M. J. CRABB

1. The spectral radius of a hermitian element of a Banach algebra, *Bull. London Math. Soc.* **2** (1970) 178–180. [114]

J. BRAM

1. Subnormal operators, *Duke Math. J.* **22** (1955) 75–94. [41]

A. BUCHHEIM

1. An extension of a theorem of Professor Sylvester relating to matrices, *Phil. Mag.* **22** (1886) 173–174. [39]

J. W. CALKIN

1. Two sided ideals and congruences in the ring of bounded operators in Hilbert space, *Ann. of Math.* (2) **42** (1941) 839–873. [96]

S. R. CARADUS, W. E. PFAFFENBERGER AND B. YOOD

1. *Calkin algebras and algebras of operators on Banach spaces* (Marcel Dekker Inc., 1974). [97]

S. H. CHANG

1. On the distribution of the characteristic values and singular values of linear integral equations, *Trans. Amer. Math. Soc.* **67** (1949) 351-367. [96]

L. COLLATZ

1. *Eigenwertprobleme und ihre numerische Behandlung* (Akademischer Verlag, Leipzig, 1945). [96]

A. M. DAVIE

1. The approximation problem for Banach spaces, *Bull. London Math. Soc.* **5** (1973) 261-266. [49, 356]

M. M. DAY

1. Means for the bounded functions and ergodicity of the bounded representation of semi-groups, *Trans. Amer. Math. Soc.* **69** (1950) 276-291. [277]

D. W. DEAN

1. Schauder decompositions in (m), *Proc. Amer. Math. Soc.* **18** (1967) 619-623. [28]

K. DE LEEUW

1. On $\mathbf{L}^p$-multipliers, *Ann. of Math.* **81** (1965) 364-379. [375, 380]

J. DIEUDONNÉ

1. Sur la bicommutante d'une algèbre d'opérateurs, *Portugal. Math.* **14** (1955) 33-38. [279]
2. Champs de vecteurs non localement triviaux, *Arch. Math. (Basel)* **7** (1956) 6-10. [279]
3. *Foundations of modern analysis* (Academic Press, 1960). [79, 96]

J. DIXMIER

1. Les moyennes invariantes dans les semi-groupes et leurs applications, *Acta Sci. Math. (Szeged)* **12** Pars A (1950) 213-227. [277]

W. F. DONOGHUE

1. The lattice of invariant subspaces of a completely continuous quasinilpotent transformation, *Pacific J. Math.* **7** (1957) 1031-1035. [57]

H. R. DOWSON [114, 144, 275, 276]

1. Restrictions of spectral operators, *Proc. London Math. Soc.* (3) **15** (1965) 437–457. [227, 278]
2. On some algebras of operators generated by a scalar-type spectral operator, *J. London Math. Soc.* **40** (1965) 589–593. [279]
3. Operators induced on quotient spaces by spectral operators, *J. London Math. Soc.* **42** (1967) 666–671. [41, 276, 278]
4. On the commutant of a complete Boolean algebra of projections, *Proc. Amer. Math Soc.* **19** (1968) 1448–1452. [279]

5. On a Boolean algebra of projections constructed by Dieudonné, *Proc. Edinburgh Math. Soc.* (2) **16** (1969) 258–262. [279]
6. Restrictions of prespectral operators, *J. London Math. Soc.* (2) **1** (1969) 633–642. [276, 278]
7. On an unstarred operator algebra, *J. London Math. Soc.* (2) **5** (1972) 489–492. [278]
8. On the algebra generated by a hermitian operator, *Proc. Edinburgh Math. Soc.* (2) **18** (1972) 89–91. [114]
9. A commutativity theorem for prespectral operators, *Illinois J. Math.* **17** (1973) 525–532. [275, 276]
10. Logarithms of prespectral operators, *J. London Math. Soc.* (2) **9** (1974) 57–64. [277]
11. Some properties of prespectral operators, *Proc. Roy. Irish Acad. Sect A* **74** (1974) (Spectral Theory Symposium) 207–221. [278]
12. m^{th} roots of prespectral operators, *J. London Math. Soc.* (2) **12** (1975) 49–52. [277]
13. On power-bounded prespectral operators, *Proc. Edinburgh Math. Soc.* (2) **20** (1976) 173–175. [277]

H. R. Dowson, T. A. Gillespie and P. G. Spain
1. A commutativity theorem for hermitian operators, *Math. Ann.* **220** (1976) 215-217. [114, 276]

H. R. Dowson and G. L. R. Moeti
1. Property (P) for normal operators, *Proc. Roy. Irish Acad. Sect. A* **73** (1973) 159-167. [278]

H. R. Dowson and P. G. Spain
1. An example in the theory of well-bounded operators, *Proc. Amer. Math. Soc.* **32** (1972) 205-208. [402]

N. Dunford
1. Spectral theory I, Convergence to projections, *Trans. Amer. Math. Soc.* **54** (1943) 185-217. [41, 95, 275]
2. Spectral theory II, Resolutions of the identity, *Pacific J. Math.* **2** (1952) 559–614. [275]
3. Spectral operators, *Pacific J. Math.* **4** (1954) 321–354. [275, 276, 380]
4. A survey of the theory of spectral operators, *Bull. Amer. Math. Soc.* **64** (1958) 217–274. [275, 276]

J. Dyer and P. Porcelli
1. Concerning the invariant subspace problem, *Notices Amer. Math. Soc.* **17** (1970) 788. [278]

P. Enflo
1. A counter-example to the approximation problem in Banach spaces, *Acta Math.* **130** (1973) 309-317. [49, 356]

K. Fan
1. Maximum properties and inequalities for the eigenvalues of completely continuous operators, *Proc. Nat. Acad. Sci. U.S.A.* **37** (1951) 760–766. [96]
2. On a theorem of Weyl concerning eigenvalues of linear transformation I, II, *Proc. Nat. Acad. Sci. U.S.A.* **35** (1949) 652–655 *ibid* **36** (1950) 31–35. [96]

L. FANTAPPIÈ
1. Le calcul des matrices, *C. R. Acad. Sci. Paris* **186** (1928) 619-621. [39]

U. FIXMAN [117, 233, 265]
1. Problems in spectral operators, *Pacific J. Math.* **9** (1959) 1029-1051. [144, 214, 227, 275, 277, 278, 402]

S. R. FOGUEL [215]
1. Sums and products of commuting spectral operators, *Ark. Mat.* **3** (1958) 449-461. [277, 280]
2. The relations between a spectral operator and its scalar part, *Pacific J. Math.* **8** (1958) 51-65. [214, 276, 277, 278, 280]
3. Boolean algebras of projections of finite multiplicity, *Pacific J. Math.* **9** (1959) 681-693. [279]

I. FREDHOLM [95, 96]
1. Sur une classe d'équations fonctionnelles, *Acta Math.* **27** (1903) 365-390. [95]

G. FROBENIUS
1. Über lineare Substitutionen und bilineare Formen, *J. Reine Angew. Math.* **84** (1878) 1-63. [39]
2. Über die schiefe Invariante einer bilinearen oder quadratischen Formen, *J. Reine Angew. Math.* **86** (1879) 44-71. [39]
3. Über die cogredienten Transformation der bilinearen Formen, *Sitzungsberichte der K. Preuss. Akad. der Wiss zu Berlin* (1896) 7-16. [39]

B. FUGLEDE [175, 184, 191, 253, 277]
1. A commutativity theorem for normal operators, *Proc. Nat. Acad. Sci. U.S.A.* **36** (1950) 35-40. [277]

D. J. H. GARLING
1. On ideals of operators in Hilbert space, *Proc. London. Math Soc.* (3) **17** (1967) 115-138. [96]

I. M. GELFAND
1. Normierte Ringe. *Mat. Sbornik N.S.* **9** (51) (1941) 3-24. [40, 41]

T. A. GILLESPIE [114, 276, 286, 356, 373]
1. Logarithms of L^p translations, *Indiana Univ. Math. J.* **24** (1975) 1037-1045. [402]
2. A spectral theorem for L^p translations, *J. London Math. Soc.* (2) **11** (1975) 499-508. [402]
3. Commuting well-bounded operators on Hilbert spaces, *Proc. Edinburgh Math. Soc.* (2) **20** (1976) 167-172. [402]
4. Cyclic Banach spaces and reflexive operator algebras (preprint). [279]

T. A. GILLESPIE AND T. T. WEST
1. A characterization and two examples of Riesz operators, *Glasgow Math. J.* **9**

(1968) 106-110. [96]
2. Weakly compact groups of operators, *Proc. Amer. Math. Soc.* **49** (1975) 78-82. [374]

G. GIORGI
1. Nuovo osservazioni sulle funzioni delle matrici, *Atti. Accad. Naz. Lincei. Rend. Cl. Sci. Fis. Mat. Natur.* (6) 8 (1928) 3-8. [40]

B. W. GLICKFELD [276]
1. A metric characterization of $C(X)$ and its generalizations to C^*-algebras, *Illinois J. Math.* **10** (1966) 547-566. [114]

I. C. GOHBERG AND M. G. KREIN
1. *Introduction to the theory of linear nonselfadjoint operators,* Transl. Math. Monographs, Vol. 18 (Amer. Math. Soc., Providence, R. I., 1969). [96]

I. C. GOHBERG, A. S. MARKUS AND I. A. FEL'DMAN
1. Normally solvable operators and ideals associated with them, *Bul. Akad. Stiince Rss. Moldoven* **10 (76)** (1960) 51-69 (Russian). *Amer. Math. Soc. Transl.* (2) **61** (1967) 63-84. [97]

R. L. GRAVES
1. *The Fredholm theory in Banach spaces,* Dissertation (Harvard University 1951). [96]

A. GROTHENDIECK
1. *Produits tensoriels topologiques et espaces nucléaires,* Memoirs Amer. Math. Soc. No. 16 (1955). [96]

P. R. HALMOS
1. Normal dilations and extensions of operators, *Summa Brasil. Math.* **2** (1950) 125-134. [230, 278]
2. Spectra and spectral manifolds, *Ann. Soc. Polon. Math.* **25** (1952) **43**–49.[233, 278]

F. HAUSDORFF
1. Der Wertvorrat einer Bilinearform, *Math. Z.* **3** (1919) 314-316. [113]

E. HELLINGER AND O. TOEPLITZ
1. Integralgleichungen und Gleichungen mit unendlichvielen Unbekannten, *Encyclopädie der Mathematischen Wissenschaften* II C.13 (1928) 1335-1616. [95]

R. H. HERMAN
1. On the uniqueness of the ideals of compact and strictly singular operators, *Studia Math.* **29** (1968) 161-165. [97]

E. HEWITT
1. A problem concerning finitely additive measures, *Mat. Tidsskr. B* 1951 (1951) 81-94. [327, 341]

E. Hilb

1. Über die Auflösung von Gleichungen mit unendlich vielen Unbekannten, *S.-B. Phys.-Med. Soz. Erlangen* (1908) 84-89. [40]

D. Hilbert

1. *Grundzüge einer allgemeinen Theorie der linearen Integralgleichungen* I-VI (Published in book form by Teubner, Leipzig, 1912). [40, 277]

T. H. Hildebrandt

1. Über vollstetige lineare Transformationen, *Acta Math.* **51** (1928) 311-318. [95]

E. Hille

1. Notes on linear transformations II. Analyticity of semi-groups, *Ann. of Math.* (2) **40** (1939) 1–47. [40]

E. Hille and J. D. Tamarkin

1. On the characteristic values of linear integral equations, *Acta Math.* **57** (1931) 1-76. [96]
2. On the theory of Fourier transforms, *Bull. Amer. Math. Soc.* **39** (1933) 768-774. [399]

A. Horn

1. On the singular values of a product of completely continuous operators, *Proc. Nat. Acad. Sci. U.S.A.* **36** (1950) 374-375. [96]

W. A. Howard [41]

B. E. Johnson

1. Continuity of linear operators commuting with continuous linear operations, *Trans. Amer. Math. Soc.* **128** (1967) 88-102. [276]
2. The uniqueness of the (complete) norm topology, *Bull. Amer. Math. Soc.* **73** (1967) 537-539. [276]

B. E. Johnson and A. M. Sinclair

1. Continuity of linear operators commuting with continuous linear operators II, *Trans. Amer. Math. Soc.* **146** (1969) 533-540. [276]

S. Kakutani [96, 189, 276, 277]

1. An example concerning uniform boundedness of spectral measures, *Pacific J. Math.* **4** (1954) 363-372. [192, 277]

R. H. Kelly

1. On commutative non-self-adjoint operator algebras, *Glasgow Math. J.* **15** (1974) 54-59. [261, 278]

D. C. Kleinecke

1. *A generalization of complete continuity,* Technical Report No. 3 to the Office of Ordinance Research, University of California, Berkeley (1954). [97]
2. Almost-finite, compact and inessential operators, *Proc. Amer. Math. Soc.* **14** (1963) 863-868. [97]

G. L, KRABBE
1. Convolution operators that satisfy the spectral theorem, *Math. Z.* **70** (1958/59) 446-462. [401, 402]
2. Stieltjes integration, spectral analysis and the locally convex algebra (BV) *Bull. Amer. Math. Soc.* **71** (1965) 184-189. [401, 402]
3. A Helly convergence theorem for Stieltjes integrals, *Indigationes Math.* **27** (1965) 52-69. [332]

S. KUREPA
1. Logarithms of spectral type operators, *Glasnik Mat.-Fiz. Astronom. Ser II* **18** (1963) 53-57. [227]

E. N. LAGUERRE
1. Sur le calcul des systèmes linéaires, Reprinted in *Oeuvres,* t.l (1898) 221-267. [39]

R. LARSEN
1. *An introduction to the theory of multipliers* (Springer-Verlag, 1971). [375, 381, 382]

M. A. LAVRENTIEFF
1. Sur les fonctions d'une variable complexe representables par les series de polynomes, *Act. Sci. Ind.* 441 (Paris, 1936). [153, 234]

T. LEZAŃSKI
1. The Fredholm theory of linear equations in Banach spaces, *Studia Math.* **13** (1953) 244-276. [96]
2. Sur les fonctionnelles multiplicatives, *Studia Math.* **14** (1953) 13-23. [96]

BOON-HEE LIM
1. *Well-bounded operators*, M.Sc. Dissertation, University of Glasgow (1974). [402]

J. LINDENSTRAUSS AND A. PELCZYNSKI
1. Absolutely summing operators in L^p-spaces and their applications, *Studia Math.* **29** (1968) 275-326. [280]

V. I. LOMONOSOV AND H. M. HILDEN [45, 54, 96]

L. H. LOOMIS
1. *An introduction to abstract harmonic analysis* (Van Nostrand, 1953). [277]

E. R. LORCH
1. Bicontinuous linear transformations in certain vector spaces, *Bull. Amer. Math. Soc.* **45** (1939) 564–569. [277]
2. The spectrum of linear transformation, *Trans. Amer. Math. Soc.* **52** (1942) 238–248. [40]
3. *Spectral theory* (Oxford University Press, 1962). [387]

G. LUMER [157, 171]
1. Semi-inner product spaces, *Trans. Amer. Math, Soc.* **100** (1961) 29–43. [113, 114]
2. Spectral operators, hermitian operators and bounded groups, *Acta Sci. Math. (Szeged)* **25** (1964) 75–85. [113, 276, 277]

YU. I. LYUBIČ
1. Almost periodic functions in the spectral analysis of operators, *Dokl. Akad. Nauk.* **132** (1960) 518–520. [114]
2. On conditions for complete systems of eigenvectors of correct operators, *Usp. Mat. Nauk.* **18** (1963) 165–171. [114]
3. Conservative operators, *Usp. Mat. Nauk.* **20** (1965) 221–225. [114]

C. A. MCCARTHY [277]
1. Commuting Boolean algebras of projections I, *Pacific J. Math.* **11** (1961) 295–307. [192, 277, 280]
2. Commuting Boolean algebras of projections II, *Proc. Amer. Math. Soc.* **15** (1964) 781–787. [280]

C. A. MCCARTHY AND L. TZAFRIRI
1. Projections in $\mathscr{L}_1$ and $\mathscr{L}_\infty$-spaces, *Pacific J. Math.* **26** (1968) 529–546. [280, 281]

G. W. MACKEY
1. *Commutative Banach algebras*, Mimeographed lecture notes (Harvard University, 1952). [277]

A. D. MICHAL AND R. S. MARTIN
1. Some expansions in vector space, *J. Math. Pures et Appl.* (9) **13** (1934) 69–91. [96]

E. H. MOORE
1. *Introduction to a form of general analysis*, The New Haven Math. Colloquium of the Amer. Math. Soc. (1906). [40]
2. *General analysis I, II*, Mem. Amer. Philos. Soc. (Philadelphia 1935, 1939). [40]

M. NAGUMO
1. Einige analytische Untersuchungen in lineare metrische Ringen, *Jap. J. Math.* **13** (1936) 61–80. [40]

B. NAGY
1. On the spectra of prespectral operators, *Glasgow Math. J.* **19** (1978) 57–61. [278]

C. NEUMANN
1. *Untersuchungen über das logarithmischen und Newtonsche Potential*, (Teuber, Leipzig 1877). [40]

J. VON NEUMANN [251–254, 297]

T. W. PALMER [276]
1. Characterizations of C*-algebras, *Bull. Amer. Math. Soc.* **74** (1968) 538–540. [114]

L. D. PEARLMAN
1. Riesz points of the spectrum of an element in a semisimple Banach algebra, *Trans. Amer. Math. Soc.* **193** (1974) 303–328. [97]

H. POINCARÉ
1. Sur les groups continus, *Cambridge Phil. Trans.* **18** (1899) 220–255. [39]

H. PORTA
1. Ideaux bilatères de transformations linéaires continues, *Compt. Rend. Acad. Sci. Paris* **264** (1967) 95–96. [96]

H. RADJAVI AND P. ROSENTHAL
1. *Invariant subspaces* (Springer-Verlag, 1973). [278]

H. REITER
1. *Classical harmonic analysis and locally compact groups* (Oxford University Press, 1968). [385, 386]

C. E. RICKART
1. The singular elements of a Banach algebra, *Duke Math. J.* **14** (1947) 1063–1077. [41]

F. RIESZ
1. *Les systèmes d'équations linéaires à une infinité d'inconnues* (Gauthier-Villars, Paris, 1913). [40, 277]
2. Über linear Funktionalgleichungen, *Acta Math.* **41** (1918) 71–98. [95]
3. Über quadratische Formen von unendlich vielen Veränderlichen, *Nachr. Akad. Wiss. Göttingen Math. Phys. Kl.* (1910) 190–195. [277]

F. RIESZ AND B. SZ. NAGY
1. *Functional analysis* (Blackie, 1955). [95, 277]

J. R. RINGROSE [45, 400, 401]
1. On well-bounded operators, *J. Austral. Math. Soc.* **1** (1960) 334-343. [400]
2. Super-diagonal forms for compact linear operators, *Proc. London Math. Soc.* (3) **12** (1962) 367-384. [96]
3. On well-bounded operators II, *Proc. London Math. Soc.* (3) **13** (1963) 613-638. [285, 288, 295, 297, 401]
4. *Compact non-self-adjoint operators* (Van Nostrand Reinhold, 1970). [96]

M. ROSENBLUM [175, 184]
1. On a theorem of Fuglede and Putnam, *J. London Math. Soc.* **33** (1958) 376-377. [277]

P. ROSENTHAL [278, 279]

W. RUDIN
1. Some theorems on Fourier coefficients, *Proc. Amer. Math. Soc.* **10** (1959) 855-859. [396]
2. *Fourier analysis on groups* (Interscience, 1962). [382]

A. F. RUSTON [77]

1. On the Fredholm theory of integral equations for operators belonging to the trace class of a general Banach space, *Proc. London Math. Soc.* (2) **53** (1951) 109-124. [96]
2. Direct products of Banach spaces and linear functional equations, *Proc. London Math. Soc.* (3) **1** (1951) 327-384. [96]
3. Formulae of Fredholm type for compact linear operations on a general Banach space. *Proc. London Math. Soc.* (3) **3** (1953) 368-377. [96]
4. Operators with a Fredholm theory, *J. London Math. Soc.* **29** (1954) 318-326. [96]
5. A direct proof of a theorem of West on sequences of Riesz operators, *Glasgow Math. J.* **15** (1974) 93-94. [96]

D. RUTOVITZ

1. On the L^p-convergence of eigenfunction expansions, *Quart. J. Math. Oxford* (2) **7** (1956) 24-38. [399]

S. SAEKI

1. Translation invariant operators on groups, *Tôhuku Math. J.* **22** (1970) 409-419. [375, 379, 381, 385]

D. SARASON [254, 278]

1. Invariant subspaces and unstarred operator algebras, *Pacific J. Math.* **17** (1966) 511-517. [253, 261, 278]
2. Weak-star density of polynominals, *J. Reine Angew. Math.* **252** (1972) 1-15. [261, 278]

R. SCHATTEN

1. *Norm ideals of completely continuous operators* (Springer-Verlag, 1960). [96]

J. SCHAUDER

1. Über lineare, vollstetige Funktionaloperationen, *Studia Math.* **2** (1930) 183–196. [95]

E. SCHMIDT

1. Auflösung der allgemeinen linearen Intergralgleichung, *Math. Ann.* **64** (1907) 161-174. [95]

J. T. SCHWARTZ

1. Two perturbation formulae, *Comm. Pure Appl. Math.* **8** (1955) 371-376. [41]

J. E. SCROGGS

1. Invariant subspaces of a normal operator, *Duke Math. J.* **26** (1959) 95-111. [41, 261, 278]

R. SIKORSKI

1. On multiplication of determinants in Banach spaces, *Bull. Acad. Polon. Sci. Cl. III* **1** (1953) 219–221. [96]
2. On Lezański's determinants of linear equations in Banach spaces, *Studia Math.* **14** (1953) 24-48. [96]

J. P. O. SILBERSTEIN
1. On eigenvalues and singular values of compact linear operators in Hilbert space, *Proc. Cambridge Philos. Soc.* **49** (1953) 201-212. [96]

W. H. SILLS [401]
1. On absolutely continuous functions and the well-bounded operator, *Pacific J. Math.* **17** (1966) 349-366. [400]

A. M. SINCLAIR [105, 111, 155, 157, 213, 276]
1. The norm of a hermitian element in a Banach algebra, *Proc. Amer. Math. Soc.* **28** (1971) 446-450. [114]
2. The Banach algebra generated by a hermitian operator, *Proc. London Math. Soc.* (3) **24** (1972) 681-691. [114]

I. SINGER
1. *Bases in Banach spaces I* (Springer-Verlag, 1970). [402]

D. R. SMART
1. Eigenfunction expansions in L^p and *C*, *Illinois J. Math.* **3** (1959) 82-97. [399]
2. Conditionally convergent spectral expansions, *J. Austral. Math. Soc.* **1** (1960) 319-333. [287, 400]

F. SMITHIES
1. The Fredholm theory of integral equations, *Duke Math. J.* **8** (1941) 107-130. [96]

M. R. F. SMYTH
1. Riesz theory in Banach algebras, *Math. Z.* **145** (1975) 145-155. [97]

M. R. F. SMYTH AND T. T. WEST
1. The spectral radius formula in quotient algebras, *Math. Z.* **145** (1975) 157–161. [97]

P. G. SPAIN [114, 275, 402]
1. On scalar-type spectral operators, *Proc. Cambridge Philos. Soc.* **69** (1971) 409-410. [171, 277]
2. On well-bounded operators of type (B), *Proc. Edinburgh Math. Soc.* (2) **18** (1972) 35-48. [285, 332, 401]

J. G. STAMPFLI
1. Roots of scalar operators, *Proc. Amer. Math. Soc.* **13** (1962) 796-798. [277]

S. B. STEČKIN [378, 380]
1. On bilinear forms, *Dokl. Akad. Nauk. SSSR* **71** (1950) 237–240. [376]

M. H. STONE
1. *Linear transformations in Hilbert space,* Amer. Math. Soc. Coll. Publ.XV (1932). [113, 277]

B. SZ. NAGY [95, 188, 277]
1. *Spektraldarstellung linearer Transformationen des Hilbertschen Raumes* (Springer, 1942). [277]
2. On uniformly bounded linear transformations in Hilbert space, *Acta Sci. Math. (Szeged)* **11** (1947) 152–157. [277]

J. J. SYLVESTER
1. On the equation to the secular inequalities in the planetary theory, *Phil. Mag.* **16** (1883) 267-269. [39]
2. Sur les puissances et les racines de substitutions linéaires, *C. R. Acad. Sci. Paris* **94** (1882) 55-59. [39]

A. E. TAYLOR
1. The resolvent of a closed transformation, *Bull. Amer. Math. Soc.* **44** (1938) 70–74. [40]
2. Linear operations which depend analytically upon a parameter, *Ann. of Math.* (2) **39** (1938) 574–593. [40]
3. Spectral theory of closed distributive operators, *Acta Math.* **84** (1951) 189-224. [41]

E. C. TITCHMARSH
1. Reciprocal formulae involving series and integrals, *Math. Z.* **25** (1926) 321-341. [376]

O. TOEPLITZ
1. Das algebraische Analogon zu einem Satze von Fejèr, *Math. Z.* **2** (1918) 187-197. [113]

J. K. TURNER
1. *Continuity of a linear operator satisfying a commutator condition, and related topics*, Ph.D. Dissertation, University of Newcastle upon Tyne (1972). [276]

L. TZAFRIRI [280, 281]
1. On multiplicity theory for Boolean algebras of projections, *Israel J. Math.* **4** (1966) 217-224. [279]

K. VALA
1. Sur les éléments compacts d'une algèbre normée, *Ann. Acad. Sci. Fenn. Ser. A* 1 No. 407 (1967). [97]

I. VIDAV [276]
1. Eine metrische Kennzeichnung der selbstadjungierten Operatoren, *Math. Z.* **66** (1956) 121-128. [114]

C. VISSER AND A. C. ZAANEN
1. On the eigenvalues of compact linear transformations, *Nederl. Akad. Wetensch. Proc. Ser. A* **55** (1952) 71-78. [96]

H. F. WEINBERGER
1. An optimum problem in the Weinstein method for eigenvalues, *Pacific J. Math.* **2** (1952) 413-418. [96]
2. Error estimation in the Weinstein method for eigenvalues, *Proc. Amer. Math. Soc.* **3** (1952) 643-646. [96]

J. WERMER [188, 240, 246]
1. Invariant subspaces of normal operators, *Proc. Amer. Math. Soc.* **3** (1952) 270-277. [248, 278]
2. Commuting spectral operators in Hilbert space, *Pacific J. Math.* **4** (1954) 355-361. [277]

T. T. WEST [67, 77, 78, 90, 92, 96, 97, 374]
1. Riesz operators in Banach spaces, *Proc. London Math. Soc.* (3) **16** (1966) 131-140. [96]
2. The decomposition of Riesz operators, *Proc. London Math. Soc.* (3) **16** (1966) 737-752. [85, 96]

H. WEYL
1. Inequalities between the two kinds of eigenvalues of a linear transformation, *Proc. Nat. Acad. Sci. U.S.A.* **35** (1949) 408-411. [96]

R. J. WHITLEY
1. The spectral theorem for a normal operator, *Amer. Math. Monthly* **75** (1968) 858-861. [175, 277]

N. WIENER
1. Note on a paper of M. Banach, *Fund. Math.* **4** (1923) 136-143. [40]

A. WINTNER
1. *Spektraltheorie der unendlichen Matrizen* (Hirzel, Leipzig, 1929). [277]

A. C. ZAANEN
1. *Linear analysis* (Noordhoff, Groningen, 1953). [95, 96, 97]

A. ZYGMUND
1. *Trigonometrical series* (Warsaw, 1935). [399, 402]

Subject Index

Index of Symbols

The page reference gives the point in the text at which the symbol is first introduced and the notation explained.